The Origin of Adaptations

The Origin of Adaptations

VERNE GRANT

1963   *New York and London*

COLUMBIA   UNIVERSITY   PRESS

Preface

THE PURPOSE OF THIS BOOK is to set forth the causal theory of evolution as applied to diploid sexual organisms. I have tried to present this theory as a continuous logical argument supported by a sample of the relevant evidence. In a word, I have attempted to provide a general framework in which we can organize our present knowledge concerning the evolutionary process in higher plants and animals.

In building the theoretical framework in these pages my main task has been to fit together the contributions of previous authors, but I have also had to add some new contributions of my own. To be sure, the task of review and synthesis has outweighed the task of development and extension at nearly every point. But the proportion between review and development varies in different parts of the book.

The classical population geneticists described elementary evolutionary changes within populations in terms of a limited number of causes, with definite directions of action, forces of action, and modes of interaction. In Part 3 I have tried to sum up the genetical theory of evolution within populations. Parts 1 and 2 provide the general and the organismic foundations for Part 3. In Part 4, on the basis of the taxonomic and ecological foundations laid down in the introductory chapters, I have attempted to extend the causal approach of the population geneticists to the more complex phenomena of speciation, and some readers may feel as I do that we are now close to having a satisfactory general theory of speciation. In Part 5, finally, we consider very briefly the applicability of our system of evolutionary mechanics to changes on a macro-evolutionary scale.

Since our treatment of evolutionary theory starts at the beginning and develops progressively, the book discusses many elementary subjects if they form an essential link in the chain of argument, whereas, on the other hand, many subjects tangential to the main theme have been omitted, however important they may be in other contexts. As with concepts, so with technical terms, I have introduced only those which I

v

found, pragmatically, to be indispensable for the presentation. The particular combination of inclusions and exclusions found in the following pages will make it difficult to categorize the book as either elementary or advanced. In fact it is both and neither. What it attempts to be is fundamental. Following the fundamental concepts of evolutionary theory takes us sometimes into the territory of classical biology and sometimes to the advanced frontiers of this science.

It has not been possible to support the various theoretical statements by reference to all the available and often voluminous evidence. Instead, an exposition of principles is usually followed by a presentation of selected examples. I have selected from among the examples known to me the one or few which seem best suited to illustrate the point. And I have omitted many complicating, but presumably unessential, aspects of the case studies that are presented. *The Origin of Adaptations* was not designed to be a complete literature review or a compendium of examples on the various topics considered. The reader interested in more detailed discussions of these topics is referred in the footnotes and bibliography to key review articles and research papers.

Where the evidence seems at present to point consistently in one direction, the task of presenting it has of course been easiest. In such cases, after giving an example or two for purposes of illustration, it has been possible to state some general conclusion. Where the evidence is conflicting, and diverse interpretations are drawn by different students, I have attempted to sum up the case pro and con, and where the evidence is incomplete, so that the prevailing conclusions are tentative, I have tried to describe the state of our knowledge, or rather the state of our ignorance, as it really is. My own statements of opinion on some controversial questions will, I hope, be plainly evident as such from the context.

Needless to say, a clear-cut distinction cannot always be drawn between live controversial issues fed by prominent discrepancies in the scientific evidence, and non-controversial conclusions which are accepted by most students but disputed by a few; but in deciding how to present each separate case I have preferred to spare the reader the trouble of fighting all over again the scientific battles of the past in their original detail, while saving his energies for the issues of the present day.

The literature cited represents the actual sources used in writing this book. If I drew an example or concept from a secondary source—a general book or review article—as was frequently the case in dealing with fields of research other than my own, the secondary source is cited and the name of the original worker is merely mentioned for historical purposes. The interested reader can track the material down to its original statement via the secondary source. In many cases, a review article will be cited in lieu of several research papers in order to economize on literature references. Where the same point has been made independently in the text of this book and in a publication by another author, I have cited the latter in the form "also John Doe, 1960."

The literature used in the preparation of this book is almost entirely that which I was able to read and digest before commencing to write in 1960. In a few instances it has been possible to include references to papers of special relevance appearing in the years 1960 to 1962.

In treating various topics I have taken some pains to indicate, textually or in the footnotes, the historical sequence in the development of ideas. These references to the historical background of our subject, incidental and inadequate as they are, may still help us see the accomplishments of our present generation in truer perspective.

Although the literature sources for this book are given in the text, there are in addition several general influences which have been so pervading as to warrant special mention here. The first of these are Darwin's *Origin of Species*, *Descent of Man*, *Voyage of the Beagle*, and other works, which kindled and directed my interest in evolutionary biology so many years ago. That interest was kept alive during a later period of semi-isolation in the tropics by J. Huxley's writings. Still later in 1946 and 1947 Professor G. L. Stebbins introduced me to the field of experimental evolution. At that time I first read Dobzhansky's *Genetics and the Origin of Species*, which has had a profound influence on my thinking, and I also learned a great deal from Mayr's *Systematics and the Origin of Species* and Stebbins' *Variation and Evolution in Plants*.

It is a pleasure to acknowledge the help received from various persons in the preparation of the manuscript for this book. Professor Th. Dobzhansky took an interest in the book from the start and provided encouragement and advice during the long period of composition. In addition, I had opportunities at various times to discuss the general outline with Professors G. G. Simpson, E. Mayr, and G. L. Stebbins. My wife, Mrs. Karen Grant, critically read each section and chapter as it came off the composer's typewriter, with an unfailing eye for ambiguities. Professor Will Jones of the Philosophy Department of Pomona College had the kindness to read and criticize the first draft of Chapters 1 and 2, and Professor G. G. Simpson read Chapter 18 in manuscript, making a number of particularly valuable suggestions.

Many other colleagues, too numerous to mention here, provided information on specific questions, or furnished illustrations, and many publishers generously granted permission to use copyrighted material, especially figures, as noted in the text. Finally, Mr. Robert J. Tilley and his staff of editors at Columbia University Press have been co-operative in many ways.

VERNE GRANT

Rancho Santa Ana Botanic Garden
Claremont, California
August, 1962

Contents

Part *1* *Introduction*

ONE

Organization and Causation

Two things are requisite to science—Facts and Ideas.   wm. whewell[1]

THE THEORY OF EVOLUTION is one of the great theories of natural science. It is an attempt to explain the historical unfolding of life on the earth as a result of the operation of natural causes, some of which reside in living organisms themselves and others in their environment. In this sense evolutionism is a biological theory; its facts are the data of biology and related fields of science, and its ideas are the theoretical constructions based on the biological data.

It might be supposed that we could expound the evolution theory strictly in terms of the relevant biological facts and ideas. The main part of this book is indeed devoted to setting forth the biological evidence and concepts necessary for an understanding of organic evolution. But the observations and deductions of biology in themselves are not quite enough.

Implicit in the theory of evolution, as in any other scientific theory, is a certain philosophical position. It is impossible to present evolutionism as a causal theory without making some assumptions, stated or unstated, regarding cause-and-effect relations. The cause-and-effect relations turn out, on examination, to presuppose the existence of units interacting upon one another in certain ways. Upon further examination these units are seen to possess properties of organization and activity, which are among the factors determining their modes of action and interaction.

These general questions are discussed at length in works dealing with the philosophy of science, and are usually passed over with little or no mention in biological studies. I have preferred in the present book to

[1] Whewell, 1859, I, 86.

take an intermediate course. It has seemed necessary to me, in writing about the theory of evolution, to preface the biology with a little bit of philosophy. We cannot hope to treat organization, action, and causation exhaustively here, but neither can we afford to take these extremely important concepts for granted, for they are basic to the causal theory of organic evolution.

The present chapter is intended as an introduction, not to evolution as such, but to the causal theory of evolution. In this chapter we shall examine the scientific method and some of the fruits of that method, particularly the concepts of organization and causation, which are pertinent to the theme of this book. Moreover, approaching these questions as biologists rather than as philosophers, we shall probe only to a depth that seems sufficient for our purposes as students of evolution.

THE SCIENTIFIC APPROACH TO NATURE

One of the tasks that science has set for itself since its earliest stages is to observe and describe nature. The historical beginnings of modern science some 350 years ago stemmed from a revolutionary change in human attitudes concerning the sources of truth. That change in attitudes, which was a part of the intellectual stirrings of the Renaissance, involved the rejection of the formerly widespread assumption that truth lies buried in the writings of traditional authorities, and the assertion of the belief that reliable knowledge about nature can be gained only by the direct observation of nature.

In describing nature scientists have found it necessary not only to collect facts but also to systematize them. When sufficient facts are gathered and are arranged systematically, the individual observations can often be seen to constitute particular cases of some more general rule. The construction of theoretical generalizations, by means of which the raw facts in their endless diversity can be reduced to some general form, to the lowest common denominator, is another task of science.

There is orderliness and pattern in nature. The pattern is not apparent to casual observers, nor is it clearly evident in the raw data of science, but it can be discovered in those data. Basic similarities and regularities may lie hidden in a mass of diverse phenomena, to be revealed when the diverse observations are analyzed and compared.

Different observations can then be collected under one concept, and different concepts can be unified in one broader generalization. The purpose of theoretical generalizations in science is to give expression to the pattern discovered in nature.

A generalization may be likened to a map or ground plan drawn from the original landscape. A certain pattern of land and sea, or hills and valleys, is present in the unmapped area, waiting to be discovered. The earliest maps, like the earliest generalizations in science, are crude and inaccurate, but they do represent a first step even though they are destined to be replaced by more accurate versions at a later date. One of the theoretical goals of science is to formulate valid generalizations, maps that summarize the pattern in nature in an accurate way.

Theoretical generalizations in biology often take the form of classification systems, a number of which will be introduced in later chapters of this book. It must be realized that no classification of biological phenomena, no set of general concepts applied to biological data—like species and race; habitat and niche; gene, subgene, and supergene—can fully represent all the varied and diverse facts, just as no map can show every detail in the terrain. And this inherent limitation of generalizations gives rise to endless controversies in biology whenever facts turn up that do not fit into the existing set of concepts.

The process of intellectual criticism may or may not produce fruitful results depending upon how much is expected of the generalizations under criticism. Some critiques lead to improvements in the system of classification and in the set of concepts, which are then embodied in the theoretical framework of a later day. But many other criticisms of the biological generalizations that we have inherited from the past, and especially those which emphasize the inadequacies of existing concepts without offering constructive suggestions for their improvement, seem to be predicated on the assumption that theoretical generalizations should account for every detail, which is like expecting a map to show the position of every stone on the ground.

When the scientist has observed the phenomena of nature, and has classified his observations, he is ready to inquire about the causes of the phenomena. The logical analysis and description of the cause-and-effect relations between phenomena constitute a third major task of science. This task is carried out by the formulation of causal theories. A causal theory is an exposition of the regular sequential relations between phenomena, and an explanation of how the earlier steps in the sequence lead to the later steps.

If generalizations are like maps, causal theories are like recipes. A map reduces the infinite variety of features in a natural landscape to a few general land forms—mountains and rivers, land and sea—and shows their relative positions; it reveals the pattern in nature in its broad outlines. A recipe prescribes a series of ingredients and actions which cause some final product to be formed. It sets forth the steps leading to the production of a certain result. A successful recipe lists all the materials and conditions which are essential to bring about a particular effect, and only the essential ones.

The theoretical phase of scientific activity has as its goal the explanation of the phenomena observed and recorded in the fact-finding phase. Scientific explanations are of two general sorts. A phenomenon can be considered to be explained, in one sense, when it is grouped with like phenomena in a more general category, and when it can be seen as a particular instance of some general regularity. Or a phenomenon can be considered as explained when its origin and development are set forth in terms of definite causes and effects.

Scientific theories, then, express the relations between phenomena: either the systematic relations, as in theoretical generalizations, or the sequential relations, as in causal theories. The so-called cell theory is an example of a theoretical generalization in biology, and the chromosome theory of heredity exemplifies a causal theory.

The formulation of generalizations is usually a necessary preliminary to the formulation of causal theories. The data must first be classified before the sequential relations between them can be investigated. The generalization that plant and animal bodies are constructed of microscopic units known as cells—the cell theory, first announced in 1839—preceded by half a century the knowledge of the steps whereby the cells divide and multiply during the growth of the body. The problem of the nature of species had to be clarified before it was possible to make significant progress toward an understanding of the processes involved in the origin of species.

The scientist does not approach nature with an empty mind. As we have seen, he makes certain basic assumptions about the world before he even gathers his facts. He assumes, namely, that the world is real and that his direct or instrumental perceptions of it are correct. A further subjective element is involved in the uses the scientist makes of his facts after he has collected them. The classification of data, the construction of generalizations, and the formulation of causal explanations are obviously mental activities. Science without logic is impossible.

The scientific process is sometimes depicted as a straightforward sequence beginning with fact-finding and ending with interpretation. In actual practice the sequence is more apt to be a cyclical one. The scientist does not usually gather facts for the sake of gathering facts. His interest in the laborious work of fact-finding is, however, likely to be whetted and directed by the presence of some idea which he would like to confirm or discard by reference to the relevant facts. It is too simple to say that facts precede ideas in a linear series; usually some ideas precede the facts, other ideas follow after the facts, and the derived ideas stimulate a search for still more facts, in a never-ending cycle.

The ideas must at any rate be relevant to the facts, supported by the facts, and in agreement with the facts, or they will be discarded from the conceptual scheme of science. The ideas that have stood the test of time, those that have been verified by reference to the observational data, are embodied in the scientific world view as it stands at any moment.

Modern science is a unique combination of Facts and Ideas, of data and philosophy, of fact-finding and thinking, of perspiration and inspiration. Compendia of information are not scientific in the best tradition if they do not reveal the meaning of the facts. A system of pure ideas about nature is likewise not a scientific work when its connection with the facts is too tenuous. Without philosophy the facts cannot attain to a scientific position; diluted by too much philosophy they also lose their scientific character. A balanced combination of the two elements, though difficult to achieve, is attainable, and is exemplified in all the great classics of scientific literature.

The world view of science is not the only possible system of ideas. Other alternative interpretations of the world have been advanced from premises that appear to be as sound philosophically as the realistic and naturalistic assumptions of science. The scientific world view does, however, claim to provide a coherent and verifiable explanation of what we see around us. The world behaves as though the scientific world view were correct.

ORGANIZATION

The universe as seen by a physical scientist in the eighteenth or nineteenth century was a system of moving bodies, a mechanical machine composed of stars and planets, of falling apples and colliding billiard balls, or ultimately of molecules and atoms. These bodies were believed to move through an imponderable ether in accordance with certain natural laws which were amenable to the laws of rational thought. The universe consisted variously of Matter, Energy, Ether, and Mind.

Biology grew up in an age dominated by the conceptions of the physical scientist, and attempted for a long time to reduce the phenomena of life to the categories recognized by the older and more mature branches of science. This long-continued attempt has been only partly successful.

Biologists began their search for an understanding of the nature of life by dissecting animal and plant bodies to find out how they are constructed. By 1839 these explorations had led to the generalization that living organisms are composed of numerous building blocks known as cells. The so-called cell theory did not, however, solve the original problem of the nature of the living state. It did not tell the biologist anything about the difference between a live animal or plant and a dead one. It merely shifted the focus from a higher unit, the individual organism, to a lower level, the cell.

In the rise of organic chemistry during the nineteenth century and later in the development of biochemistry many biologists saw new opportunities to solve the riddle of life by studying the processes that occur within the cell. Chemical and biochemical probings at levels below the cell might bring about the long-sought unification of the life phenomena with established phenomena in the physical world. Now, knowledge of the chemical structure of living matter, including that most vital substance, the genetic or hereditary material, and an understanding of the biochemical reactions that go on in living organisms are worthwhile for their own sake. There is a biochemical and a chemical, as well as a cellular, basis for life, and it is highly desirable from the standpoint of scientific analysis to find out as much as possible about these structural and material foundations.

A living organism, say a rabbit or coyote, is, however, more than a mixture of water, salt, nucleic acid, protein, sugar, and fat, and more than a bundle of biochemical reactions, just as it is more than an aggregation of cells. The rabbit or coyote lives and breathes, grows and behaves, searches for food, fights, and repairs its wounds, *as a whole.*

A concept of life which accepts the living organism as a whole unit in itself, the organismic concept, began to take shape in the early decades of this century following the failures of previous attempts to explain life completely in mechanistic, chemical, or vitalistic terms.[2]

The whole organism maintains itself by a series of coordinated activities in relation to its surroundings. These coordinated activities are made possible by a particular internal organization of the constituent parts and processes. Not only is a certain array of building blocks present in a living organism, but those building blocks are put together in a certain working relationship to one another. A coyote or a rabbit is the sum total of its biochemical and cellular parts *plus* the integration of those parts into an organized working system.

The integration is achieved through the subordination of the constituent parts to the functioning of the whole. In the growth of a woody plant certain cells forming a layer in the stem normally secrete a waxy substance into their own cell walls. This waxy substance becomes cork, which being impervious to liquids protects the plant as a whole from desiccation. The same imperviousness also results in the death by suffocation and starvation of the cork-secreting cells. The formation of corky tissue is suicidal to certain living cells in the stem, but is essential to the life of the individual plant. Individual organisms are frequently sacrificed in the interest of a still higher organizational unit, the breeding population, in which the individuals exist as subordinate members. Thus in many insects the adults die in the process of reproduction; male honeybees, for example, die as individuals as a result of the act of copulation which ensures the future life of the population.

Organization is a fundamental component of life. Since living organisms obviously exist in the universe, the concept of organization

[2] Woodger, 1929; J. S. Haldane, 1931, 1935; Needham, 1936; Bertalanffy, 1952, and his earlier works cited therein.

must be included in any set of generalizations about the structure of the universe. The organismic concept has indeed been carried over successfully from biology into physical and chemical systems. As the physicist Max Planck has stated, "The conception of wholeness must . . . be introduced in physics as in biology."[3]

A chemical molecule is not simply a collection of atoms. Each kind of molecule is composed of certain kinds of atoms present in definite numbers and proportions and hooked together in a specific way. A water molecule consists of one oxygen atom linked by electron bonds to two hydrogen atoms, but the water molecule has unique properties of its own which are not the sum total of the properties of its constituent atoms. Water is a chemically stable liquid at room temperatures, while oxygen and hydrogen are explosive gases under the same conditions. Oxygen and hydrogen lose their identity as units and their individual properties when they combine chemically to form water molecules. The characteristics of water stem from the particular structural organization of the water molecule itself.

LEVELS OF ORGANIZATION

The analysis of the structure of the world has led to the discovery of an extensive series of organizational units, which have taken the place of matter in the scientist's conceptual scheme. The units which enter into the constitution of the universe range in size from the elementary particles of physics (the electrons, protons, and neutrons) to the astronomical objects (the planets, stars, solar systems, and galaxies).

The units found on our planet also range widely in size and complexity. At the lower limit of this range are the physical particles (electrons, protons, etc.). Larger and more complex are the chemical units (atoms, molecules, macromolecules, colloidal particles). The elementary biological entities (genes, chromosomes, cells) are still more complex. At the upper limit of the range we find the macroscopic biological units (individuals, populations, species, biotic communities), and the complex organization created by man (human societies) (as in Fig. 1).

It has been found that the larger units are composed of smaller units. An animal body is made up of organs which in turn are composed of

---

[3] Quoted by J. S. Haldane, 1935, 71.

cells. The cells are highly intricate units in themselves, consisting of various primary parts (i.e., nucleus, cytoplasm, cell wall) and their even smaller subdivisions (chromosomes, chloroplasts, mitochondria, etc.). The cellular body can be reduced as a whole to colloidal droplets, and these to a series of chemical compounds (water, proteins, etc.). The

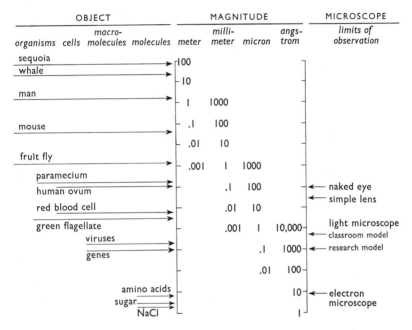

Fig. 1. The relative sizes of various chemical and biological units

Based on G. G. Simpson, C. S. Pittendrigh, and L. H. Tiffany, *Life: An Introduction to Biology* (New York, Harcourt, Brace and World, Inc., 1957).

compounds are reducible to a limited variety of chemical elements (carbon, hydrogen, oxygen, nitrogen, sulphur, phosphorus, and others). The atoms making up the animal's body are just like those in the rocks, the sea, and the atmosphere. And these atoms, like all else on earth, can be resolved into smaller components, chiefly electrons, protons, and neutrons.

The recognition that smaller units are the building blocks of more, complex units has frequently led to the conclusion that either the former or the latter has no real existence as an organizational unit. Genes are composed of smaller units and some students would discard the gene theory for that reason. Other students have suggested that "no particulate genes exist," i.e., that genes do not exist as organizational units, for the opposite reason that groups of genes are assembled into larger units, chromosomes or chromosome segments, which act as integrated wholes.[4] Some nineteenth-century biologists, having discovered the almost universal existence of cells in living organisms, took the position that the multicellular individual is an aggregation of cells. We now know that the individual is a physiological unit and that the cells composing it are units too.

The facts of modern population biology reveal that the individual in turn is an elementary component of a still more complex unit, the breeding population; yet the individual remains a real unit even though it is a subordinate member of another unit.[5] Species are composed of lower units, individuals, populations and races, and are sometimes the constituents of higher units, groups of hybridizing species or syngameons, from which it has been concluded that species are "mental units rather than biological units" having "no foundation in reality."[6] Similar opinions have been stated regarding the biotic community, composed as it is of different interacting species.[7]

These arguments are based on the fallacious assumption that the pattern in nature is simple; that we must make an either-or choice between a limited array of organizational units; and that if A is demonstrably a discrete organizational unit, then B cannot be a real unit too.[8] The fact that atoms have been subdivided into smaller particles does not force us to deny the existence of either atoms or electrons and protons.

In fact a hierarchy of organizational complexity exists in the world. The smaller and simpler units are the building blocks which make up a larger and more complex unit. This latter is in its turn one of the building blocks of a still more complex unit. A series of successively

[4] Goldschmidt, 1940, 203; also 1955, 42.

[5] Allee, Emerson, Park, Park, and Schmidt, 1949, 6, 263 *passim*; Dobzhansky, 1950; Cordeiro and Dobzhansky, 1954.

[6] Davidson, 1954; Mason, 1950; see Grant, 1957, for a critique of this view.

[7] Gleason, 1926; Mason, 1947. For an opposing view of the reality of the community see Allee, Emerson, Park, Park, and Schmidt, 1949, 10, 436, 721 *passim*.

[8] Grant, 1957.

higher levels of organization can be discerned in the structure of the universe. The different kinds of units occupy positions on the different levels of this stepwise series.[9]

ACTIVITY AND FIELD

We can observe that the various organizational units have motions and actions. Therefore activity is another fundamental feature in the universe.

The actions of the units always take place in a framework of space and time. Planets move in their orbits; biochemical reactions occur in the bodies of organisms; organisms live in their environment; the human mind functions in the brain of social man. These actions require time as well as space. The relevant frame of reference is large and long in the case of astronomical happenings, microscopic and short for biochemical processes, and of intermediate proportions for many biological events; but, whether large or small, long or short, the space-time framework is an indispensable component of every happening in the world. A third fundamental component in nature is the continuum of space and time, or field.[10]

Shakespeare's contention that the world is a stage is applicable to the phenomena of nature as well as to the affairs of man. A dramatic play requires a combination of three elements: a company of actors, the acting of the parts, and a stage on which the performance takes place. In human affairs men act out their unrehearsed roles on a local, national, or international scene. In the world of living nature the same elements are present, but take a different form. The actors may be coyotes and rabbits; their actions are their natural inborn and learned responses and movements; and their stage is the territory in which they live. Similarly, the happenings of inanimate nature consist of units, actions, and field. The units in this case may be the heavenly bodies; the actions are their motions; and the field is the large-scale continuum of space and time in which their paths and orbits exist.

[9] The concept of different levels of organization in biology has been stated, among many others, by Woodger (1929), Needham (1936), Allee et al. (1949), Bertalanffy (1952 and earlier), and is expressed in the outlines of several symposia, i.e., Symposium (1942), and Symposium (1956).
[10] Einstein and Infeld, 1938.

The process of scientific generalization has thus led to the recognition of three basic categories. In the first place, a variety of organizational units exist in the world. In the second place, these units have activity. And, thirdly, their actions take place in a framework or field. The most general statement we can make about the universe is that it consists of various kinds of organizational units acting in a field. The field and the active units together comprise a system.

CAUSATION

Causation consists of the sequential order of events in a system composed of active units. The organizational units, whether they be moving particles, reactive chemicals, or living organisms, act according to their intrinsic properties. When the units exist together in combinations, and each acts in its characteristic way, their individual actions lead to reactions, and the primary reactions lead to secondary and tertiary reactions. Every unit in the system, and indeed the system as a whole, undergoes changes as a result of the actions and interactions. These changes, moreover, are found to be orderly and not chaotic, for the sequence of events is repeatable. (Exceptions to this statement will be considered in the final section of this chapter.)

Consider a simple mechanical system consisting of a few billiard balls sitting at rest in scattered positions on a pool table. Let one of the balls be hit by a cue. This ball will now move in a line across the table. If it strikes another ball lying in its path, its motion will be deflected into a new direction and the second ball will be impelled into a motion its own. Each ball may in the course of its new motion strike one or more other balls, sending these off in various directions. When the friction of the table surface overcomes the force of motion of the balls, they come to rest again in new positions.

The changes in the system of balls can be described throughout as an orderly series of causes and effects. The initial cause was the force imparted to the first ball by the cue. Without this event none of the subsequent events would have occurred. The direction and velocity of the first ball determined the course of motion of the second ball, and, conversely, the mass and inertia of the second ball deflected the first ball into a new path at the moment of collision. The new motions of these two balls, which are the effects of previous events, now become the causes of new events—the motions of several other balls. The final resting positions of all the balls on the table are the result of the direction and force of their previous motions, and these vectors are due in turn to a chain of causal relations going back to the start of the play.

It is appropriate to illustrate causation with the motion of balls, since the idea of cause and effect seems to have grown out of the study of change and motion, both in Aristotelian philosophy and in mechanical physics.

Prior to Newton the cause of the motion was usually attributed to some vaguely conceived property or virtue. For Aristotle the ultimate cause of motion was desire. Aristotle also related the cause of motion to the "wonderful properties" of circles. "The circle," he stated "contains the principle of the cause" of motion.[11]

The earliest students of mechanics in the modern period of science— Copernicus, Kepler, Galileo, and others in the late sixteenth and early seventeenth centuries—were too much occupied with working out the numerical and geometrical regularities in the motions of bodies, with the paths and periodicities of planets and the velocities of falling bodies, to inquire how those motions are caused.[12] Or, if they did speculate on the cause of motion, they found their answer in some imaginary virtue. Thus Kepler stated in 1609 that "the vehicle of that Virtue which urges the planets, circulates through the spaces of the universe after the manner of a river or whirlpool, moving quicker than the planets."[13]

The idea that the cause of motion is an external force acting upon a body was apparently clearly seen and stated for the first time by Newton in the *Principia* (1687). The following passage is revealing.[14]

In this sense rational mechanics will be the science of motions resulting from any forces whatsoever, and of the forces required to produce any motions, accurately proposed and demonstrated. . . . Then from these forces . . . I deduce the motions of the planets, the comets, the moon, and the sea. I wish we could derive the rest of the phenomena of Nature by the same kind of reasoning from mechanical principles, for I am induced by many reasons to suspect that they may all depend upon certain forces by which the particles of bodies, by some causes hitherto unknown, are either mutually impelled towards one another, and cohere in regular figures, or are repelled

[11] Whewell, 1859, I, 88–89.
[12] Whewell, 1859, I, 311, 322.
[13] Whewell, 1859, I, 320, 386–87, 415.
[14] Newton, 1687, Preface, quoted from the English edition reprinted by the University of California Press, Berkeley, 1934.

and recede from one another. These forces being unknown, philosophers have hitherto attempted the search of Nature in vain.

After this prefatory statement, Newton went on to propound the law of universal gravitation, the first successful causal theory in science. In the *Principia* he related particular motions to particular forces, and thus showed how particular effects followed from particular causes. Our modern concept of cause as an external force bringing about a change in a system of active units seems to have been introduced into scientific thought by Newton in the latter seventeenth century.

The notion of causality rode to its prominent position in science with the rise of classical physics in the eighteenth and nineteenth centuries, and was carried over into the other younger sciences, including biology. As we shall see later, the concept of causality was taken over in biology with varying degrees of success, depending on where and how the biologist went about the search for causal relations.

The notion of cause and effect is sometimes explained away by philosophers of science. I cannot discuss the philosophical issues involved,[15] but it is perhaps significant that the search for causal explanations has led to virtually all of the great advances in science.

THE INVESTIGATION OF CAUSATION
IN ISOLATED SYSTEMS

In practice the scientist investigating the causes of a phenomenon or process tries to work with an isolated system composed of a manageable number and variety of acting units. He deals with one segment of the complicated machinery of nature at a time. Causation can be studied best in a closed system where all the variables can be known. In the system thus isolated for special study a scientist tries to include representatives of the essential elements and factors and exclude the unessential ones. Difficult and complex processes of nature are deliberately made as simple as possible before they are subjected to a causal analysis; refractory problems are, so to speak, softened up before they are attacked.

Ideally, a scientist working out causal relations prefers to set up his own isolated system consisting of a selected group of units acting under

[15] For a critical examination of causality in science see Bunge, 1959. The present chapter was written before I discovered Bunge's excellent book.

prescribed conditions. An artificial system designed to reveal causal relations or to test causal explanations—to discover new truths or confirm preconceived theories—is an experiment. The great value of experimentation lies in the possibility this method offers of selecting, out of the great welter of interactions in nature, a limited number of elements and factors whose reactions can be observed in a closed system. The experimentalist can, in short, arrange the conditions himself so as to include only one or a few causal influences and exclude all others.

The ideal situation for a scientific investigation cannot always be realized in this imperfect world. Living organisms are intrinsically complex systems in which numerous variable factors are always at work simultaneously. The practical difficulties standing in the way of isolating these factors one at a time are often insurmountable. As a consequence, experiments in biology often yield results which, while pointing definitely to the validity of some conclusion, fall short of furnishing proof.

It is not possible to devise experiments at all bearing on some of the problems with which scientists have to deal. Such sciences as astronomy, geology, and anthropology are essentially nonexperimental. Causal theories relating to mountain-building or the advance of ice ages may have to depend on direct observation of uncontrolled natural phenomena, since the reproduction of these geological processes under experimental conditions is impossible. We can perform breeding experiments with garden peas, rats, and flies, but not with men, and our knowledge and theories of human genetics cannot be subjected to direct experimental tests.

The theoretical scientist who is unable to set up experiments owing to circumstances inherent in the nature of his problem nevertheless approaches his problem in the same general way as the experimentalist; that is, he considers an isolated system of units and actions. He circumscribes one segment of the whole universe for special study and deals with it as though it were actually isolated. He fixes his attention on one or a few significant relations within this semi-isolated system and excludes, in his thinking at least, the many other relatively insignificant factors. One of the hallmarks of genius in a scientist is the ability, often intuitive, to recognize what is essential and primary and what is incidental and secondary in a complex series of events.

A scientist working on a problem that is not amenable to direct experimentation can often extrapolate from the experimentally based conclusions obtained in a related field. The laws of motion can be tested experimentally with small material objects on earth, and can then be extended satisfactorily to the movements of planets. The laws of heredity were first worked out from the results of experiments with peas and flies. A large number of facts about human heredity can be explained if and only if it is assumed that the same laws of heredity are valid for the human species.

One of the tests of a causal theory is the ability of an observer to make accurate predictions with the aid of this theory. If the present condition of the active units in a system is known, and if the causal relations between these units are understood, it should be possible to foresee the future condition of the system. In a system of billiard balls on a billiard table it is possible, knowing the laws of motion, to predict that a certain force and direction imparted by a cue to a ball of certain weight will cause a second ball to move in the direction of a pocket. This prediction can be put to the test of experiment. Predictions are also made in systems that are not susceptible to experimentation. Thus the future paths of the planets around the sun can be predicted exactly.

THE LIMITED APPLICABILITY OF PHYSICS AND CHEMISTRY IN BIOLOGICAL THEORY

The task of scientific explanation of our universe began of necessity with the investigation of the more simple phenomena. Natural science cut its eyeteeth on the gross inanimate material objects on earth and in the heavens, and, as a result, the first sciences to develop a satisfactory body of causal explanations were classical physics and astronomy. Their first great success was the formulation of the laws of motion of material bodies. The causes of the actions and reactions of billiard balls on a table, and of the motions of the planets around the sun, were reduced to simple laws. The same mechanical concepts of matter and motion were later extended successfully to the explanation of other physical phenomena, such as heat, gas pressure, surface tension, and Brownian movement.

The great successes of classical mechanics encouraged many eighteenth- and nineteenth-century thinkers to believe that a mechanical type of cause and effect is universally valid. The phenomena of light and electricity, of chemistry, and of life were thought to be especially complicated systems of mechanically interacting molecules. If one knew enough about the movements of the large numbers of molecules in an electrical machine or living organism one could predict its future state with the same reliability as applies to billiard games and planetary motions. Admittedly the practical difficulties involved in obtaining and assessing the information about billions of mechanical particles needed to make these predictions were very great, perhaps insurmountable, but in theory at least the most complex phenomena of chemistry and biology could be reduced to causation on a mechanical level of organization.

This thesis was stated in 1812 by the French astronomer Laplace[16]:

Let us imagine an Intelligence who would know at a given instant of time all forces acting in nature and the position of all things of which the world consists; let us assume, further, that this Intelligence would be capable of subjecting all these data to mathematical analysis. Then it could derive a result that would embrace in one and the same formula the motion of the largest bodies in the universe and of the lightest atoms. Nothing would be uncertain for this Intelligence. The past and the future would be present to its eyes.

Helmholtz in 1847 stated: "The task of physical science is finally to reduce all phenomena of nature to forces of attraction and repulsion."[17] In 1862, in the *First Principles*, Herbert Spencer applied mechanistic principles consistently to the whole range of evolutionary phenomena, characterizing evolution as "an integration of Matter and dissipation of Motion."[18]

It was partly the desire to bridge the gap between physical systems and life that led many biologists to investigate the fine structure of living matter. Among the results of such investigations were an endless number of observations concerning the structure of living organisms, along with a growing realization that these facts could be fitted together into a map but not a recipe for life. For the ultimate properties of life are found to reside in a particular organization, which is lost when the living organism is analyzed into its parts. Mechanical theories proved

[16] Quoted by Frank, 1957, 263.
[17] Quoted by Dobzhansky, 1960.
[18] Spencer, 1884, ch. 13 *passim*.

inadequate to account for many biological phenomena, and for many physical phenomena too, like electromagnetism and radioactivity, and were abandoned in these fields in favor of organismic conceptions.

It should be made clear that mechanical explanations of cause and effect have not been invalidated generally. Those explanations are fully valid in their own sphere. Nor do living systems violate mechanical laws in any way. They transcend them. The bones and muscles in the legs of a rabbit or coyote are hooked together as an efficient mechanical system of levers and pulleys. The work involved in moving these legs requires energy, which the rabbit or coyote must obtain by eating food. These and other processes of the living organism are explicable in mechanical terms.

Still other processes are explicable in biochemical terms. Modern biochemistry, having achieved enormous results in its own realm, is regarded in many quarters today as being fundamental for the development of biological theory in general.[19] Biochemistry is by way of assuming a responsibility for the formulation of causal explanations in biology like that assumed by mechanics a century ago, and thus history is repeating itself. But biochemistry is only able to explain the causes of *some* biological processes. A rabbit or coyote is no more a chemical machine in the final analysis than it is a mechanical machine or an electrical machine. No one ever saw a chemical machine, a mechanical machine, or an electrical machine run away from coyotes or chase rabbits.

Just as mechanical cause-effect relations are valid for mechanical systems and chemical causal relations are valid in chemical systems, so must the causes of many life phenomena be sought within biological systems and not at some lower, sub-biological level. The scientist who sets out to investigate the causes of biological phenomena by analyzing living organisms into their structural components is likely to end up by making contributions in biochemistry or some other field while failing to discover causal relations in biology. But the scientist who takes the biological units for granted, as Newton took the planets, and studies their actions and interactions within systems of their own, is more apt to make progress in the formulation of causal explanations. A scientist must first recognize the organizational units with which he has to deal

[19] This view is stated explicitly by Commoner, 1961, among others.

before he can hope to explain the cause and effect relations between them.

This conclusion follows from the nature of causality, which as we have seen consists of the sequential order of events in a system composed of active units. The type of organizational units making up the system determines their characteristic mode of action. Causal explanations, being descriptions of the way these actions bring about reactions, must correspond to the level of organization in the system under investigation. If the units are living organisms, and if the object of the inquiry is to discover cause-and-effect relations between them, it is fruitless to attempt to reduce their actions to mechanical or biochemical levels.

The presence of coyotes is a factor determining the numbers of rabbits living in a territory; an increase in the numbers of coyotes will normally cause a decrease in the rabbit population. Eventually the expansion of the coyote population will level off when it reaches the limits set by the supply of rabbits available for food, and so the numbers of rabbits are retroactively a determinant as well as an effect of the numbers of coyotes. Similar causal relations exist between mountain lions and deer, and it is well known that the killing off of mountain lions by man has resulted in the excessive multiplication of deer, with pronounced secondary effects on the plant life which the deer browse on for food. No amount of research in physics or biochemistry would reveal these causal relations.

Biochemical and physical information is thus capable of furnishing an insight into some, but only some, of the causal relations in living systems. In other words, biological phenomena are reducible in part, but not entirely, to the laws governing sub-biological systems.

As we shall see in a later discussion in Chapter 3, living organisms have developed historically out of chemical systems. The kinship between the two levels of organization is apparent when we seek, in vain, for absolute distinctions between the simplest forms of life and the more complex chemical systems. (See Chapter 3.) The gap between the living and non-living states becomes narrow in comparisons of the simplest known organisms with some kinds of chemical systems. That gap was bridged in the past during the course of cosmic evolution, and may be bridged again in future research. In any case the simplest organisms can be described very largely in biochemical terms.

One of the consequences of the long-continued evolution process, however, has been the emergence of life out of the biochemical stage in which it originated, and the attainment of new and higher levels of organization. A system composed of interacting organisms, of rabbits and coyotes, is far removed from its biochemical antecedents, like the superstructure of a bridge from its foundations. For the description of the interaction between rabbits and rabbits, and between rabbits and coyotes, biochemical laws will no longer suffice, and strictly biological laws must be formulated.

THE FORMS OF CAUSAL RELATIONSHIPS IN BIOLOGICAL SYSTEMS

Three different kinds of causal explanations can be given for any biological phenomenon. One can explain how a given structure or process works in a physiological and biochemical sense. Or the adaptive role of the structure or function in the life of the organism may be elucidated. Or the historical steps leading to the development of the phenomenon may be described.[20]

Take the rabbit's ear for example. First, how does it function as an organ? The eardrum vibrates when sound waves are received, and these vibrations are transmitted by small bones in the middle ear to a fluid in the inner ear, which in turn communicates the sense impression to the brain. Secondly, what is the functional significance of the ear? It enables the rabbit to hear sounds, including those made by its enemy, the coyote. And, thirdly, how did the ear develop? Historically the upper and lower jaws and certain gill arches of fishes became progressively changed in form and function during a period hundreds of millions of years long into the various parts of the mammalian ear.[21]

Thus in one sense the functioning of the rabbit ear can be explained as the effect of a specific structural mechanism. In another sense the ear is explained as the effect of evolutionary processes which have endowed rabbits with the ability to detect and escape from their predators. And in still another sense the ear is the end result of a particular series of evolutionary transformations.

[20] Huxley, 1943, 40.
[21] Huxley, 1943, 40.

Cause-and-effect relations in biological systems exhibit certain characteristic features, which we may briefly consider. In the first place, biological events are usually determined by a multiplicity of causes. Rarely if ever can we trace an effect back to a simple cause in biology.

In a purely mechanical system of billiard balls the motion of ball B might be determined simply by the impact of ball A, and ball B might go on to hit ball C. Or a simple cause might have multiple effects, as when ball B hits two other balls, each of which then hits two more balls. Single causes generally do produce multiple effects, as Herbert Spencer emphasized long ago.[22] The pathway of events for a series of simple causes and simple effects can be diagrammed as in Fig. 2a. The case of a simple cause with multiple effects can be represented as a series of branching pathways (as in Fig. 2b).

Now let us suppose that three balls collide with a fourth; a single effect is then determined by the joint action of multiple causes (as in Fig. 2c). This is a very common, in fact a normal, situation in biological systems. Coyotes are not the only factor controlling the numbers of rabbits; hawks, man, diseases, vegetation, and weather also have an influence on the abundance of rabbits. These multiple causes, moreover, are not independent of one another but are interacting. Thus complex interactions exist between the causal agents, man, hawks, and coyotes, between weather and vegetation, and sometimes between man and weather. The case of multiple and interacting causes, which is diagrammed in its simplest form in Fig. 2d, is also very common in biology and, we might add, is difficult to deal with quantitatively.

Billiard ball B, after being set in motion by ball A, might glance off the side of the billiard table and hit ball A on the rebound, thus initiating a cycle in which the effect reacts upon the causal agent (as in Fig. 2e). Circular or feedback relations are another widespread aspect of causality in biological systems. Unlike the example of the billiard balls, where balls A and B might bounce back and hit one another twice or even three or four times before they stop, biological cause-effect cycles often keep going indefinitely. Thus the abundance of coyotes determines, among other factors, the number of rabbits, and the rabbits in turn determine to some extent the abundance of coyotes, through an unlimited number of cycles which is best described as a

[22] Spencer, 1884, ch. 20.

continuous equilibrium. Many physiological processes conform to the
cyclical type of causal relation, in that an end product is fed back into

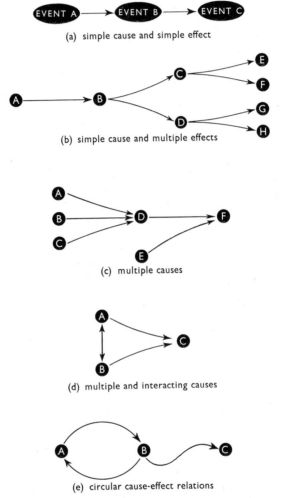

(a) simple cause and simple effect

(b) simple cause and multiple effects

(c) multiple causes

(d) multiple and interacting causes

(e) circular cause-effect relations

Fig. 2. Different types of cause and effect relations

the system whereupon it has some effect on the formation of the same
product during the next cycle.

Not uncommonly the product fed back at the end of cycle I has a
boosting effect on cycle II, as when a growth-promoting substance is

formed which speeds up the formation of more growth substance. Later we shall consider some examples of the boosting type of circular cause-effect relation in evolutionary systems.

Needless to say, the causal relations in actual living systems are composites of the various separate aspects mentioned here and diagrammed in Fig. 2.

When the series of events reaches a certain degree of complexity it becomes necessary to give up the search for particular causes of particular effects in favor of a statistical treatment of the whole series as a process. We can see each link in the chain of causes involved in the movements of a few billiard balls, but not in the interactions of millions of gas molecules, the behavior of which must be described as a statistical process.

The fact that nature must often be described in terms of statistical laws has led Schrödinger to question the fundamental reality of cause-effect relations.[23] It is not necessary, however, to throw the baby out with the bath water. Causality is not negated, but merely placed beyond our practical reach, in cases of complex sequences involving very large numbers of units.

A process occurring in a short time span and involving a limited number of sequential steps may be repeated again and again in nature or in experiments. Physiological processes have the regularity of heartbeats. Chemical reactions can be repeated exactly in the laboratory. But processes taking place over very long periods of time, and requiring very large numbers of sequential steps, are not repeatable in the same way and are not exactly predictable.[24]

When a new mountain range rises up and a new river flows down to the sea, the river may happen to be diverted at one point by a fallen log and will therefore cut its future channel in a new direction. The indeterminant event of a tree falling in one spot at one moment causes the youthful river to flow around it and take a different course from the one it otherwise would have followed. If the whole historical sequence could be allowed to occur over again, it is highly unlikely that this tree would fall over where and when it did. The main trend of the process would be the same, insofar as the river would always seek the

---

[23] Schrödinger, 1935, ch. 6 *passim*.
[24] Dobzhansky, 1954.

lowest ground and eat its way preferentially through the softest rocks, but many details determined by chance happenings would be expected to turn out differently in each repetition.

Causality is present in historical processes as well as in short-term processes, but because of the infinitely large number of causal steps involved in historical processes, and the improbability of those steps occurring in exactly the same order twice in a row, historical processes lack the repeatability and determinism of simpler, short-term changes.[25] Causal theories attempting to explain changes on a historical time scale must therefore restrict themselves to the main trends, leaving out of account the details. Earlier in this chapter we stated that changes in nature are orderly; now to complete this statement by its opposite we must add that there is real spontaneity in nature.

[25] Dobzhansky, 1954.

TWO

The Establishment of Causal Theories in Evolutionary Biology

A CAUSAL THEORY is acceptable only if it is logically sound and supported by the available evidence; the different parts of the theory must be consistent both internally with one another and externally with the observable facts. Furthermore, a causal theory should be framed so as to dispense with unnecessary assumptions. The simplest explanation that fits the facts is most apt to be accepted by scientists as the true explanation. Newton in 1687 laid down the rule that "we are to admit no more causes of natural things than such as are both true and sufficient to explain their appearances."[1] This was no doubt an extension of Occam's Razor of the early fourteenth century that "entities must not be unnecessarily multiplied."

Even when a theory fulfills these conditions it is not regarded as proven, for alternative explanations must be considered. It is frequently possible to offer two or more equally logical and plausible explanations for the same phenomenon. The alternative explanations are embodied in different hypotheses, a hypothesis being a provisional causal explanation which is verifiable but not yet verified. Many controversies in science are due, not to any disagreement between rival scientists regarding the facts, but to the causal interpretations placed on those facts.

The only real solution in such cases is to obtain more facts bearing on the point of difference between the rival hypotheses. The relevant evidence when obtained may point in any one of several directions. (1) It may turn out that the opposing schools of thought are equally wrong, in which case the process of formulating a valid theory must be started all over again. (2) The new facts may be in utter disagreement

[1] Newton, 1687, 1934 reprint, 398.

27

with the predictions made from one hypothesis, which is then dropped, and in good agreement with another hypothesis, which has to be accepted. (3) Quite frequently one hypothesis, while not decisively victorious, is seen to be much more probably correct than any other in the light of the new evidence. (4) Or, again, correct as well as incorrect features may be discovered in both or all contending hypotheses, and the true solution of the problem is provided by an essentially new theory compounded out of the acceptable elements in the various older hypotheses.

It may be instructive to consider some examples from evolutionary biology of the different ways in which a theory becomes established. The examples are numbered in the following sections to correspond with the various possible outcomes of theoretical investigations listed in the above paragraph.

THE SEVERAL LINES OF EVIDENCE POINTING
TO ORGANIC EVOLUTION (3)

The great majority of biologists in the seventeenth and eighteenth centuries believed that the existing species of plants and animals had originated in separate acts of creation about 6,000 years ago and had remained unchanged in form down to the present time. The belief in the fixity of species was stated, among others, by John Ray in the *Historia Plantarum* (1704) and Carl Linnaeus in the *Systema Natura* (1735). A few biologists in the eighteenth and early nineteenth centuries expressed the opposite view that the existing species have developed by a process of branching evolution from common ancestors that existed far back in earth history. The idea of evolution was ventured by Buffon in the *Histoire Naturelle des Animaux* (1760), and formally stated as a consistent scientific hypothesis by J. B. de Lamarck in the *Philosophie Zoologique* (1809). The nineteenth-century naturalist Charles Darwin then applied himself for some two decades to the gathering and collating of the available evidence bearing on this question, which he first published in *The Origin of Species* (1859). In this great book Darwin showed that a world of facts about living organisms and their fossil remains in the rocks makes sense on the evolution hypothesis and on no other view. A large amount of further evidence accumulated after 1859, by Darwin himself in later years, by his successors in the latter nineteenth-century, and by numerous workers during this century, has greatly strengthened the conclusion that the existing forms of life have evolved by series of progressive changes from pre-existing forms. This evidence can be briefly summarized as follows:

(a). Fossil plants and animals are found in rocks estimated to be millions or even hundreds of millions of years old on the basis of geological and chemical methods of dating. Some modern species, like the opossum, have fossil histories stretching far back into geological time. Therefore the act-of-creation hypothesis is wrong as far as the time of origin of some species is concerned. Most fossil plants and animals are unlike the modern forms of life in greater or lesser degree, and, furthermore, a series of fossils of the same plant or animal group collected in rocks of successively younger geological age often exhibits a graded series of changes in morphological characters, as would be expected on the evolution hypothesis but not necessarily on the act-of-creation hypothesis.

The evolution hypothesis presupposes that ancestral species once existed from which various modern groups have diverged. Species with characteristics intermediate or transitional between those of later, more divergent forms have indeed been discovered in a number of cases in the fossil record. Some of the distinctions between modern birds and reptiles break down in the fossil Archaeopteryx of Jurassic age (Fig. 3); most of the distinctions between modern carnivorous mammals and hoofed mammals, the carnivores and ungulates, respectively, are lost in the transitional creodonts and condylarths of Early Tertiary age.

(p). Different living members of the same group tend to resemble one another in their general plan of structural organization (as in Fig. 4). As Darwin put it[2]:

What can be more curious than that the hand of a man, formed for grasping, that of a mole for digging, the leg of the horse, the paddle of the porpoise, and the wing of the bat, should include similar bones, in the same relative positions? ... Nothing can be more hopeless than to attempt to explain this similarity of pattern in members of the same class, by utility or by doctrine of final causes. ... On the ordinary view of the independent creation of each being, we can only say that so it is;—that it has pleased the Creator to construct all the animals and plants in each great class on a uniform plan; but this is not a scientific explanation. ... The explanation is to a large extent simple on the theory of [descent]. ... If we suppose that an early progenitor ... had its limbs constructed on the existing general pattern, for whatever

[2] Darwin, 1872, ch. 14.

Fig. 3. Archaeopteryx, restored

Romanes, 1896.

purpose they served, we can at once perceive the plain signification of the homologous construction of the limbs throughout the class.

(c). An interesting and significant aspect of the existence of corresponding or homologous structures in different species of the same group is the occurrence of vestigial organs (Figs. 5 and 6). A well-developed and functional organ in one member of a group of plants or animals may be represented in another member by a homologous organ which is atrophied and nonfunctioning. In Darwin's words[3]:

> Organs or parts in [a rudimentary and atrophied] condition, bearing the plain stamp of inutility, are extremely common, or even general, throughout nature. It would be

[3] Darwin, 1872, ch. 14.

Fig. 4. Homologous bones in the manus (distal part of the limb) of several types of mammals

(a) Generalized limb.
(b) Hand of tenric (Insectivore).
(c) Digging hand of mole (Insectivore).
(d) Foot of man (Primate).
(e) Paddle of whale (Cetacean).
(f) Foot of dog (Carnivore).
(g) Hoof of deer (Ungulate).
(h) Hoof of horse (Ungulate).
(i) Wing of bat (Bat).

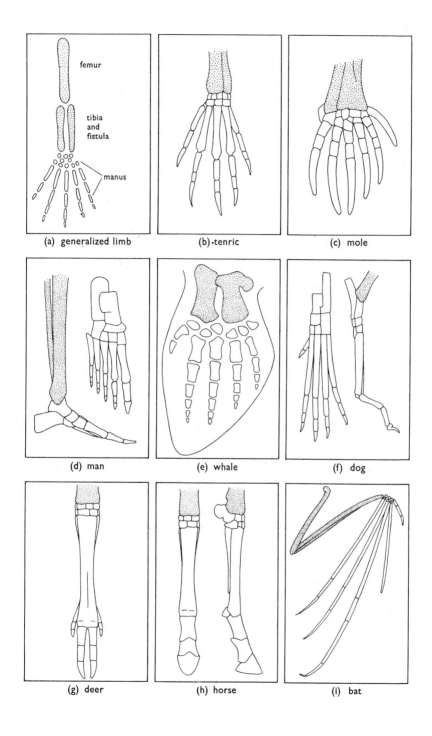

(a) generalized limb

(b) ·tenric

(c) mole

(d) man

(e) whale

(f) dog

(g) deer

(h) horse

(i) bat

impossible to name one of the higher animals in which some part or other is not in a rudimentary condition. In the mammalia, for instance, the males possess rudimentary mammae; in snakes one lobe of the lungs is rudimentary; in birds the "bastard-wing" may safely be considered as a rudimentary digit, and in some species the whole wing is so far rudimentary that it cannot be used for flight. What can be more curious than the presence of teeth in foetal whales, which when grown up have not a tooth in their heads; or the teeth, which never cut through the gums, in the upper jaws of unborn calves?

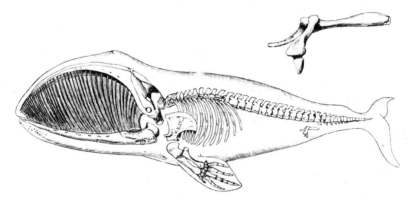

Fig. 5. Vestigial pelvis bones in the Greenland whale

The pelvis is shown enlarged above.

Romanes, 1896.

In reflecting on [rudimentary organs], every one must be struck with astonishment; for the same reasoning power which tells us that most parts and organs are exquisitely adapted for certain purposes, tells us with equal plainness that these rudimentary or atrophied organs are imperfect and useless. In works on natural history, rudimentary organs are generally said to have been created "for the sake of symmetry," or in order "to complete the scheme of nature." But this is not an explanation, merely a re-statement of the fact. Nor is it consistent with itself; thus the boa-constrictor has rudiments of hind-limbs and of a pelvis, and if it be said that these bones have been retained "to complete the scheme of nature," why, as Professor Weismann asks, have they not been retained by other snakes, which do not possess even a vestige of these same bones? ... An eminent physiologist accounts for the presence of rudimentary organs, by supposing that they serve to excrete matter in excess, or matter injurious to the system; but can we suppose that the minute papilla, which often represents the pistil in male flowers, and which is formed of mere cellular tissue, can thus act? Can we suppose that rudimentary teeth, which are subsequently absorbed, are beneficial to the rapidly growing embryonic calf by removing matter so precious as phosphate of lime?

But, he adds, an organ useful under certain conditions in the life of one species, might become injurious under the different conditions of life in another species, which has evolved to a state in which that organ is reduced. The organ persists in a vestigial form, however, due to the conservatism of heredity, and bears witness to a former stage in the history of the species when the organ was fully developed and functional. "On the view of descent with modifications, we may conclude that the

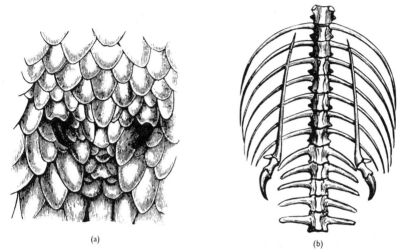

(a)   (b)

Fig. 6. Vestigial hind-limbs of the python

(a) Scaly ventral surface around the vent showing the protruding horny ends of hind limbs (marked B).

(b) Section of the skeleton.

Romanes, 1896.

existence of organs in a rudimentary, imperfect, and useless condition, or quite aborted, far from presenting a strange difficulty, as they assuredly do on the old doctrine of creation, might even have been anticipated in accordance with the views here explained."[4]

(d). When biologists use structural similarities and differences between contemporary organisms as a basis for classifying them into groups, they find that small groups with various features in common, functional as well as vestigial, can be arranged with other such groups into more inclusive circles of resemblance. Thus all races of men have,

[4] Darwin, 1872, ch. 14.

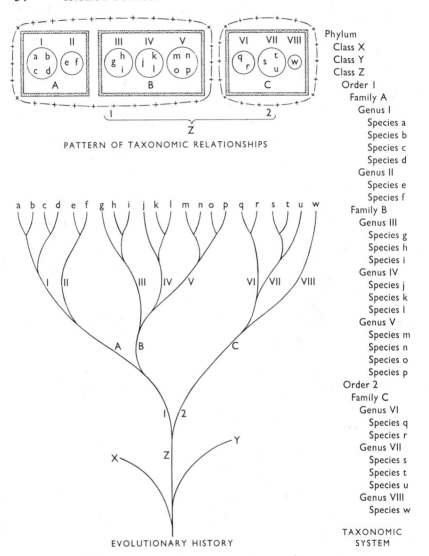

Phylum
Class X
Class Y
Class Z
Order I
  Family A
    Genus I
      Species a
      Species b
      Species c
      Species d
    Genus II
      Species e
      Species f
  Family B
    Genus III
      Species g
      Species h
      Species i
    Genus IV
      Species j
      Species k
      Species l
    Genus V
      Species m
      Species n
      Species o
      Species p
Order 2
  Family C
    Genus VI
      Species q
      Species r
    Genus VII
      Species s
      Species t
      Species u
    Genus VIII
      Species w

TAXONOMIC
SYSTEM

PATTERN OF TAXONOMIC RELATIONSHIPS

EVOLUTIONARY HISTORY

Fig. 7. Three perspectives of the same evolutionary pattern

Below: The evolutionary development of a group of organisms during a period of time.

Above: The pattern of relationships between the living members of the group.

Right: The system of classification of the same group.

despite their differences, numerous characteristics in common. So do all chimpanzees. And so do all gorillas, all orangs, and all gibbons. There are other characters common to man and the great apes collectively. The assemblage of man and the great apes is in turn a subordinate group along with the monkeys, tarsiers, and lemurs within the primates. The primates, together with the carnivores, ungulates, moles, porpoises, bats, rabbits, etc., possess common characters uniting them into a still more inclusive grouping, the mammals.

Darwin noted: "This classification is not arbitrary like the grouping of the stars in constellations." The subordination of organic groups within organic groups is, on the contrary, a reality of nature.[5] This reality is expressed in our taxonomic systems of classification of organisms by the hierarchy of groups, from races and species at the lower level, through genus, family, and order, to class, phylum, and kingdom as the most inclusive categories. Different species with characters in common are united into one genus; different genera with morphological similarities can be grouped naturally in the same family; similar families collectively form a single order; similar orders belong to the same class; and so on.

The "natural subordination of organic beings in groups under groups" (as Darwin phrased it) can be readily explained on the hypothesis that the modern forms of life have developed by a process of divergence from common ancestors existing at successively more remote points of time. In the branching pattern of evolution, the species of one genus represent the outermost twigs which have diverged from their progenitor at a relatively recent time in the history of the group; the genera of a family are branches going back to an earlier point of divergence; the families of an order represent still older, main branches; and so on until we reach the most basic dichotomy in the evolutionary history of life manifested to us today in the division of organisms into kingdoms (Fig. 7). Why the modern species of organisms should resemble one another in descending degrees if they originated in separate and independent acts of creation has never been explained.

(e). If the present-day species had been created separately and independently in the beginning and had remained unchanged in their characters since, they would be expected to be distinct from one another.

[5] Darwin, 1872, ch. 14.

If, on the other hand, species arose by divergence from common ancestral forms, and if the process of gradual change were going on continually, we would expect to find the present-day species at all stages of separation. We would expect to find cases of well-distinguished species along with cases of slightly different species and borderline cases where it is impossible to decide whether the differentiated forms are members of two species or of one. And this is what biologists do find when they survey the world of life.

(f). We now have direct observational evidence of evolutionary changes in successive generations of flies, corn plants, moths, hamsters and other animals and plants. Some of these changes have been made to happen in experiments; others have been witnessed in nature during the lifetime of one or more human observers. A few of the numerous recorded case histories will be reviewed in later chapters.

It is true that the evolutionary changes which have been directly observed are very minor compared with some of the transformations required by the evolution hypothesis. The evolution of a black moth from a gray moth, or of a corn plant with highly proteinaceous kernels from one with slightly proteinaceous kernels, during a period of years leaves the final product unchanged as far as its basic characteristics are concerned. The evolution hypothesis embraces, in addition to small evolutionary changes from one kind of moth or corn plant into another, also such far-reaching transformations as are involved in the differentiation of the carnivores, ungulates, primates, and bats from some common mammalian ancestor.

Evolutionists reason that if small changes can occur in a short time, large-scale changes can take place during the many millions of years of earth history. But adherents of the act-of-creation hypothesis argue that we do not have direct evidence that changes of this magnitude have been brought about by evolution. A series of fossils taken out of successive geological strata may reveal what we interpret to be an evolutionary trend, but since many gaps exist in the geological record we cannot be sure that the different kinds of fossils were ever connected geneologically. The sequence of fossils can be interpreted as a series of separate creations as well as an evolutionary series. The evolutionary biologist has to admit that this is a possible interpretation and one that has not been disproved.

Indeed it is not clear how the act-of-creation hypothesis can be rigorously disproved, inasmuch as it refers to alleged historical events which were not witnessed; nor for the same reasons can we expect to have direct observational evidence of large-scale evolutionary changes that took place during eons of time, unless the "man from Mars" comes forward with time-lapse motion pictures. We can, however, state that the act of creation hypothesis is highly improbable in the light of all the relevant facts. It is also blasphemous, as Dobzhansky has pointed out, since it implies that the Creator has rigged a great bulk of facts belonging to several independent lines of evidence in such a way as to mislead honest-minded students of His works.[6]

THE SYNTHESIS OF A NEW THEORY
OUT OF THE VALID ELEMENTS
IN VARIOUS OLDER HYPOTHESES (4)

In the establishment of causal theories in science, the formation of a new doctrine out of the acceptable features in two or more older rival hypotheses is a more common occurrence than is often realized. We have seen that the controversy over the origin of species ended in the defeat of the act-of-creation hypothesis and the triumph of the evolution theory. This being the case, the question arises as to what forces bring about evolutionary changes. Different controlling factors have been proposed by different students of evolution and embodied in different causal theories: the theory of the direct effects of the environment (Lamarck); the selection theory (Darwin), the mutation theory (De Vries); the theory of isolation (Wagner); the theory of the germplasm (Weismann); and others.

From our present vantage point none of the older explanations of the evolutionary mechanism can be said to be adequate; indeed they are all one-sided and incomplete to some degree, and yet there are important grains of truth in nearly all of the older causal theories. During the period from 1930 to the present, various evolutionary biologists put these grains of truth together into an essentially new theory of evolutionary mechanisms. The currently accepted theory is at once synthetic, coherent, and new; it has borrowed freely from many earlier ideas, but it has worked those ideas over into an internally consistent whole, and it

[6] Dobzhansky, 1955a, 228.

thus represents a product of thinking that did not exist until recent times.

THE EXPERIMENTAL VERIFICATION OF HYPOTHESES (2)

The separate parts of the modern theory of evolutionary mechanisms can be tested experimentally. In many cases we are thus able to obtain clear-cut decisions between alternative explanations of evolutionary processes. The study of organic evolution has in fact become an experimental science to an extent which is little appreciated by laymen generally, by many biologists specializing in other fields, and even by some writers dealing with evolution. We will have occasion to examine some examples of evolution experiments in later chapters.

CONCLUSIONS

Causal theories in evolutionary biology, as in other branches of science, do not claim to have an absolute finality. They are on the contrary subject to constant revision and modification as new facts are discovered and new insights gained. The theory of natural selection held by evolutionists at the present time is so different from the selection theory of Darwin's time as to represent essentially a new doctrine. It would be rash to assume that evolutionary theory a century hence will not differ profoundly from our present-day concepts.

Some critics regard the changing theories of science as a sign of weakness in the whole scientific approach to nature. This criticism reflects a personal attitude. No one wishing to find absolutely certain answers to his questions about the world should be seeking them within the open-minded framework of science in the first place. The person in quest of certainty and finality should consider one of the numerous authoritarian dogmas available, some of which have come down to us unchanged since the time of barbarism.

The feeling held by some persons that the answers of science should somehow possess a permanence and immutability that they obviously lack, and that science therefore has not quite measured up to its task of explaining the world, is perhaps due to a vestige of medieval habits of thinking. During the Middle Ages, when cultural advance had come to a virtual standstill and ideas were stagnant, it was natural to believe in

Eternal Truth.[7] In the rapidly changing world of today, the best we can hope for, from science at least, is Relative Truth.

Relatively accurate answers can be intellectually satisfying to a mature mind; at least they are better than relatively inaccurate answers or downright false ones. The long-range trend in the growth of scientific theories is toward the elimination of such inaccuracies as can be detected. With each successive revision the theory comes closer to giving an accurate account of a natural process, but it never reaches perfection and the task of revising is never done. This is as true of the causal theory of organic evolution as of any other theory in the natural sciences. And so we have to expound that theory in the present book, not as an Eternal Truth, but as an explanation valid in the light of our present information, though undoubtedly including misconceptions that will have to be corrected in the future.

[7] Jones, 1952, 997.

THREE

The Nature and Origin of Life

ORGANIC EVOLUTION AS A STAGE
IN COSMIC EVOLUTION

THE PRESENT STATE of the universe, the earth, and life on earth is the result of a long period of historical development. Cosmologists have divided this historical development into four main stages. The time scale for the successive stages of cosmic evolution is reckoned in billions of years or in eons (an eon being defined as a unit of time 1 billion or 10^9 years long).[1]

The earth is estimated to be about 4.5 eons or possibly 5 eons old. This estimate is based on radioactive methods of dating, which are becoming increasingly accurate. It is well known that radioactive elements change spontaneously into new forms, as uranium to lead. The time required for these transmutations is also known. The proportion between the original radioactive substance and its derived products in any rock is therefore a reliable measure of the age of that rock.[2] Some of the rocks in the earth's crust contain minerals more than 4 eons old.[3]

The universe is generally assumed to be older than the earth, but how much older is a moot point. Some cosmologists believe that the present system of galaxies and stars was formed as a result of a tremendous explosion which took place shortly (1 eon or even much less) before the origin of the earth.[4] Other students believe that the original explosion might have occurred as long as 15 eons ago and hence 10 or 10.5 eons before the earth was formed.[5] Still other cosmologists maintain that the

---

[1] Urey, 1960.
[2] These methods are described by Knopf, 1957.
[3] Knopf, 1957; Briggs, 1959.
[4] Gamov, 1955a, b.
[5] Hoyle, 1961.

galaxies have had a continuous existence through infinite time.[6] If so, the time previous to the origin of the earth is infinitely long.

However this may be, the *matter* of which the universe is composed has been formed at some time in cosmic history. Whether matter was formed suddenly from a pre-existing radiant soup consisting of un-associated protons, neutrons, and electrons, as Gamov suggests, or is being formed continuously, as Hoyle believes, remains to be decided.

In any case, the original and basic form of matter seems to be the hydrogen atom. Hydrogen atoms, once present, have undergone nuclear reactions inside the stars to form the other more complex types of atoms, such as helium, carbon, nitrogen, oxygen, sulphur, iron, lead, etc.[7] The formation of the existing array of atomic species, the elements, was a gradual process which may have continued into the earliest period of earth history. We may call this largely preterrestrial phase in the development of the universe the stage of atomic evolution.

This first phase of cosmic evolution was followed by a second phase during which the different kinds of atoms—hydrogen atoms, oxygen atoms, carbon atoms, iron atoms, etc.—became combined into chemical compounds of various degrees of complexity. The phase of chemical evolution lasted from an early period in earth history until about 1 eon ago.[8]

It is believed that conditions existed in the latter part of the period of chemical evolution which permitted the formation and accumulation of very complex carbon compounds. Such compounds could have served as the food materials for the simplest and most primitive forms of life, which are assumed to have originated at this time.

The origin of life at some unknown time when the earth was still fairly young, perhaps about 3 eons ago, initiated the third phase of cosmic evolution. This is the phase of organic evolution, which has continued at an ever-increasing tempo up to the present time. The time span of organic evolution has encompassed immense transformations in the world of life, transformations from submicroscopic organisms to giant trees 200 feet tall, and from simple biochemical systems to complex animal bodies. Life has multiplied from a small number of individuals

[6] Hoyle, 1956.
[7] See Fowler, 1956.
[8] Calvin, 1956.

to countless billions of individuals filling nearly every inhabitable spot to saturation, and diversified from a few similar kinds of organisms to over four million species fitted for life in different parts of the land, sea, and air. (See Chapter 4.)

The ascendancy of life did not put an end to chemical evolution in an absolute sense. Inorganic chemical reactions which bring about the formation of complex compounds occur under natural conditions even today. But at some point in history the forces of organic evolution gained a predominating role over physical forces in shaping the natural productions on the earth. Organic evolution became so much more effective than chemical evolution an eon or two ago that the former may be said to have replaced the latter in a relative sense.

One of the products of organic evolution was an erect two-legged apelike mammal who roamed the savannahs of Africa and Asia about a million years ago. This creature was the ancestor of modern man, who by virtue of a peculiar combination of factors—tool-making, social grouping, learning ability, and language—was destined to enter upon a fourth phase in cosmic evolution. The phase of cultural evolution is marked by the growth and the transmission from generation to generation of a cultural heritage. This heritage, consisting of cumulative knowledge and understanding of the world, has made man the master of his outer environment and may yet give him mastery of himself. There is nothing like this progressive accumulation and transmittal of knowledge in the animal kingdom.

Man's control over nature is bringing the period of organic evolution to a close. The plants and animals useful to man are evolving under domestication and by artificial breeding practices. Their camp-followers, the weeds and pests, evolve naturally in man-made environments. The fate of many wild species is either extinction or preservation in wild-life refuges and national parks. Although the processes responsible for organic evolution will continue to operate as before, their influence on future developments in man's world will be relatively slight by comparison with the effects of man himself.

Each of the stages of cosmic evolution is an outgrowth of the preceding stage. Human cultural evolution grew out of organic evolution; organic evolution is an extension of chemical evolution; and chemical evolution is an offshoot of atomic evolution. One stage passes by gradual transitions into the next. In this respect it would be correct to regard cosmic evolution as a single process of change from simple forms of matter to ever more complex organizations. It is equally true that some important changes of state have occurred in the gradual historical development of complex structures.

If one builds a fire under a pot of water, the water will gradually rise in temperature up to a certain point and then turn into steam. The statement that steam is merely heated water contains only part of the truth. Steam in this case is derived from liquid water and retains the same chemical composition as the parental material, but has risen to a new and different state.

In a similar way the statements that man is an animal, and that life is a chemical system, are undeniably true, yet do not tell the whole story. It is true enough that man's body and part of his mind belong to the animal kingdom and have been shaped by organic evolution; yet other parts of the human mind are the product of social institutions and culture. Similarly, it cannot be denied that life is a special kind of chemical system. The statement is academic, however, and does not advance our understanding very much, because in the next breath we must add that life is a chemical system possessing the special properties of life. The whole truth is that chemical evolution transcended itself when the first living systems arose, just as organic evolution transcended itself in man.

We can recognize in the course of cosmic evolution three significant changes of state from a lower to a higher level of complexity. Corresponding to the successive levels are the evolutionary stages—atomic, chemical, organic, and cultural—into which we can divide time.

Figure 8 is an attempt to show the *approximate* duration of the four grand periods of cosmic evolution on the time scale of earth history. We will take as our starting point the origin of the earth in the Year Zero, and will assume that the earth is 4.5 eons old, so that we are now living in Year 4.5 Billion. Atomic evolution was going on in the universe for an unknown length of time before the earth existed and may have continued into the earliest part of terrestrial history. Chemical evolution probably began sometime before Year 1 Billion and prevailed until about Year 3 Billion.[9] For reasons which will be explained later, organic evolution is assumed to have begun about Year 1.5 Billion, and has

[9] Calvin, 1956.

continued until the present time. The time span of cultural evolution is too short to show in accurate proportions on the scale in Fig. 8, for this phase began less than one million years (.001 eon) ago and would have to be represented by a line too thin to be seen with the naked eye.

There are good reasons for believing that the processes involved in all phases of cosmic evolution have been under the control of natural causes, some of which have already been discovered and explained, and others of which will be discovered in the future. An exposition of the causes of atomic evolution, chemical evolution, the origin of life, and

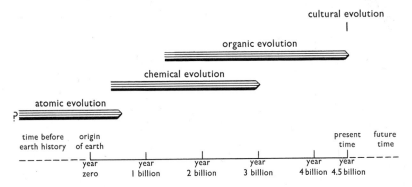

Fig. 8. *The four main stages of cosmic evolution on the time scale of earth history*

cultural evolution is beyond the scope of this book. The present book is concerned with the stage of organic evolution. We will understand our subject better, however, if we can view it in the framework of cosmic development, and to this end we mention briefly the stages of chemical and cultural evolution which delimit organic evolution at both ends.

The starting point of organic evolution cannot be defined with exactness for two reasons. In the first place, the transition from non-living chemical systems to primitive living systems took place about 3 eons ago. We have no direct observations of the events that occurred in that remote period of time. But even if scientists did have any means of observing the chemical and biochemical processes that took place so long ago, they still could not be sure exactly where chemical evolution left off and biological evolution began. The change of state between the two evolutionary levels was probably gradual.

The problem of the origin of life is inseparable from the problem of the nature and distinguishing characteristics of life. Everyone, biologist and layman alike, recognizes that a rabbit is alive. It moves and breathes, responds to stimuli, adjusts to its surroundings, eats, grows, and reproduces.

The blue-green algae lack most of these characteristics in the everyday sense. These microscopic organisms consist of single cells or colonies of cells which float in water or lie sedentarily on soil or snow. They do not eat food; they have no body organs or nervous system; they lack mobility, irritability, and responsiveness in the ordinary sense of the words. On the other hand, the blue-green algae possess a cellular organization which, though simple, permits them to grow, maintain themselves, and reproduce their kind. The layman is apt to regard blue-green algae as inanimate scum, but the biologist, taking their ability to grow and reproduce into account, assigns them to the world of living things.

Smaller and simpler even than the blue-green algae are the viruses, which exist as parasites on various kinds of living organisms. They are too small to be seen through any microscope with the aid of visible light waves, but their characteristics are revealed in other ways. Photographs taken through an electron microscope, which uses a stream of electrons instead of visible light rays, show them to be tiny spheres, rods, or key-shaped particles. Chemical analysis indicates that they consist of a nucleic-acid core surrounded by a protein shell. Although viruses cannot be seen directly, the results of their infective activities are visible enough. Some viruses cause influenza and other diseases in man and animals; others cause various plant diseases such as tobacco mosaic; and still others parasitize bacteria.

They multiply rapidly. A single virus particle attacking a bacterium breeds 200 new virus particles exactly like itself in 24 minutes. Each daughter particle can then infect another bacterial cell and produce 200 more viruses 24 minutes later.[10] Occasionally a strain of virus mutates or changes spontaneously to a new form possessing different characteristics, which reproduces itself true to type thereafter.

[10] Stent, 1955.

Viruses do not exist as organized cells. This is shown by the fact that they can be reduced in the chemical laboratory to a pure chemical form without being killed. The protein and nucleic-acid fractions of tobacco mosaic virus have even been separated chemically; yet these fractions if reassembled on the leaves of tobacco plants become infective and grow and multiply again.[11] The organization of the virus particle resides in its nucleic acid and protein macromolecules.

Biologists are not agreed as to whether or not viruses are living. This is very interesting, because it suggests that the viruses may occupy a position near the borderline between living and non-living systems. Chemical compounds capable of producing harmful effects on bacteria, plants, or animals are clearly not living. If therefore we wish to stress the fact that viruses can be obtained in crystalline form without permanent loss of virulence, we can categorize them as poisonous chemicals and exclude them from the world of life. Or if we adopt the premise that all life exists in organized cells, the viruses would have to be considered non-living. On the other hand, the viruses share with living organisms the properties of growth, self-maintenance, reproduction, and mutability. By these important criteria they have to be regarded as living.

Let us examine these fundamental characteristics of life—self-maintenance, growth, reproduction, and mutability—in more detail.

A living organism maintains itself by taking in raw materials and food from the environment, converting them to new usable chemical forms, and assimilating them as a part of the cell or body. The identity and individuality of the organism are preserved through time and changing external conditions by its ability to make various substances a part of itself. Out of the random array of chemical substances in the environment, the organism selects certain ones in certain proportions which it can fit into its pattern of organization. Those substances which cannot be fitted in are either rejected or eliminated.

In short, life creates order out of disorder, and maintains organization in the midst of chaos. Life and man are, in fact, the main agents in our world which build and extend organization, and thereby work against the general trend toward increasing disorder. Now the maintenance of organization requires energy. The source of energy for the order-creating activities of living systems is food.

[11] Fraenkel-Conrat, 1956.

Many of the chemical reactions involved in life processes take place on the surfaces of colloidal droplets and particles. These small particles represent aggregations of molecules and macromolecules. Living protoplasm has a gelatinous colloidal structure.

The functions and characteristics of an organism by means of which it achieves inner stability and maintains itself in a disorderly environment can be traced ultimately to the genetic material located in the cells. This material consists chemically of long chains of nucleic acid (deoxyribose nucleic acid or DNA in virtually all organisms) associated with protein. The orderly molecular structure of the nucleic-acid molecules is the basic form of organization in living systems.

The genes are centers of biological activity arranged in a linear sequence on the DNA chains. Each gene directs a specific chemical reaction or governs the formation of a particular kind of enzyme or other protein or polysaccharide.[12] In this way the gene controls the functions and characteristics of the organism at their source. The orderliness inherent in the structure of the genetic material itself is conferred through successive biochemical steps to the protein molecules it causes to be synthesized. And these in turn translate the order into specific functions and characteristics of the organism.

The DNA of a complex higher organism like a rabbit contains thousands of genes, each of which has a different specific structure and hence a different primary biochemical action. The live animal, the mature rabbit, is then a mosaic of numerous functions and characteristics determined by different centers of activity on the nucleic acid macromolecules.

The organism may increase in size by consuming and assimilating more raw materials and food than it eliminates as waste products. It grows from within. It adds new gene centers of biochemical activity. The growth process may transcend the limits of a single individual, as when one individual gives rise to two or more individuals like itself. Reproduction is basically a special aspect of growth.

Growth and reproduction are also the result of gene action. The activity of genes is twofold. Not only does each gene direct the formation of specific proteins or other chemical substances, but it also governs the formation of copies of itself. The copying activity of genes leads to the perpetuation of the gene-controlled characteristics during the

[12] Beadle, 1955.

lifetime of one individual and in its descendants. In other words, the self-duplication of genes is the basis of both growth and reproduction.

Three of the basic properties of life—self-maintenance, growth, and reproduction—are thus results of the activity of the genes. What of the fourth, mutability? In the process of duplication, genes usually form exact copies of themselves. Rarely, however, a daughter gene is formed which differs in some way from the parental gene. The altered gene or mutation produces a somewhat different biochemical substance in the cell and hence brings about some change in the characteristics of the organism. The mutant gene now duplicates itself accurately in its altered form. The change is hereditary. Mutation is an infrequent but nonetheless a constant and regular aspect of gene duplication, and mutability is therefore one of the characteristics of life.

The genes have, on the one hand, a tendency to multiply, and on the other hand a tendency to mutate and give rise to altered copies. In terms of living organisms and their populations, these two tendencies can be expressed as follows. First, each kind of organism tends to increase in numbers. Secondly, the individuals are not all alike in their genic constitution.

Now the ability of an organism to multiply in numbers is unlimited, whereas the environmental resources from which it can obtain the necessary means of life are limited. In consequence of the increase in numbers, a competition eventually sets in between the individuals for food, raw materials, space, and other factors necessary for life. Since the competing individuals are not all alike in respect to their gene-controlled characteristics, some individuals will prove more successful than others in the so-called struggle for existence, and will leave more progeny. In this way some forms of each gene will increase in frequency, while other forms of the same gene will decrease in frequency. A change in the relative frequencies of the different forms of a gene is organic evolution. One of the inevitable properties of life in a world of limited resources is therefore the tendency to evolve.

Most of what has been said in the foregoing paragraphs can be summarized schematically (Fig. 9). We have found the basic properties of life to be *organization, self-maintenance, growth, reproduction, mutation,* and *evolution*. The ultimate carriers of these properties are order-creating, self-duplicating centers, or genes, on macromolecules composed (usually) of the nucleic acid DNA. Given a supply of raw materials and food, the genes perpetuate their specific molecular structure, control specific enzyme reactions, govern the synthesis of particular kinds of proteins and polysaccharides, direct the formation of identical copies of themselves, and produce some altered copies. The

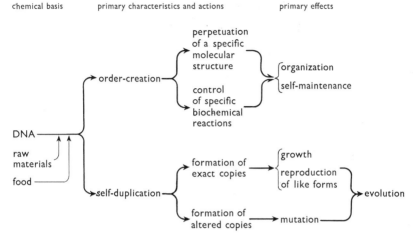

Fig. 9. A scheme relating the basic characteristics of life to the intrinsic properties and modes of action of DNA macromolecules

diagram shows that the several properties of life mentioned above are the result of different aspects of gene behavior.

SIMILARITIES BETWEEN LIVING AND
NON-LIVING SYSTEMS

These various properties are found among non-living systems. The power of self-duplication can be built into mechanical models. Robots are being designed to perform a variety of functions, including thinking, and there is no reason why a robot cannot be made which would assemble other robots like itself out of a given supply of nuts and bolts.[13] A self-duplicating robot would probably make mistakes in the copying process and hence "mutate" from time to time. Automobiles have evolved tremendously during a relatively short period. Robots and automobiles thus have some features in common with living organisms.

[13] For recent experiments along these lines see Penrose, 1959.

The essential distinction seems to be that robots and automobiles are mechanical machines, whereas life is an organic machine.

Organic chemicals as such are not living. Some of the complex carbon compounds found in living organisms can be synthesized in the laboratory. This does not elevate them above the chemical level of complexity. Nor would an aggregation of all the chemical constituents found in the body, including nucleic acids and proteins, constitute life. Life exists when these chemical substances form an organized and self-perpetuating system.

It is easy to tell the difference between a mechanical machine or chemical mixture and a living organism. The difference between a chemical reaction system and a self-perpetuating system of nucleic acids, the one non-living and the other living, is not so clear-cut. Chemical processes are known which perpetuate themselves as long as a supply of raw materials is available (autocatalysis), create an orderly arrangement of parts out of a random distribution of molecules (crystallization), store and release energy slowly in the maintenance of the system, mutate occasionally, and change with time. Why is a chemical system with one or more of these characteristics said to be non-living? Why is a system of nucleic acids and proteins with the same characteristics considered to be more than chemical?

Most, if not all, of the processes of living organisms—irritability, movement, respiration, growth, self-perpetuation, order-creation, mutation, and evolution—can be found in various non-living chemical systems.

For example, a suspension of carbon tetrachloride in a concentrated solution of certain sugars has surface-tension relationships causing it to move and behave like an amoeba. Many chemical reactions yield energy, and some such energy-yielding reactions are coupled with an energy-absorbing reaction, so that the energy is stored and released slowly as in respiration. The orderliness so characteristic of life has its counterpart in the intrinsic tendency of many organic molecules to pile up in an orderly way during crystallization as a result of their surface configurations. Growth is another aspect of crystallization.[14]

Chemical reactions can be self-perpetuating as long as the supply of raw materials lasts if one of the products catalyzes or speeds up the

[14] These examples are taken from Calvin, 1956.

formation of more products like itself. Thus cuprous ion catalyzes the reaction between hydrogen and cupric ion to give cuprous ion plus hydrogen ion, and the reaction will continue as long as cupric ion and hydrogen are present. A mixture of hydrogen and cupric ion will remain together without reacting indefinitely if cuprous ion is absent. Occasionally, however, a cupric ion changes spontaneously or "mutates" to cuprous ion; then the reaction described above goes to completion.[15]

Where alternative products are formed by a reaction, and one of these products catalyzes its own formation more efficiently than do the alternative products, the former will increase in abundance. This change corresponds to evolution on a purely chemical plane, as Calvin has noted. It may be illustrated by the change from simple iron ion to iron porphyrin. Bare iron atoms in water solution catalyze the reaction: hydrogen peroxide → water + oxygen. Iron surrounded by porphyrin, a ringlike organic molecule, catalyzes this same reaction a thousand times more effectively. Furthermore, porphyrin with iron, in a solution of hydrogen peroxide, catalyzes the formation of more porphyrin. Therefore, in a mixture consisting of hydrogen peroxide, simple iron (in abundance), and iron in porphyrin (in traces), an evolutionary change will take place from a low to a high frequency of the porphyrin, as the simple chemicals having a weakly developed catalytic ability are replaced by complex molecules with a more efficient catalytic ability.[16]

The various properties of life, not excepting self-perpetuation, mutation, and evolution, are thus found individually in chemical systems. The question posed earlier regarding the distinction between living and non-living chemical systems is therefore very difficult to answer in a definitive way. For life is related to chemical systems in much the same way that steam is related to liquid water: it is the same and yet not the same. Perhaps our conclusion must be that life is a unique type of chemical system; one in which all of the special attributes— organization, self-maintenance, growth, reproduction, mutation, and evolution—are present together.[17]

[15] Calvin, 1956.
[16] Calvin, 1956, 1959.
[17] Calvin, 1956.

THE CHEMICAL PRELUDE TO THE ORIGIN OF LIFE

As we have seen, it is not altogether clear just where the evolution of matter left off being chemical and began to be living. This is as we would expect if life developed out of and made use of pre-existing chemical-reaction systems.

In the early period of earth history, the different chemical elements reacted and combined with one another to produce molecules of various kinds. Among the chemical compounds thus formed were those destined to be used at a later date as building blocks and food sources by the first living systems. Chemical evolution paved the way and prepared the materials for the origin of life.

It is believed that the atmosphere before the appearance of life differed markedly in its composition from the atmosphere of the present time. Whereas our modern atmosphere contains much oxygen but little hydrogen, the primitive atmosphere is supposed to have contained little or no free oxygen but much hydrogen.[18] In addition to hydrogen, the earth's primitive atmosphere probably contained water, ammonia, and methane, and was thus like the atmospheres of several other planets in our solar system at the present time.[19]

Various ways are known by which these compounds could react to form large and complex organic molecules. The action of lightning bolts or ultraviolet light in the primitive atmosphere of hydrogen, water, ammonia, and methane would lead to the formation of numerous organic compounds. Miller subjected such an atmosphere to electrical discharges in an experiment and obtained acetic acid, formic acid, three other organic acids including propionic, the amino acid glycine, 24 other amino acids, and several other organic compounds.[20] (See Fig. 10 for the structural formulas of some of these compounds.)

A solution of carbon dioxide in water, exposed to high-energy radiation, will yield formic acid. Further irradiation will build formic acid up to oxalic acid and eventually to succinic acid (Fig. 10). These syntheses have been carried out experimentally in a cyclotron. Under natural conditions the radiations necessary to bring about the same

[18] Oparin, 1938.
[19] Urey, 1952.
[20] Calvin, 1956; Gaffron, 1960.

results could be emissions from radioactive materials or cosmic rays from outer space.[21]

A water solution of formic acid (or formaldehyde) and ammonia (or nitric acid or nitrate), if irradiated by ultraviolet light from the sun, gives rise to amino acids, as has also been demonstrated experimentally.[22]

Under present-day conditions complex organic compounds such as these would either be oxidized by the free oxygen in the atmosphere or

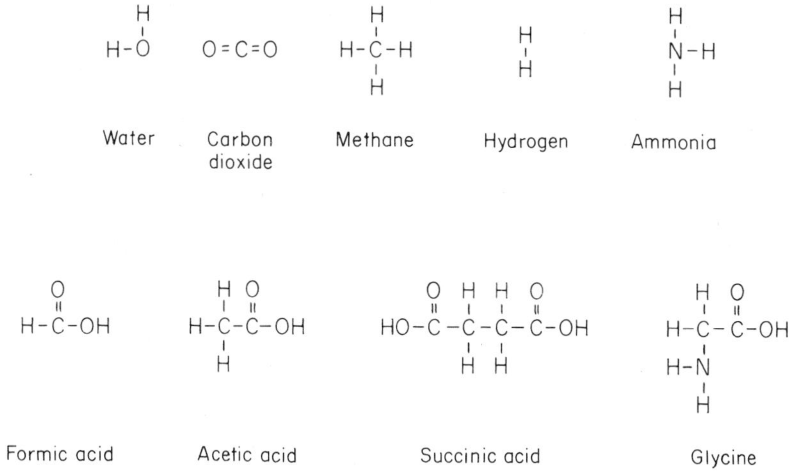

```
      H                      H                        H
      |                      |                        |
  H-O      O=C=O        H-C-H          H           N-H
                            |          H           |
                            H                      H

  Water    Carbon       Methane      Hydrogen     Ammonia
           dioxide
```

```
     O               H  O                O  H  H  O              H  O
     ||              |  ||                ||  |  |  ||            |  ||
  H-C-OH        H-C-C-OH           HO-C -C -C- C-OH        H-C - C-OH
                     |                       |  |                 |
                     H                       H  H               H-N
                                                                  |
                                                                  H

  Formic acid    Acetic acid        Succinic acid            Glycine
```

Fig. 10. Structural formulas of several inorganic and simple organic molecules which play important roles in living systems

From Calvin, *Am. Scientist*, 1956. By courtesy of Dr. Calvin.

eaten up by living organisms. Neither of these methods of chemical change existed in a lifeless earth with an atmosphere devoid of free oxygen. Under primitive conditions large carbon and hydrogen molecules would not only arise, but would endure and accumulate.[23]

In the course of millions of years the organic compounds would accumulate in the waters of the earth until these attained the consistency of broth. Such a broth would provide, as Urey pointed out, a favorable medium in which living systems capable of utilizing all this stored energy might arise. As a matter of fact, the materials synthesized in Miller's

[21] Calvin, 1956.
[22] Calvin, 1956.
[23] Oparin, 1938; Wald, 1954; Calvin, 1956; Symposium, 1959; Gaffron, 1960.

experiment make a very good growth medium for many modern micro-organisms.[24]

POSSIBLE STAGES IN THE STRUCTURAL DEVELOPMENT OF NUCLEIC ACID

As noted earlier, life is a system of nucleic acids which is able to extract energy from complex organic molecules, and raw materials from these and other sources, to maintain a certain pattern of organization and extend that organization through reproduction. The characteristic properties of life are properties of the genes. A naked gene, capable of maintaining itself by means of biochemical reactions and duplicating itself, would represent the simplest kind of living system. The origin of life would therefore be the origin of a naked gene, as suggested independently by Muller and Darlington about 1930.[25] Of course the term "naked gene" has acquired a different meaning today after spectacular advances in the field of genetical biochemistry. By naked gene we mean a macromolecule of DNA or of some simpler form of nucleic acid possessing the genetic properties necessary for self-maintenance.

The DNA macromolecule in contemporary organisms consists of long chains of alternating phosphate and sugar molecules to which bases are attached as side groups. The chains occur in pairs. The sugar-phosphate chains of a pair lie parallel to one another and are linked together by transverse base side groups. The sugar and phosphate molecules are identical throughout the DNA macromolecule, but the bases may be of four or more types (symbolized as A, C, G, and T), and the linear sequence of the different bases along the length of the macromolecule may vary. Thus in any given DNA macromolecule the pattern of the bases might be AAAGAG in one region, TAGGCA in another, CTTGA at another site, and so on (Fig. 11d).[26]

The variety and diverse specializations of life may derive from the manifold sequences of the bases. A site on the DNA chain with the base order AAAGAG may guide one biochemical reaction and lead to the formation of one type of protein or other substance which enhances the survival ability of the organism under certain environmental conditions;

[24] Gaffron, 1960.
[25] Muller, 1929; Darlington, 1932, 450–51.
[26] Watson and Crick, 1953; Crick, 1954.

a second site with the base order TAGGCA will determine some other specific reaction advantageous to the perpetuation of the organism; and the third site with the pattern CTTGA will perform still another function.

A multiplicity of specialized responses and reactions is required in any complex organism which has to cope with conditions in a heterogeneous environment. This multiplicity of reactions and fine adjustments to environmental conditions is a universal feature of all known organisms, from viruses to rabbits, all of which can be said to be complex in some degree. Complex organisms endowed with specialized reaction systems may, however, have been preceded in earlier times by very simple forms with simple and undiversified reactions.

The chemical steps that led to the formation of DNA as we now find it are not yet fully understood. We can speculate that sugar-phosphate chains possessing a single type of base side group might have developed in the course of chemical evolution. We can speculate further that such chains might have been able to maintain themselves in an appropriate chemical solution. Such a macromolecule, being uniform along its length, could grow by the addition of new terminal groups, and reproduce by simple fragmentation (Fig. 11a). A nucleic acid macromolecule consisting of single types of base, sugar, and phosphate groups linked together in a regular repetitive pattern in chains of variable length would perhaps be the simplest kind of naked gene.

More specialized biochemical reactions could be carried out by a primitive nucleic acid macromolecule containing not one but two types of bases. If the more complex macromolecules containing two types of bases were able to guide the biochemical reactions leading to their own survival and reproduction in more effective ways than the simple nucleic acid chains, as might easily be the case, they would tend to supplant the simpler forms. And if the two-base macromolecule could function with a wide variety of arrangements of its two bases, so that the linear order of the bases could be unspecific and indefinite, the primitive gene could continue to grow by elongation and reproduce by fission (Fig. 11b).

A primitive two-base species of nucleic acid macromolecules would have been relatively successful in self-maintenance and fecundity as long as it did not have to face the competition of more efficient and better

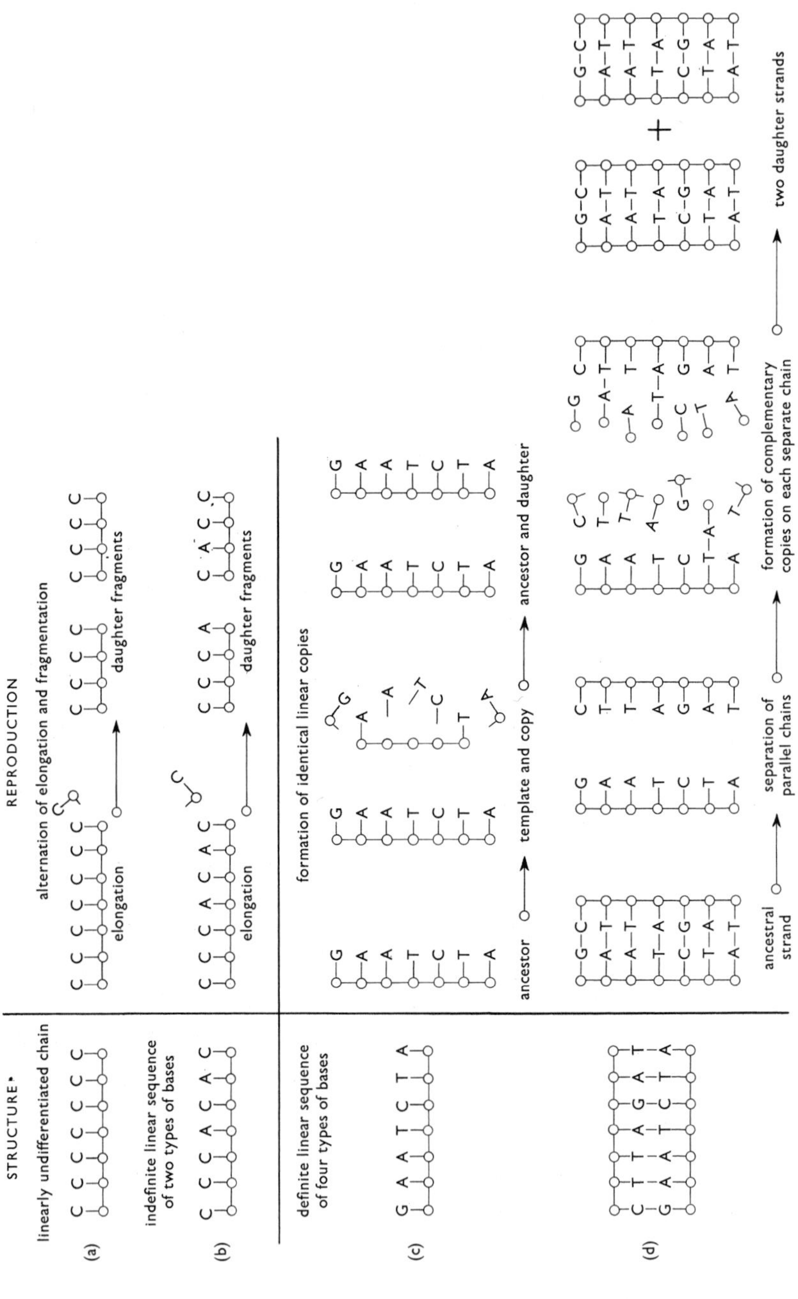

STRUCTURE·

REPRODUCTION

(a) linearly undifferentiated chain

alternation of elongation and fragmentation

elongation → daughter fragments

(b) indefinite linear sequence of two types of bases

elongation → daughter fragments

(c) definite linear sequence of four types of bases

formation of identical linear copies

ancestor → template and copy → ancestor and daughter

(d)

ancestral strand → separation of parallel chains → formation of complementary copies on each separate chain → two daughter strands

specialized forms of nucleic acid. Superior species of nucleic acid were, however, possible and did obviously arise.

The DNA macromolecules that we find in the world today have a diversity of specific reaction sites spaced along the length of the chain. They consist of strings of different genes. The order and arrangement of the base side groups is believed to be specific in each genic site on the string. A string of genes cannot grow by elongation or reproduce by fragmentation. It must build up new macromolecules from the parental ones by a copying process which takes place along the whole length of the chain (Fig. 11c and d).

A single sugar-phosphate chain bearing four types of bases arranged in a definite sequence, and hence differentiated linearly in a specific way, cannot multiply by an alternation of elongation and fission. Instead the ancestral macromolecule must act somehow as a template to guide the synthesis of an identical linear copy (Fig. 11c). The stage exemplified diagrammatically by Fig. 11c is not entirely hypothetical. It has recently been found that the DNA of two small bacterial viruses, S13 and ϕX174, is probably in the form of simple chains of sugars and phosphates with attached bases.[27]

The problem of exact replication of linearly differentiated nucleic acid macromolecules is met in most modern organisms by a more sophisticated method based on a more complex structure. The DNA consists of parallel chains with a complementary linear order of bases. Replication involves the separation of the paired chains and the formation of complementary copies on each separate strand. Where one pair of

[27] Sinsheimer, Tessman; see Beadle, 1960, 50.

Fig. 11. Hypothetical stages in the evolution of DNA

The various structural stages have different methods of reproduction as shown.

(a) Nucleic acid consisting of a main chain of sugar and phosphate molecules (represented by circles and bars, respectively) with bases of a single type (C as shown here) attached as short side chains.

(b) Nucleic acid with two types of bases (C and A here), the linear order of which is not definite.

(c) Nucleic acid with a definite linear sequence of four types of bases (A, C, G, T).

(d) Nucleic acid consisting of a double chain of sugar and phosphate groups linked together by the base side chains. The four types of bases have a definite linear order, and can be paired in certain combinations only (A—T, C—G).

parallel chains with a specific linear order of bases occurred before, two such paired chains with the same base order now exist (Fig. 11d).

The paired structure of the DNA macromolecules found in contemporary organisms, and the ability of these macromolecules to duplicate themselves by a linear copying process, may be looked upon as adaptations to secure the most efficient reproduction, not of the sugar-phosphate backbone of DNA, or of single types of base side groups, but of the particular sequences of several different kinds of bases existing at different sites along the linear axis of the chain.

The historical steps by which DNA as we now know it evolved from simpler molecular forms several eons ago may never be retraced today. It may, however, be possible some day to synthesize simple types of DNA in the laboratory and observe their subsequent evolutionary changes under laboratory conditions. In this way man could gain a better understanding of the steps that are actually possible, and, by extension, of the approximate sequence of events that might have occurred during the threshold period between chemical evolution and organic evolution in the buried past.[28] Whatever the details regarding the stepwise evolution of nucleic acid turn out to be, it seems improbable that the complex form of DNA found in modern life could have arisen in one step. It is more likely that the build-up of molecular complexity was a gradual process, and that simpler molecular species, such as sugar-phosphate esters and linearly undifferentiated nucleic acid chains, preceded the formation of the type of DNA macromolecule found in modern living organisms. In short, simple crude naked genes probably came before strings of differentiated genes.

Among all the countless molecular aggregations and combinations that occurred during the long period of chemical evolution, a simple naked gene might have arisen. We do not know whether this event occurred once or repeatedly, but once was enough. Owing to the gene's power of self-duplication, the processes of life and organic evolution, once started, could continue indefinitely.

[28] Since the above was written, Kornberg and his co-workers have found that a solution of raw nucleotides without any DNA to begin with can spontaneously form a simple type of DNA under favorable conditions in the laboratory. The spontaneously formed DNA is then capable of self-duplication, as is shown by its production of more DNA like itself when introduced as a primer into a fresh mixture of appropriate raw materials. These workers also find that single-stranded DNA is more efficient than double-stranded DNA as a primer in such solutions. See Beadle, 1962.

In nearly all contemporary ogranisms the genes are borne in micro-scopically visible structures, chromosomes, inside the cells of the body. The genes are surrounded by colloidal protoplasm in varying quantities. The body, the cell, the colloidal protoplasm and the chromosome may be viewed as superstructure erected during the course of organic evolution to ensure the better survival of the genes. Among contempo-rary organisms this superstructure is reduced to a minimum in the viruses, which lack true cells and consist of a string of genes enclosed within a protein sheath. That the genetic material of viruses has a threadlike form is indicated by genetic evidence in bacteriophage[29] and by morpho-logical structure in tobacco mosaic virus.[30] The virus particle may not be a naked gene, but it is a scantily clad gene string.

POSSIBLE TIME AND PLACE OF THE ORIGIN OF LIFE

The first forms of life must have lived in the organic broth from which they gained their nourishment and hence in some water body on the earth. It is usually assumed that this water body was the sea. This is not a necessary assumption. The site of the origin of life might equally well have been a sun-warmed pond on the land.[31] In fact, the organic molecules would have had a better chance of becoming concentrated to the point where they could aggregate and form living macromolecules and colloidal droplets in ponds and pools than in the infinite waters of the sea.[32]

Furthermore, there are theoretical reasons for expecting that more efficient types of naked genes would evolve more rapidly in ponds and pools than in the oceans. Assume that relatively inefficient genes came into being independently in the ocean and in a pond or lake. The great oceans would remain unsaturated with the primitive living particles for a very long period of time. An unsaturated environment is tolerant of inefficient and ill-adapted organisms and permits such forms to survive. The smaller permanent water bodies, by contrast, would reach a point of overcrowding and competition for a limited food supply in a relatively

[29] Benzer, 1955.
[30] Fraenkel-Conrat, 1956.
[31] Wherry, 1936; Sagan, 1957; Zirkle, 1960.
[32] Sagan, 1957.

short time. These conditions would stimulate evolutionary changes leading to the emergence of new types of organisms with more efficient means of self-maintenance and reproduction. The improved types of primitive organisms originating in small water bodies could then go on to colonize the sea. In the course of time they would replace the pre-existing inefficient forms of life in the sea.

Wherry has pointed out that the type of metallic element present in plant protoplasm corresponds more closely to the chemical features of a lacustrine environment than to those of a marine environment. Of the two widely available alkali metals, sodium and potassium, the former is abundant in the ocean, while the latter is held in vast excess in the clays underlying fresh water lakes and ponds. If plant life had originated in the sea, he argues, it would probably have made use of the principal metallic element present there, namely, sodium. The alkali metal which enters almost universally into the composition of plant protoplasm is not sodium, however, but potassium. The universal presence of potassium in protoplasm can be accounted for readily enough on the hypothesis that the protoplasm of plants originated in ponds or lakes, but is less easily explained by the alternative hypothesis that plant life arose in the ocean.[33]

When did the origin of life take place? We can no more assign a date for this event than we can specify the site. Various lines of evidence indicate that life existed on the earth at least 2 eons ago. The evidence consists partly of fossils in very ancient rocks and partly of theoretical considerations relating to the composition of the atmosphere.

The primitive atmosphere contained much hydrogen, as we have already seen, whereas the present atmosphere with much (21 percent) free oxygen is a product of the photosynthetic or food-manufacturing activities of green plants. The age of the oxidizing atmosphere thus gives us an estimate of the age of photosynthetic organisms. Different methods of estimating the age of an oxidizing atmosphere from the chemical composition of minerals in the earth's crust point to an age of 2 eons.[34]

The oldest known fossils are simple algae found in limestone in Rhodesia, and blue-green algae, fungi, and flagellates in the Canadian

[33] Wherry, 1936.
[34] Briggs, 1959.

shield. The Rhodesian rocks are estimated to be 2.6 to 2.7 eons old and the Canadian rocks between 1.3 and 2.0 eons old by radioactive methods of dating. Carbon of a chemical type normally associated with living organisms has been found in rocks from the Canadian shield 2.5 eons old.[35]

Fungi, flagellates, and photosynthetic algae stand low on today's scale of life, but are much more complex and highly organized than virus particles and naked genes. The first appearance of such organisms 2 eons or more ago implies a previous period of organic evolution of unknown duration. Although we do not know when life originated, we can assume that this event occurred a long time before the algae, fungi, and flagellates had evolved. It is for this reason that we have taken a date about 3 eons ago (around the Year 1.5 Billion) as the starting point of life and organic evolution. Blum suggests that the origin of life occurred between 2.7 and 3.8 eons ago and probably occurred closer to the latter than to the former figure.[36]

[35] Tyler and Barghoorn, 1954; Briggs, 1959.
[36] Blum, 1955, 157.

FOUR

The Course of Evolution

The long pageant of evolution extending over one billion years appears to have been brought about by fundamental causes which are still in operation and which can be experimented with today. TH. DOBZHANSKY[1]

NUTRITIONAL STAGES IN THE EARLY PERIOD OF EVOLUTION

ALL FORMS OF LIFE derive the energy necessary for the maintenance of their organization and for their vital processes from carbon compounds. Green plants, both the lower algae and the higher land plants, are able to manufacture their own supply of carbon compounds out of carbon dioxide and water by means of the energy in sunlight. This process of food manufacture is photosynthesis. Those organisms such as animals, fungi, viruses, and most bacteria that live on the energy-rich carbon compounds which they find in their environment are said to be heterotrophic, and heterotrophs dependent on dead organic matter are called saprophytes.

In the modern world the energy-rich carbon compounds that sustain life are contributed almost exclusively by photosynthetic green plants. In the early period of organic evolution, before photosynthetic plants existed, the organic compounds were furnished by non-living chemical syntheses. The primordial forms of life, whether they were naked genes or nearly naked gene strings, had to live on a supply of organic chemicals accumulated during preceding eras of chemical evolution. They were necessarily saprophytes.

It will be recalled that free oxygen was virtually or entirely absent from the earth's atmosphere before and at the time life originated. The only way in which the primordial saprophytic organisms could derive

---

[1] Dobzhansky, 1951a, ix.

energy from organic molecules under these anaerobic or oxygenless conditions is by fermentation. Fermentation follows the course: sugar → alcohol (or some other by-product) + carbon dioxide + energy.[2]

In using up the existing supply of organic compounds by fermentation, the primitive heterotrophic organisms were living on borrowed time, as man is doing today in the combustion of oil reserves. If the organic broth had been used up before photosynthesis or some similar method of food manufacture had been developed, life would have died out.[3]

The ability to synthesize complex organic substances was probably developed gradually by a series of mutational steps in the evolution of primitive organisms.[4] The earliest organisms depended on complex compounds already in existence. As these became used up, some organisms may have mutated to new forms capable of carrying out the final steps in the synthesis of complex organic compounds from pre-existing simpler organic compounds. With the gradual depletion of the reserves of organic broth, it became necessary for organisms to evolve metabolic processes of their own rendering them more and more independent of previously formed organic compounds. The photosynthetic organism, able to start with raw inorganic compounds and build up from them sugars and other complex organic molecules, represented an advanced stage in the evolution of synthesizing ability.

The process of photosynthesis may be represented by the reaction: carbon dioxide + water + sunlight → sugar + oxygen. The source of the carbon in organic compounds built up by photosynthesis is thus carbon dioxide.[5] This gas, however, was either absent from the primitive waters and atmosphere or present in only minute traces.[6] Photosynthesis would have been impossible during the earliest phase of organic evolution owing to the absence of one of the necessary raw materials, carbon dioxide.

But carbon dioxide is one of the waste products of fermentation.

[2] Oparin, 1938, ch. 8; Wald, 1954.
[3] Oparin, 1938; Wald, 1954.
[4] Horowitz, 1945.
[5] For a recent review of photosynthesis, summarizing the chemical steps and energy relations involved, see Arnon, 1960.
[6] Urey, 1952; Gaffron, 1960.

The primitive saprophytes in their fermentation of the organic soup released this gas into the water and air in ever-increasing quantities. The rise in abundance of carbon dioxide made photosynthesis possible. The exhaustion or near exhaustion of the reserves of organic compounds in the waters placed a premium on the development of photosynthetic methods of manufacturing organic compounds. By a series of steps as yet unknown, living organisms evolved which possessed the fairly complex apparatus necessary to carry out photosynthesis.[7] That apparatus consists of a cell containing specific pigments (especially chlorophyll) and specific enzyme systems. The simplest modern organisms capable of photosynthesis are the blue-green algae and certain bacteria.

The evolutionary development of photosynthesis had other important consequences in the early history of life. The primordial saprophyte living in a world devoid of free oxygen had to gain its energy from carbon compounds by fermentation. Now fermentation is a form of incomplete oxidation which leaves most of the energy potential of the carbon compounds unexploited. Where free oxygen is available the same materials can be completely oxidized in the process of respiration.

Respiration, consisting of the reaction: sugar + oxygen → carbon dioxide + water + energy, is a far more efficient source of energy than fermentation. One hundred and eighty grams of sugar will yield about 700,000 calories by respiration, but only about 20,000 calories in fermentation. Therefore a fermenting organism would have to go through about 6,300 grams of sugar to obtain the same amount of energy that a respiring organism could get from 180 grams of sugar.[8]

Under the anaerobic atmospheric conditions that prevailed in the early period of earth history, respiration was not possible. Photosynthesis, however, by giving off oxygen as a by-product, created an atmosphere which made the development of respiration possible.[9]

It is difficult to overestimate the degree to which the invention of cellular respiration released the forces of living organisms. No organism that relies wholly upon fermentation has ever amounted to much. Even after the advent of photosynthesis, organisms could have led only a marginal existence. They could indeed produce

[7] Some possible steps are discussed by Calvin, 1959.
[8] Oparin, 1938; Wald, 1954.
[9] Oparin, 1938; Wald, 1954.

their own organic materials, but only in quantities sufficient to survive. Fermentation is so profligate a way of life that photosynthesis could do little more than keep up with it. Respiration used the material of organisms with such enormously greater efficiency as for the first time to leave something over. Coupled with fermentation, photosynthesis made organisms self-sustaining; coupled with respiration, it provided a surplus. To use an economic analogy, photosynthesis brought organisms to the subsistence level; respiration provided them with capital. It is mainly this capital that they invested in the great enterprise of organic evolution.[10]

In the early period of organic evolution before free oxygen had accumulated in quantity, life was restricted to an aqueous environment. The emergence of life from the water onto the land was made possible by photosynthesis and the availability of oxygen in two ways.

In the first place, fermentation breaks sugar down into carbon dioxide plus various organic compounds such as alcohol, lactic acid, formic acid, or acetic acid. These partially decomposed compounds are poisonous to living organisms. In an aqueous environment they can be washed away. An organism living outside the water, in a terrestrial or aerial environment, will not be able to get rid of the poisonous waste products of fermentation so easily. Respiration, however, breaks sugar down all the way to carbon dioxide plus water. The end products of respiration are not poisonous and in any case land-living organisms can easily get rid of them.[11]

The entry of oxygen into the atmosphere also liberated organisms in another sense. The sun's radiation contains ultraviolet components which no living cell can tolerate. We are sometimes told that if this radiation were to reach the earth's surface, life must cease. That is not quite true. Water absorbs ultraviolet radiation very effectively, and one must conclude that as long as these rays penetrated in quantity to the surface of the earth, life had to remain under water. With the appearance of oxygen, however, a layer of ozone formed high in the atmosphere and absorbed this radiation. Now organisms could for the first time emerge from the water and begin to populate the earth and air. Oxygen provided not only the means of obtaining adequate energy for evolution but the protective blanket of ozone which alone made possible terrestrial life.[12]

STAGES OF STRUCTURAL COMPLEXITY

The ever-increasing efficiency of living systems, as they successively developed photosynthesis, respiration, and land life, called for and was

[10] Wald, in *Scientific American*, August, 1954.
[11] Wald, 1954.
[12] Wald, in *Scientific American*, August, 1954.

accompanied by increases in the complexity of the structural organization. A bacterium with relatively few functions to perform can lead
a successful life with the simplest kind of cellular structure. A photosynthetic alga carries out a greater variety of functions for which it
requires a more complex cellular apparatus corresponding to a more
efficient division of labor. A rabbit must surmount numerous difficulties and carry out numerous and often conflicting functions in order

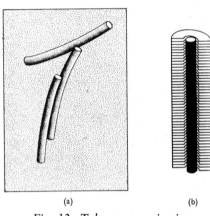

(a)   (b)

Fig. 12. Tobacco mosaic virus

(a) A virus particle as revealed by the electron microscope.
(b)–(c) Its structure consists of a strand of nucleic acid surrounded by a helix of
    protein. The length of the particle is about 3,000 A or about 1/100,000 of an
    inch.

From Fraenkel-Conrat, *Scientific American*, 1956.

to survive on the land; it has, however, in its complex body composed of
numerous differentiated cells and organs the necessary equipment for
living and surviving in a difficult terrestrial environment.

We may recognize several significant steps in the evolution of
structural complexity in living organisms.

The naked gene. This hypothetical unit is supposed to have consisted
of a short segment of nucleic acid, which lived in and fed saprophyti-
cally on the primordial organic broth.

The nucleoprotein particle. As represented by the viruses, the nucleoprotein particle consists of a string of separate genes surrounded by a
protein mantle (Fig. 12). The modern viruses may not be the living
descendants of a primordial form of life. The very fact that they live

3000 A

(c)

as parasites on such advanced organisms as mammals and flowering plants proves that they are themselves the products of relatively recent evolutionary changes. Nevertheless they possess a simple structural organization comparable to that which primordial saprophytic organisms now extinct may have had during the early stages of organic evolution. The saprophytic nucleoprotein particle is in the last analysis

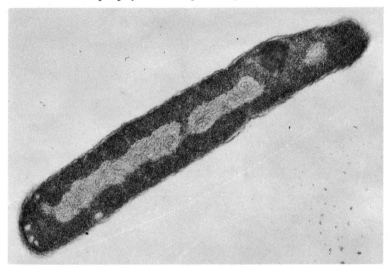

Fig. 13. Colon bacterium, Escherichia coli

The DNA is contained in a central (light-colored) region, which is not separated by a membrane from the rest of the cell.

From Kellenberger, *J. Biophys. Biochem. Cytology,* 1958. Photomicrograph by courtesy of Dr. Kellenberger.

a hypothetical unit, like the naked gene, but one which can reasonably be inferred to have existed on the analogy of known virus particles.

The simple cell. This unit, consisting of DNA strings diffused through cytoplasm, and bounded as a whole by a membrane, is exemplified in the bacteria and blue-green algae. Such cells possess only a rudimentary cellular organization, and reproduce by simple fission (Figs. 13 and 14).[13]

[13] Recent findings on the internal structure of simple cells are reviewed by Dillon, 1962.

The nuclear cell. Here we have a much more elaborate organization. The DNA is grouped into organized chromosomes which possess differentiated parts (the centromere, surrounding membrane, and the like). The chromosomes in turn are grouped into a special region, the nucleus, which is separated by a nuclear membrane from the cytoplasm. The cytoplasm contains various organelles specialized for the performance of different functions: mitochondria for respiration, chloroplasts

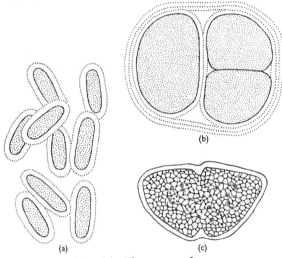

Fig. 14. Blue-green algae

(a) A single-celled form (*Gloeothece linearis*).
(b) A colonial form in which several cells are enclosed within a gelatinous sheath (*Chroococcus turgidus*).
(c) Structure of a cell of *Chroococcus turgidus*. There is a reticulum of protoplasm with genetic material distributed through the central portion.

From G. M. Smith, *Cryptogamic Botany* (New York, McGraw-Hill Book Company, Inc., 1955).

for photosynthesis, vacuoles for the storage of raw and waste materials, eye-spots for the reception of light stimuli, flagellae for locomotion, and so on. The entire structure is bounded by a membrane or cell wall. The differentiation of parts within the cell, and especially the separation of nucleus and cytoplasm by a membrane, makes possible the division of labor which enables the cell to carry out a variety of physiological processes (Figs. 15 and 16).[14]

[14] Lwoff; see Stebbins, 1960.

In reproduction the chromosomes in one nucleus divide precisely and separate to two daughter nuclei (mitosis). In addition, the nuclei of different individuals occasionally combine to form one collective nucleus (sexual fusion). The resulting nucleus possesses two sets of chromosomes (and is said to be diploid). It returns to the condition

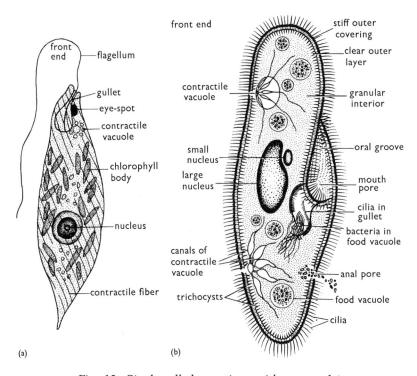

Fig. 15. Single-celled organisms with true nuclei

(a) The green flagellate, Euglena.   (b) Paramecium.

From R. Buchsbaum, *Animals without Backbones* (Chicago, University of Chicago Press, 1948).

of one set (the haploid condition) at some time in the life cycle by an orderly segregation of each member of a pair of chromosomes to a different nucleus (meiosis).

The life cycle of cellular organisms with sexual reproduction thus consists of sexual fusions, which establish a diploid chromosomal condition, and meiotic divisions, which reduce the number of chromosomes to the haploid condition. In some single-celled organisms like the plant flagellates and unicellular green algae, meiosis occurs immediately after sexual fusion, so that the organism during the greater part of its life cycle is haploid. In the ciliated protozoans, on the other

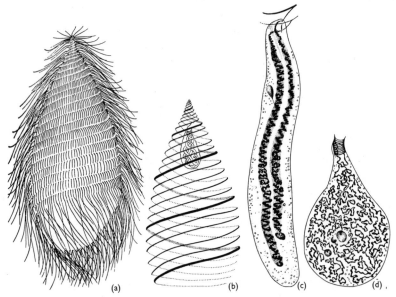

Fig. 16. The single-celled nuclear organism, Holomastigotoides tusitala, *and its nuclei and chromosomes*

(a) Surface view of organism showing the covering of hairlike flagellae.
(b) Anterior end of same with flagellae removed so as to reveal the internal nucleus.
(c) Nucleus enlarged showing two chromosomes. Each chromosome consists of a coiled strand of darkly stained genetic material surrounded by a matrix and attached to a fiber above by a centromere.
(d) Nucleus at a different stage when the chromosomal strands are uncoiled.

Cleveland, 1949

hand, meiosis is postponed until sex cells are formed; the latter, consequently, are haploid, but the organism itself is diploid.

The simple multicellular body. The aggregation of cells into a body paves the way for a differentiation between cells, some specializing in food manufacture, others in reproduction, still others in the attachment of the body to its substrate, and so on. The multicellular body can thus perform more complex functions than any single cell alone. The simplest multicellular bodies are colonies and filaments composed of a few or scores of cells, as exemplified by various small algae. Relatively simple multicellular bodies consisting of hundreds or thousands of cells with a greater differentiation of parts are illustrated by the liverworts and sponges (Fig. 17).

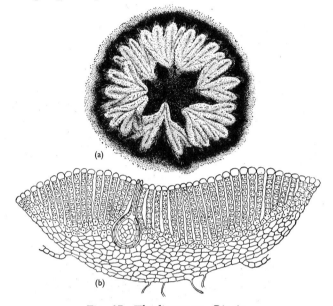

Fig. 17. The liverwort, Riccia

(a) Plant as seen from above.
(b) Longitudinal section through the plant showing the cellular structure and differentiation of tissues. The uppermost cell layer forms a protective surface. Below this is a green photosynthetic tissue containing air spaces which facilitate the exchange of gases in photosynthesis. The lower portion of the plant body is colorless and serves for food storage. Filamentous rhizoids project from the lower surface into the soil, anchoring the plant to its substrate and absorbing water. Embedded in the upper plant body is a female reproductive organ.

From G. M. Smith, *Cryptogamic Botany* (New York, McGraw-Hill Book Company, Inc., 1955).

The complex multicellular body. In the gradual evolution of structural complexity we pass at some undefined point from the relatively simple to the relatively complex body. The latter consists of millions of cells grouped into well-differentiated tissues and organs. Thus in a fish or crab well-marked differences are found between the cells and organs composing the muscular system, the skeletal system, the nervous system, the digestive system, the reproductive system, and so on (Fig. 18).

Complex multicellular organisms, as exemplified by molluscs, crabs and other arthropods, and fish, evolved in the oceans, rivers, and lakes. The invasion of the land called for an even greater degree of structural

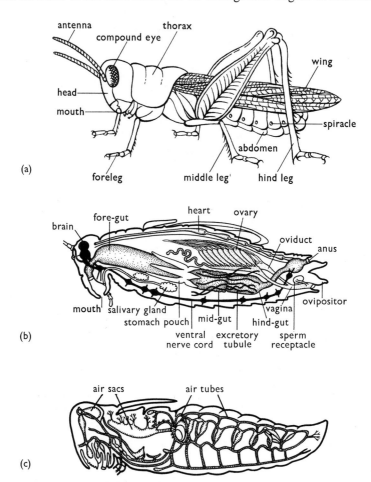

Fig. 18. Anatomy of a grasshopper

(a) Main external organs.   (c) Respiratory system.
(b) Main internal organs.

From R. Buchsbaum, *Animals without Backbones* (Chicago, University of Chicago Press, 1948), with modifications.

complexity. Water is the universal solvent of living systems. In the primitive aqueous environment, water is constantly available and can be continuously passed through the organism. Life on land must have a tough skin to protect the organism from excessive loss of water, and must in addition possess specialized structures for renewing its supply of water. Water is a buoyant medium which holds aquatic plants and animals up, but a plant or animal living on the land must have a strong skeleton to hold itself up. Water has a moderating influence on the climate; not so the land, which heats up by day and during summer, and cools off at night and in winter, to a much greater extent than water bodies. Life on land must therefore develop means of maintaining a constant body temperature in spite of weather fluctuations, as in mammals and birds, or the ability to become dormant during periods of extreme cold or heat, as in many plants and insects.

The terrestrial and aerial environments presented these and other challenges to life, which could only be met by the evolutionary development of new specialized structures able to perform the many new functions required of a land-living organism. The higher vertebrate animals, the insects and other higher arthropods, and the vascular plants met the challenge successfully. The long evolutionary process by which they achieved the ability to live successfully on land involved the perfection of many new and special adaptations, which are reflected in a very complex body structure.

The society. The multicellular organisms normally exist in reproductive groups, or breeding populations, consisting of several or many individuals. A division of labor often develops between the individuals of a population. All the higher animals and some of the higher plants have two kinds of individuals, males and females, specialized for performing different roles in reproduction, the males for fertilizing the eggs and contributing foreign genes to the new individuals, and the females for nourishing the eggs and the young for varying lengths of time.

The differentiation between the individuals of a reproductive community is carried to an extreme degree in insect societies. Among honeybees, for example, the female functions are subdivided between two kinds of females; the queens receive the sperms from the males and lay the eggs, and the workers nourish and tend the developing young. Some ant societies (like bee societies) consist of males, queens,

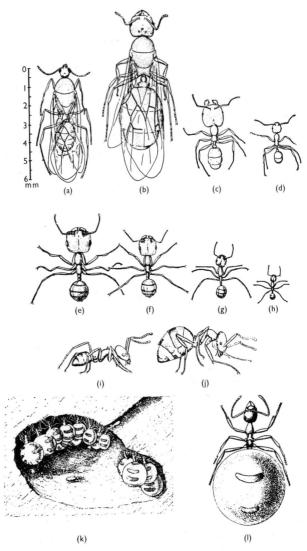

Fig. 19. *Different specialized types of individuals, or castes, in ants*

(a)–(d) Italian ant, *Pheidole pallidula;* male (a), queen (b), soldier (c), and worker (d).

(e)–(h) Harvester ant, *Messor barbarus;* large soldier (e), medium-sized soldier (f), smaller form (g), and worker (h).

(i)–(j) Chilean desert ant, *Tapinoma antarcticum;* worker (i) and water carrier (j).

(k)–(l) Desert ant, *Myrmecocystus;* hanging honeypots in a special chamber of the ant nest (k), and honeypot filled with sweet juices (l).

Not to same scale.

From W. Goetsch, *Die Staaten der Ameisen* (Göttingen, Springer-Verlag, 1953), rearranged.

and workers, and have in addition one or more kinds of soldiers, and even individuals with greatly distended abdomens for the storage of water and honey (Fig. 19).

The ability of a population of social bees or ants to gather food, protect itself from its enemies, survive in deserts and other extreme habitats, and reproduce itself is undoubtedly enhanced by its supra-individual division of labor. Here, as in previous levels of evolution, complexity of organization is the price of success in an environment beset with numerous and diverse challenges.

HISTORY OF LIFE

We may next attempt to relate the various stages in the evolution of life to the geological time scale. The fossil record, our only direct indication of the life of the past, becomes increasingly fragmentary as we go back in history. The Cambrian period, which began about 600 million years ago,[15] is the earliest period for which abundant fossils are available. Enough fossils are known from the vast pre-Cambrian era to make it certain that simple forms of life were in existence then, but the pre-Cambrian fossil record is exceedingly scanty and poorly preserved. Consequently we must proceed by inferences from indirect evidence and be content with uncertain approximations in our attempts to reconstruct the earliest chapters in the history of life.

The existence of fossil algae and of an oxidizing atmosphere 2 or more eons ago makes it necessary to assume that primitive saprophytic particles existed even earlier. We have assumed on this basis that life in its most elementary forms was present about 3 eons ago, or in the Year 1.5 Billion.

Figure 20a shows the 3-eon span of life on the 4.5-eon scale of earth history. The era of life is enlarged and divided into stages in the next diagram (Fig. 20b). We have only a few fixed points on this scale prior to the Cambrian, namely, the archaic fossil algae, fungi, and flagellates found in rocks about 2 eons (1.3 to 2.7 eons) old.[16] From the Cambrian on, the history of life is increasingly well documented by fossils. The reasons for this are not far to seek. Until living organisms had reached a stage of complexity which included the formation of hard shells and

[15] Kulp, 1961.
[16] Tyler and Barghoorn, 1954; Briggs, 1959.

skeletons they were unlikely to be preserved as fossils. Obviously a great deal of organic evolution had to occur before the emergence of complex multicellular bodies with hard parts. Figure 20b shows indeed that the stage of complex multicellular plants and animals from the Cambrian to the present comprises only the last one-sixth of the estimated time span of organic evolution.

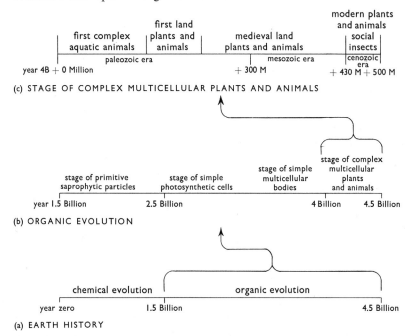

(c) STAGE OF COMPLEX MULTICELLULAR PLANTS AND ANIMALS

(b) ORGANIC EVOLUTION

(a) EARTH HISTORY

Fig. 20. The main stages of organic evolution on the geological time scale

The stage of complex multicellular organisms is shown on an enlarged scale in the next diagram (Fig. 20c), which encompasses geological time from the Cambrian period to the present. (The geological periods are listed for reference in Table 1.) The first complex animals in Cambrian and Ordovician times were aquatic molluscs, arthropods, and fish. The first land life consisted of primitive vascular plants in the Silurian period. Animals began to invade the land later when plants had become established and could furnish a food source in the new terrestrial habitat. The first animals to invade the land were arthropods

Table 1. The geological periods and their ages in millions of years[a]

Era	Period	Epoch	Time since beginning of period	Duration of period and era
Cenozoic	Quaternary	Recent		.011
		Pleistocene	1	1.0
	Tertiary	Pliocene	13	12
		Miocene	25	12
		Oligocene	36	11
		Eocene	58	22
		Paleocene	63	5
				63
Mesozoic	Cretaceous		135	72
	Jurassic		181	46
	Triassic		230	49
				167
Paleozoic	Permian		280	50
	Carboniferous		345	65
	Devonian		405	60
	Silurian		425	20
	Ordovician		500	75
	Cambrian		600?	100
				370

[a] Kulp, 1961.

represented by scorpions in late Silurian time and spiders in the Devonian. These were followed by the amphibians in late Devonian time and by land snails in the Carboniferous.[17]

The adaptations for life on dry land were not perfected all at once. In fact a great part of the evolutionary changes in the vascular plants, terrestrial arthropods, and land vertebrates during the next 200 million years were concerned with completing the emancipation of these organisms from the ancestral waters.

Plants and animals with a modern aspect, the flowering plants, mammals, birds, moths, butterflies, bees, ants, etc., did not appear in abundance until the Cretaceous and Tertiary periods. Modern plants belong essentially to the last 135 million years of earth history, and

[17] Simpson, Pittendrigh, and Tiffany, 1957, ch. 31.

modern types of animals to the last 75 million years. Man appeared
on the scene about 1 million years ago. These general sequences are
shown graphically in Fig. 21.

We can recognize perhaps two aspects in the modernization of life.
A great many of the modern families and genera of plants and animals
were in existence by the Cretaceous period or the early Tertiary period.
Most of the species belonging to those families and genera date back
to the latter part of the Tertiary period or to the Pleistocene. The horse

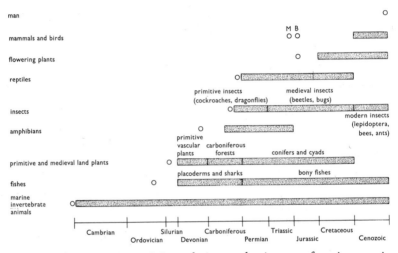

*Fig. 21. The successive origin and rise to dominance of various major
groups of animals and plants*

The lengths of the geological periods are shown in proportion to their duration in
absolute time. The first fossil record of a group is indicated by a circle; the period of
expansion of each group is shown by a wavy line. A group may have existed before
the time of our earliest fossil record, however, and usually persisted as a subordinate
element in the fauna or flora after the period of maximum expansion. Ages of the
various groups from Simpson (1949), Simpson, Pittendrigh, and Tiffany (1957),
Ross (1956), and other works.

family, for example, is about 58 million years old; the average age for
eight contemporaneous and extinct genera in this family is 7.5 million
years; while the modern species of horses are less than 1 million years
old.[18] Many, if not most, species of mammals now living can be traced
back in the fossil record to the early Pleistocene and are thus 100,000

[18] Simpson, 1953a, 32, 261.

or more years old; however, very few living mammalian species are as much as 1 million years old, and some are younger than the average, a case in point being various species of land mammals in Great Britain which are 80,000 years old.[19] The ages of many modern species of flowering plants and insects, though not so well documented paleontologically, are probably comparable.

One hundred thousand years represents 1/630 of the Cenozoic era; 1/6,000 of the stage of complex multicellular organisms since Cambrian time; and 1/30,000 of the whole history of life. This period of time is about the width of a pencil line on the scale in Fig. 20c and could not be shown at all on the scale in Fig. 20b. Most, though not all, of the species of plants and animals in the world today are the productions of the last moment of time in the long process of organic evolution.

THE DIVERSITY OF LIFE

The diversity of organisms generated in the course of evolution is very great. There are 8,600 species of birds and 3,200 species of mammals. Some 20,000 species of fishes have been described. The number of described species of insects is about 850,000. There are 80,000 described species of molluscs, 4,500 of sponges, and over 18,000 of wormlike animals. The lower vascular plants run to 10,000 known species, the algae to 8,700, and the fungi to 40,400.[20]

The number of described species of flowering plants can be estimated as 286,000 from the following considerations. In 1941 Jones estimated on the basis of a careful survey that 195,000 species of flowering plants had been described. He also pointed out that an average of 4,800 new species are being described each year in this group.[21] We have multiplied this average annual increment of 4,800 by 19 years to bring the estimate up to date for 1960. The resulting estimate of 286,000 agrees fairly well with that obtained by two other methods, and is in line with the figures given by several other authors.

The foregoing and other estimates are summarized in Table 2. As shown there, the total number of species of modern organisms now known to science is close to 1.5 million. This figure agrees well with

[19] Simpson, 1953a, 37.
[20] References given in Table 2.
[21] Jones, 1941.

Table. 2. Estimated numbers of described species of organisms[a]

Animals		
Chordates		39,500[b]
Fishes	20,000	
Reptiles and amphibians	6,000	
Birds	8,600	
Mammals	3,200	
Echinoderms (starfish and sea urchins)		4,000[b]
Arthropods		923,000[b]
Crustaceans	25,000	
Insects	850,000	
Segmented worms		7,000[b]
Molluscs (snails, clams, octopuses)		80,000[b]
Nematodes		10,000[b]
Flat-worms		6,000[b]
Coelenterates (jellyfish and corals)		9,000[b]
Sponges		4,500[b]
Miscellaneous smaller groups		7,300[b]
Total animals		1,090,300
Plants		
Flowering plants		286,000[c]
Gymnosperms (conifers and cycads)		640[d]
Ferns and fern allies		10,000[d]
Bryophytes (mosses and liverworts)		23,000[d]
Algae		8,700[e]
Green	5,275	
Red	2,500	
Brown	900	
Total plants		328,300
Fungi		
True fungi		40,000
Slime molds		400
Total fungi		40,400[f]
Protists		
Protozoans, diatoms, plant flagellates		30,000[g]
Monerans		
Blue-green algae		1,400
Bacteria		1,630
Viruses		200
Total monerans		3,200[h]
Grand total		1,492,200

[a] Subgroups not all listed; numbers in the subtotals and totals rounded off to nearest hundred; sources given below.

[b] Mayr, Linsley, and Usinger, 1953, p. 4. [c] See text.
[d] Jones, 1941. [e] Smith, 1938, vol. 1.
[f] Ainsworth and Bisby, 1954, pp. 138, 235.
[g] Mayr *et al.*, 1953; Ross, 1956. [h] Smith, 1938; Ainsworth and Bisby, 1954.

an earlier estimate of 1.5 million known species, made by Dobzhansky on the basis of somewhat different figures for the various groups.[22]

The task of describing the species of animals, plants, and micro-organisms is far from completed as of today. Some groups like the birds and mammals are now well known taxonomically, and few if any species remain to be described. In other groups, such as the flowering plants, marine animals, and insects, many species remain to be discovered and described scientifically. Whereas the number of described species of fishes is 20,000, the estimated total number of fishes is closer to 40,000.[23] The amount of unfinished business in insect taxonomy is best indicated by the following quotation.

Approximately three-fourths of a million species of insects have so far been described and named, and their number is being gradually increased from year to year. So far as those competent to judge are able to estimate, it seems probable that this number represents perhaps one-fifth or one-tenth of those which actually exist upon our planet at the present time.[24]

How many species, both described and undescribed, exist in the world today? Obviously we cannot say definitely until the many undescribed species are described and a new census is taken. In the meantime we can only make a rough "guesstimate" of the minimal numbers of species on the basis of some arbitrary assumptions. We will assume that the numbers of species in the well-studied land verte-brates will not be increased significantly by future research. For the enormous group of insects, on the other hand, we will take the most conservative assumption possible, namely, that the total number is $5 \times 750,000$. We will accept the estimate that there are 40,000 species of fishes. Such groups as the flowering plants, fungi, microbes, marine invertebrates, and spiders, in which new species are being described constantly, may increase in the future to 120 percent of their present numbers. This latter is actually a very conservative assumption; in many poorly known groups of plants and animals the total number of species is probably several times greater than the number of described species. The insects are by no means exceptional in regard to the ratio

[22] Dobzhansky, 1951a, 8.

[23] Spector, 1956, 533.

[24] Brues, Melander, and Carpenter, 1954, 6, quoted from the *Bulletin of the Museum of Comparative Zoology*, Harvard University, 1954.

between total diversity and known diversity. But even with the foregoing conservative assumptions, the estimated total number of species of modern organisms turns out to be at least 4.53 million.

Evolution has been going on for about 3 eons and in that time there has been a continual turnover of species. Only a handful of the species now living can trace their ancestry back unchanged into the remote geological past. Many modern species in such rapidly evolving groups as the mammals, insects, and flowering plants are about 100,000 years old, as noted above. Many other major groups are more conservative, and most periods in geological history have been more stable than the present. Simpson suggests an average duration of 500,000 to 5 million years for all species throughout all time.[25]

By making use of a few arbitrary but reasonable assumptions, we can obtain a very rough guesstimate of the total number of species, living and extinct, that have existed on earth since the Cambrian. We shall assume first that the total number of species in the modern world is 4.5 million; secondly, that the number of species present at the beginning of the Cambrian period was 25,000; and, thirdly, that the number of species has increased exponentially through history until it has reached its present all-time high. The rate of exponential increase is assumed to be represented by a curve similar to that given by Simpson for many major groups.[26] The length of time from the beginning of the Cambrian to the present is taken to be 600 million years. Let us assume finally that the average longevity of a species during this time has been 5 million years. Then the estimated total number of species since Cambrian time would be about 1.6 billion.

If we take the average longevity of species as 500,000 years, instead of 5 million as assumed above, the turnover in species during geological time would be 10 times greater, and the total number of species would be 16 billion. Since the assumption that there are 4.5 million recent species is minimal, the estimates of 1.6 to 16 billion species are if anything conservative.

If the number of species existing at any one time during the long pre-Cambrian era was only a few thousand, and if the longevity of the pre-Cambrian species was similar to that of post-Cambrian life, the

[25] Simpson, 1952.
[26] Simpson, 1949, chs. 3–6; Simpson, Pittendrigh, and Tiffany, 1957, chs. 30–32.

total number of pre-Cambrian species would range from 1 million to tens of millions. These figures, added to the post-Cambrian numbers expressed in billions, would not change the latter materially. In other words, in an array of more than a billion species, the gain or loss of a few million would scarcely be noticed.

Simpson has approached the problem in a similar but slightly different way.[27] He has assumed that the number of recent species is around 2 million. He has then attempted to estimate the average number of contemporaneous species at any one time level in geological history, which is a figure somewhere between the original 1 species and the postulated present 2 million. The total time span of organic evolution is then divided by the average duration of species, which is assumed to be 500,000 to 5 million years, and multiplied by the average number of species per time level. By this method Simpson has arrived at a guess, as he calls it, of between 50 million and 4 billion species during the whole history of life since pre-Cambrian time. For various reasons the figure 50 million can be considered too low. In view of the numerous uncertainties involved, the other figure of 4 billion suggested by Simpson may be regarded as being in reasonably good agreement with the figures of 1.6 or 16 billion arrived at above by a partially independent method.

It should not be imagined that the separate species are uniform within themselves. Many modern species of organisms consist of many millions or billions of individuals, no two of which are identical. The fact that this individual variation is circumscribed within limits, which makes possible the classification of individuals into species, should not cause us to lose sight of the great amount of organic diversity below the species level.

THE TAXONOMIC TREATMENT OF ORGANIC DIVERSITY

The diversity of life is inconceivable. A good taxonomist, after spending a lifetime in the study of a special group or the fauna or flora of a particular geographical region, may be able to name 2 or 3 thousand species at sight. A taxonomist with an exceptional memory and long experience may be able to keep the names and distinguishing characteristics of 4 or 5 thousand species straight in his mind. A single

[27] Simpson, 1952.

taxonomist, even the best, can deal effectively with only a small fraction of the world of life.

The monumental task of describing the species of animals, plants, and microbes has occupied a small army of taxonomists since the time of John Ray and Carl Linnaeus in the seventeenth and eighteenth centuries. Their task is, as we have seen, far from completed today.

The descriptive phase of taxonomic work has been accompanied by the search for similarities between organisms as a basis for classifying them into circles of relationship. Similar and presumably related individuals are first placed together in one species; related species are then grouped in the same genus; related genera are placed in the same family; families are grouped into orders; and so on up the hierarchy through classes and phyla to kingdoms. The system of classification is the taxonomist's way of reducing the bewildering diversity of species to a limited and manageable number of categories. It is his way of treating individuals and species as particular examples of more general rules. The system of classification as it stands at any moment is the embodiment of the theoretical generalizations arrived at by generations of taxonomists up to that time.

In the process of reducing particular items to their lowest common denominator, the individuality of the particular items is necessarily lost. This is the inherent disadvantage of all generalizations in biology. The disadvantage is a serious one in certain types of studies. Studies of the composition of breeding populations, for example, may be hampered by a typological or group concept, which hides from view the individual variation. Typological approaches to population studies are quite properly rejected in such cases. Some population students have gone on to reject type concepts in general, overlooking the fact that the reduction of diversity to its lowest common denominator is as desirable and necessary for some purposes as it is undesirable for others.

The formulation of generalizations that reveal the pattern of nature in its main outlines has been one of the goals of science for 350 years, and unless it can be shown that a special exception should be made in the case of taxonomy, that goal is as legitimate for the taxonomist as for the worker in other branches of science. The chief form of theoretical generalization in taxonomy is the concept and description of taxonomic groups. Group concepts should be adopted along with a full awareness of their limitations. The taxonomist should see the diversity of organisms with one eye and the group similarities with the other; he must see both the trees and the forest.

A broad generalization, like a map drawn to a small scale, may be useful for some purposes but quite useless in others. For example, if one wants to see the irregularities on the ground in their fine detail, one might not be able to use a map at all. But the amount of ground that can be covered in this fashion is limited. If one is curious about the main course of a river from its headwaters to its outlet in the sea, it will be necessary to draw a general plan from which many details are omitted. If the object is to show the shape and position of the continents and oceans, the map of the larger area will have to be drawn on a still smaller scale and most of the rivers will have to be left out entirely. This loss of detail is inevitable, however, in the outlining of the major features of the earth's face.

Similarly, in the mapping of organic diversity, we may want to see the raw terrain in one case, while in another case we may want to survey a somewhat broader sector with a map drawn to a correspondingly small scale. Or we may wish to see the broad pattern of groupings and relationships in the world of life. If we are to generalize successfully at this broadest level, we must be prepared to dispense with the fine details altogether.

THE KINGDOMS OF ORGANISMS

It has been found that the multiplicity of organisms can be reduced to a few kingdoms: (1) *the animal kingdom,* consisting of the sponges, coelenterates (jellyfish and corals), several phyla of worms, molluscs (snails, clams, etc.), arthropods (crabs, spiders, insects), echinoderms (starfish), and chordates (tunicates, fish, land vertebrates); (2) *the plant kingdom,* which includes several main groups of algae, the bryophytes (mosses and liverworts), ferns, several other groups of lower vascular plants, the cycads, conifers, and flowering plants; (3) *the fungus kingdom,* containing the slimemolds and several groups of true fungi; (4) *the protistan kingdom,* containing the nucleated single-celled organisms, namely, the diatoms, green flagellates (including euglenas), dinoflagellates, animal-flagellates, rhizopods (amoebas and radiolarians), and ciliates; (5) *the moneran kingdom*, composed of organisms without true nuclei, that is, without a chromosome apparatus separated by a nuclear membrane from the cytoplasm of the cell, as in the blue-green algae, bacteria, spirochaetes, and viruses.

The foregoing classification follows (with certain modifications to be mentioned later) a system proposed recently by Whittaker.[28] It represents an advance over the traditional classification of all organisms into just two kingdoms, plants and animals. Yet it is still unnatural in several respects. The red algae and brown algae are placed here in the plant kingdom, the sponges in the animal kingdom, and the slime-molds in the fungus kingdom, for convenience until the true relationships of these and other lower groups become better understood. Whittaker assigned the monerans and the nuclear protistans to separate subkingdoms of the same kingdom, whereas the two groups are treated as separate kingdoms here. The importance of the distinction between nuclear cells and non-nuclear cells or particles is recognized in either classification, and the difference between the four-kingdom system of Whittaker and the present modification of it is therefore one of emphasis. As will be shown below, the plants and animals are more closely alike in their basic genetical architecture than the protistans and monerans.

Logically, one might postulate a hypothetical sixth kingdom of simple precellular organisms to include the hypothetical naked genes and the hypothetical saprophytic nucleoprotein particles.

In the course of evolution, simple saprophytic particles gradually grew in complexity until they reached a cellular state of organization. Similarly, the simple non-nuclear cellular forms eventually crossed the boundary line separating the moneran kingdom from the protistan kingdom, thus attaining a more highly organized type of single-celled life. The plant and animal kingdoms have diverged from some common ancestor among the protista.

The plants are sedentary multicellular organisms which produce food in the process of photosynthesis. The animals are motile multicellular organisms which consume by ingestion the food produced by plants. The vast difference between a sedimentary green plant and a motile food-devouring animal needs no emphasis.

[28] Whittaker, 1959.

The discovery of a very similar cellular structure and a similar genetic machinery in the two kingdoms is, in the light of the many evident differences, all the more remarkable. The basic organization of the genetic material into chromosomes and of the chromosomes into nuclei, the precise and complicated sequence of events by which those chromosomes and nuclei divide in mitosis and meiosis, and the machinery of sexual reproduction, are strikingly similar in plants and animals, in angiosperms and insects, in lilies and in grasshoppers. These facts could not have been predicted; they had to be discovered repeatedly before their significance could even be realized. It is as though animals and plants are built on the same basic cellular plan and have diverged with respect to different modes of nutrition.

The same features of cellular structure and genetical machinery, the same chromosome apparatus and chromosome cycles, and the same sexual processes are found among the single-celled protista. Among the protista, however, the green flagellates such as Euglena seem to stand at the parting of the ways of the plant and animal kingdoms. They have the motility of an animal combined with the chlorophyll-bearing plastids and food-manufacturing power of a plant. Their protozoan relatives are motile non-photosynthetic food-consumers and are claimed by zoologists as members of the animal kingdom. Their closest algal relatives are photosynthetic green food-producers and are claimed by botanists for the plant kingdom. The dividing line between plants and animals is indeed hazy and perhaps undefinable where the two kingdoms converge upon the living representatives of their common protistan ancestor.

It is a curious fact, the significance of which is still not entirely clear, that diploid life cycles developed in the single-celled stage in the lines of evolution leading to the animal kingdom, but developed much more slowly in the lines leading to the plant kingdom. The animal-like protista, such as the ciliated protozoa, have diploid life cycles, as do all the multicellular animals. The plant flagellates, on the other hand, have haploid life cycles, and so do many of their algal relatives.[29] Among the simple multicellular land plants, the liverworts and mosses, the predominant phase in the life cycle is still haploid, and that phase is still present though subordinate in the ferns.

[29] Stebbins, 1950, 157; 1960, 207.

The fungi are believed by many students to constitute a third kingdom derived independently of the plants and animals from colorless motile protistan ancestors. Nutritionally they differ from both plants and animals in that they neither manufacture food nor eat it, but live by decomposing organic matter and absorbing some of the decomposition products. Correlated with their special mode of nutrition is a distinctive structural organization which is fundamentally different from that of animals or plants.[30]

The protista, then, constitute a plexus from which the multicellular organisms have been derived. Below the protista, when we turn to the monerans, we enter a strange underworld of organisms without the familiar landmarks of biology. The monerans lack nuclei and true chromosomes; consequently, they lack mitosis; they lack true sexual reproduction and meiosis; in the case of the viruses, they even lack cells. What is left is the same basic genetical material as in higher organisms, usually DNA, with associated protein carried in cells or subcellular particles.

The genetic material of the bacteria and viruses is passed on in reproduction according to rules very different from those followed by the protista and their higher derivatives.[31] Herein lies a fundamental difference between the monerans and protistans. Yet, chemically, it is the same kind of genetic material with the same kinds of gene action. These characteristics point to the kinship of the monerans with the other kingdoms of organisms. The gap between the monerans and protistans is well-marked, yet it is one which could have been crossed by evolving monerans of the past.

PHYLOGENETIC VERSUS CAUSAL APPROACHES
TO THE STUDY OF EVOLUTION

Organic evolution may be studied from two different points of view: the historical and the causal. Some biologists are concerned primarily with the sequence of evolutionary steps within a particular group of organisms. They want to reconstruct the evolutionary history of the group as it gradually unfolded in the course of geological time. The

[30] See Whittaker, 1959, for a statement of the case with numerous literature references. In separating the fungi from the true plants here I have also been influenced by discussions with my colleague, Dr. Richard K. Benjamin.

[31] Stebbins, 1960.

group whose genealogy is under investigation may be one or a few related species; it may encompass a larger number of forms such as the modern and ancient horses and their primitive and unhorselike ancestors; or one may wish to describe on a much broader scale the evolutionary development of the whole animal or plant kingdom. The evolutionary history of a group of organisms, whether small or large in scope, is known as its phylogeny.

The phylogenies of many plant and animal groups are now known with reasonable accuracy as a result of paleontological and/or comparative morphological studies. The fossil record, and in particular the sequence of fossil remains in a succession of geological deposits, provides direct evidence of the phylogeny of a group. The comparison of the morphological and embryological features of the different living members of the group may yield indirect evidence of value, particularly where some forms preserving primitive characteristics are still in existence as so-called living fossils.

The other main approach in evolutionary biology considers organic evolution as a process. The object of inquiry in this case is to discover the mechanisms which bring about evolutionary changes. Our purpose in the present book is to seek causal explanations for the process of organic evolution.

We have seen that the great majority of living species represent the latest productions of the long-continued evolution process. Insofar as we base our ideas about evolutionary mechanisms on the study of modern and recent species, such as fruit flies, mice, corn, and cotton, we have to assume that the causal forces revealed today were also in operation in previous stages of organic evolution. This assumption is justified by everything we have learned about the regularity of natural causal mechanisms in the world. But we can also assume that the same evolutionary forces may have taken a somewhat different course when operating in primitive saprophytic particles than the one followed in vinegar flies or corn plants. The nature of the organism itself is one of the factors determining the type of evolutionary pattern.

This latter assumption is also justified by the facts. To be sure, we cannot resurrect long-extinct saprophytic particles for experimental purposes, but we can study the genetics and evolution of viruses, bacteria, and protista. A great deal of research has been carried out in recent years on the genetics of the simplest living things. The results obtained to date have complemented in an important way the results obtained from the study of higher plants and animals. Moreover, new discoveries about the genetic mechanisms in microorganisms are being made rapidly today, and will inevitably affect our future concepts of evolution.

Part 2 *Organism and Environment*

FIVE

Adaptation

NEARLY EVERY PART of the earth's surface is inhabited by life. Living organisms are found in the surface of the sea and in the great ocean depths, on the land from the tropics to the polar latitudes, and thousands of feet high in the atmosphere. Bare rocks are colonized by lichens, glaciers by red algae, and hot springs by blue-green algae. No one kind of organism, however, lives in all these varied environments.

Each kind of organism normally lives in a particular zone of the earth's surface characterized by a certain range of environmental conditions. We expect to find sponges on the ocean bottom, earthworms in moist humus-containing soil, cacti in warm deserts, ducks in fresh-water ponds and the air above, and colon bacteria in the human intestine. The natural abode of an organism in which it can sustain life is its habitat. Every kind of organism occurs in a particular habitat; removed from its normal habitat it is like a fish out of water.

The reason why each kind of organism is restricted in nature to its own habitat and is not normally found elsewhere is that it is specialized or adapted for making a living under one particular set of conditions. Adaptation, or the hereditary adjustment to the environment, is one of the universal features of life. Earthworms are well fitted to live on the organic materials in soil. No other kind of organism can compete with them in their own territory. But, conversely, earthworms are unable to live in the habitats of ducks or sponges and if they are accidentally dispersed to such places they soon die.

Each kind of animal, plant, or microorganism is a complex of adaptations for performing its life functions in its normal habitat. "Can a more striking instance of adaptation be given," asked Darwin in 1859, "than that of a woodpecker for climbing trees and for seizing insects in the

93

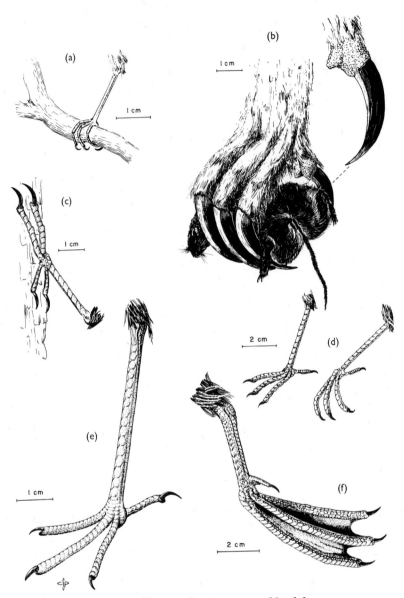

Fig. 22. Different adaptive types of bird feet

(a) The perching foot of a warbler. Audubon warbler (*Dendroica auduboni*).
(b) The grasping foot of an owl. Horned owl (*Bubo virginianus*).
(c) Climbing foot of a woodpecker. Acorn woodpecker (*Melanerpes formicivorus*).
(d) Walking and scratching foot of a quail. California quail (*Lophortyx californicus*).
(e) Wading foot of a heron. Green heron (*Butorides virescens*).
(f) Swimming foot of a duck. Pintail duck (*Anas acuta*).

94

chinks of the bark?"[1] The webbed feet of a duck set toward the rear of the body represent an adaptation for swimming; the strong sharp talons of an owl are an adaptation for clutching prey; the opposable front and rear toes of a warbler are an adaptation for perching on branches; the foot of the quail with its forward pointing toes and sturdy toenails adapts it for walking and scratching on the ground; the woodpecker's feet with their long flexible sharp-pointed toes, two pointed forward and two turned backward, are well adapted for clinging to the upright trunk of a tree (Fig. 22).

The leaves of plants contain sugar, starch, protein, and water, and hence form a desirable food for many animals. Green leaves are also essential for the life of the plant. Various characteristics which help to protect these vital organs from herbivorous animals are found in different plants. The leaves are surrounded by spines in the locust (*Robinia pseudoacacia*) and many acacias, and by sharp pointed twigs in hawthorn, plum, and ceanothus. The leaves have barbed or saw-toothed margins in agave, thistles, sedges, and some grasses. Stinging hairs are developed in the leaves of nettles (*Urtica*) and poisons in the leaves of belladonna (*Atropa*), jimson-weed (*Datura*), and poison hemlock (*Conium*).[2]

CONCEALING COLORATION IN ANIMALS

The colors of animals are often adaptive. Many animals possess concealing coloration; they resemble their background and are consequently less apt to be noticed by their natural enemies. A horned lizard is concealed by its gray, brown, or reddish color against a background of sand and rock. The speckled gray form of the British moth, *Biston betularia*, is almost invisible when perched on lichen-covered bark, and the dark form of the same species is well concealed on a soot-covered tree trunk (Fig. 23).[3]

The general situation is well summarized by that great naturalist and evolutionist, A. R. Wallace.[4]

The fact that first strikes us in our examination of the colours of animals as a whole, is the close relation that exists between these colours and the general environment.

[1] Darwin, 1859, ch. 6.
[2] Kerner, 1894–95, I, 430 ff.
[3] Kettlewell, 1956.
[4] Wallace, 1889, 190–93. A similar statement is given by Cott, 1940, 5–6.

Thus, white prevails among arctic animals; yellow or brown in desert species; while green is only a common colour in tropical evergreen forests.

In the arctic regions there are a number of animals which are wholly white all the year round, or which only turn white in winter. Among the former are the polar bear and the American polar hare, the snowy owl and the Greenland falcon; among the latter the arctic fox, the arctic hare, the ermine, and the ptarmigan.

Fig. 23. Two forms of the peppered moth (Biston betularia), *the speckled gray and the melanic forms, perched on a lichen-covered trunk, and on a soot-covered trunk*

In the left hand picture the speckled gray moth is barely visible below and slightly to the right of the melanic moth.

From Kettlewell, *Heredity*, 1956; photograph by courtesy of Dr. Kettlewell.

In the desert regions of the earth we find an even more general accordance of colour with surroundings. The lion, the camel, and all the desert antelopes have more or less the colour of the sand or rock among which they live. The Egyptian cat and the Pampas cat are sandy or earth coloured. The Australian kangaroos are of similar tints, and the original colour of the wild horse is supposed to have been sandy or clay coloured. Birds are equally well protected by assimilative hues; the larks, quails, goatsuckers, and grouse which abound in the North African and Asiatic deserts are all tinted or mottled so as closely to resemble the average colour of the soil in the districts they inhabit.

Passing on to the tropical regions, it is among their evergreen forests alone that we find whole groups of birds whose ground colour is green. Parrots are very generally green, and in the East we have an extensive group of green fruit-eating pigeons; while the barbets, bee-eaters, turacos, leaf-thrushes (Phyllornis), white-eyes (Zosterops), and many other groups, have so much green in their plumage as to tend greatly to their concealment among the dense foliage.

There can be no doubt that these colours have been acquired as a protection, when we see that in all the temperate regions, where the leaves are deciduous, the ground colour of the great majority of birds, especially on the upper surface, is a rusty brown of various shades, well corresponding with the bark, withered leaves, ferns, and bare thickets among which they live in autumn and winter, and especially in early spring when so many of them build their nests.

An additional illustration of general assimilation of colour to the surroundings of animals, is furnished by the inhabitants of the deep oceans. Professor Mosely of the Challenger Expedition . . . says: "Most characteristic of pelagic animals is the almost crystalline transparency of their bodies. So perfect is this transparency that very many of them are rendered almost entirely invisible when floating in the water, while some, even when caught and held up in a glass globe, are hardly to be seen."

Such marine organisms, however, as are of larger size, and either occasionally or habitually float on the surface, are beautifully tinged with blue above, thus harmonizing with the colour of the sea as seen by hovering birds; while they are white below, and are thus invisible against the wave-foam and clouds as seen by enemies beneath the surface.

Among the birds inhabiting the American tropical forests, concealing coloration is best developed in the species which live on or near the ground and which are consequently most vulnerable to such predators as snakes, skunks, and coatis. The wrens, ovenbirds, antbirds, wood rail, wood quail, and tinamou are brown flecked with lighter spots and blend into the pattern of light and shadow in the forest. These birds also have the habit of remaining motionless when an enemy comes near—an adaptive behavioral trait. The roof of the forest where less danger exists from predators is inhabited by many gorgeously colored birds, such as toucans, tanagers, manakins, flycatchers, woodpeckers, trogons, and motmots, which have neither dull protective colors nor quiet habits.[5]

Zonal differences in the degree and kind of concealing coloration are also found among ocean fishes. The species inhabiting the surface waters of the sea, for example, tuna, marlin, and flying fish, are characteristically blue or blue-green above and white or silvery below. Predators looking down from above are likely to miss these forms in the

[5] Wallace, 1889, 277–78; Chapman, 1929, ch. 13.

general sea of blue, while predators looking up from below see only a grayish or silvery sea. Deeper in the sea, below 200 meters where no light penetrates from above, the fishes are predominantly black or red in color.[6]

Some animals do not merely efface themselves generally in their background, but resemble objects of no interest to their enemies or their prey.

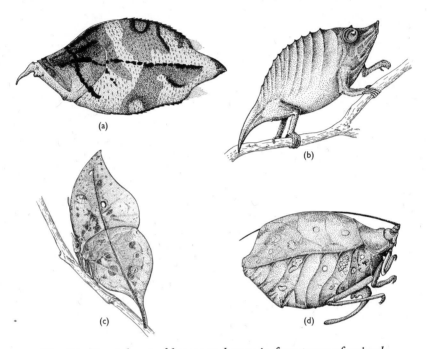

Fig. 24. Special resemblances to leaves in four types of animals

(a) The fish, *Monocirrhus polyacanthus*.   (c) Butterfly, *Kallima paralecta*.
(b) Lizard, *Rhampholeon boulengeri*.   (d) Grasshopper, *Cycloptera excellens*.

Parts (a), (b), and (d) from H. B. Cott, *Adaptive Coloration in Animals* (London, Methuen and Company, Ltd., 1940).

For example, a Brazilian fish, *Monocirrhus polyacanthus*, which lives in quiet streams and pools in the forests of the lower Amazon valley, closely resembles a dead leaf in both its mottled coloration and its flat leaflike form (Fig. 24a). In its native waters this fish lies on or near the

[6] Wallace, 1889, 193; Savage, 1963.

bottom where it is lost to view among the water-logged foliage, and even when taken up in a net it is difficult to distinguish from the dead leaves among which it lives. Not only human observers, but also the natural prey and enemies of *Monocirrhus polyacanthus*, are deceived by its leaf-like attributes.[7]

An arboreal lizard of the Belgian Congo, *Rhampholeon boulengeri*, has the curved form, irregular outline, and veinlike markings of a leaf attached to a stem (Fig. 24b).[8] Again, among amphibians, an Amazonian toad, *Bufo typhonius*, has the contour and green color of the leaves among which it lives.[9] A famous example of leaf resemblance among animals is the Indian butterfly, Kallima (Fig. 24c). Among moths, the caterpillars of various hawk moths and the adults of some others, like Miniodes and Timandra, have a leaf like appearance.[10] The grasshopper, *Cycloptera excellens*, exemplifies a leaf insect belonging to still another order of animals (Fig. 24d).

Various other insects exhibit special resemblances to bark, lichens, pebbles, twigs, flowers, or even bird droppings.[11] A great variety of such disguises is developed in the single insect order Orthoptera including the mantids and grasshoppers.[12]

The effectiveness of concealing coloration has been tested and verified in numerous experiments. Such experiments have been carried out with birds preying on caterpillars, moths, grasshoppers, mantids, mice, and fish; with lizards preying on caterpillars; and with crows preying on chicks.[13]

The results obtained in an experiment with the mantid, *Mantis religiosa*, in Italy are typical. This insect exists in two forms, a green form which inhabits green grass and a brown form which lives in dried grass. When the forms were transferred to a contrastingly colored background, as green insects in brown grass or vice versa, a much higher mortality resulted from predation by birds than when the insects were kept in their normal harmonious surroundings.[14] Similarly, the

[7] Cott, 1940, 311–13.
[8] Cott, 1940, 134.
[9] Cott, 1940, 316.
[10] Cott, 1940, 317–20.
[11] Cott, 1940, 336, 392 *passim*.
[12] Cott, 1940, 336–37.
[13] See Cott, 1940, 174 ff., for summary.
[14] Cesnola; see Cott, 1940, 179–81.

peppered moth, *Biston betularia*, exists in two forms in Great Britain, which differ strikingly in their visibility on different backgrounds. On a lichen-covered tree trunk the normal speckled gray form of the moth is well concealed, but the melanic form is conspicuous; contrarily, on a soot-covered trunk the melanic form is inconspicuous, but the gray form stands out (Fig. 23). These moths are attacked and eaten by various birds, and it has been shown that the birds find and catch the conspicuous moths more often than the cryptic ones.[15] An experiment demonstrating the effectiveness of concealing coloration of fish in relation to fish-eating birds is described in Chapter 9.

WARNING COLORATION

Many animals are protected from their enemies by possessing poisonous or distasteful properties. Bad-smelling or unpalatable bodies are found in many groups of bugs, beetles, and butterflies; in the families Pentatomidae and Reduviidae of the order Heteroptera; in the families Carabidae, Coccinellidae, Lycidae, and Lampyridae of the Coleoptera; and in the adult monarch and Heliconia butterflies as well as in the caterpillars of various other Lepidoptera.[16] Among mammals and birds, the flesh of shrews and drongos has a nauseous taste and is refused by some flesh-eating animals.[17] Irritating or repellent secretions are given off by skunks, coccinellid beetles, lycid beetles, pentatomid bugs, and the stingless tropical bees (Meliponinae).[18] The skin of the salamander, *Salamandra*, is poisonous.[19] Bees, wasps, sting rays and other fishes, and coral snakes and other venomous snakes can inflict poison stings on their enemies.

Now it is well known that many animals, far from being concealingly colored, possess bright colors arranged in a conspicuous color pattern. The red, yellow, and black bands of the coral snake, the yellow and black of bees and wasps, and the bold combination of black and white in the skunk form common examples of conspicuousness. A conspicuous appearance is frequently associated, moreover, with poisonous or distasteful attributes, as in the examples just mentioned.

[15] Kettlewell, 1956.
[16] Cott, 1940, 229, 256.
[17] Cott, 1940, 256 ff., 264.
[18] Cott, 1940, 255.
[19] Cott, 1940, 253.

Most amphibia are green, gray, or brown, and blend in with their background. But a few are boldly colored with patterns of yellow and black, or red and black, or red and blue, etc. These conspicuous species are also distasteful or poisonous in every case that has been studied. The bright-colored South American frog *Dendrobates tinctorius*, which occurs in black and red, black and yellow, and black and white forms, secretes a poison in its skin, which is used by the Colombian Indians as an arrow poison. The black and yellow South American toad Atelopus is also poisonous and unacceptable as food to predatory animals. The European salamander, *Salamandra maculosa*, has a yellow and black poisonous skin.[20]

In a comparison of the relative acceptability to insectivorous birds of 5,000 individual insects belonging to 200 species, Jones found that conspicuous markings of red, orange, or yellow occurred generally in the least acceptable insects, but were absent from the resting postures of the more palatable species. All—100 percent—of the cryptic insects were highly acceptable to the birds, while 81 percent of the conspicuous insects were relatively unacceptable.[21]

Carpenter fed a varied assortment of insects totaling 244 species to a Cercopithecus monkey in Africa. The monkey's reactions permitted the different species to be classified as edible or distasteful. Carpenter himself classified the same insects as conspicuous or cryptic. When the two classifications were combined, the following correlations became apparent:

	Edible	Distasteful
Cryptic	83	18
Conspicuous	23	120

Thus here again showiness is correlated with distasteful qualities.[22]

To explain this correlation between a showy color pattern and noxious flesh or poisonous stings in many different animals, A. R. Wallace, Fr. Müller, Poulton, and other naturalists of the late nineteenth century suggested that the showy colors serve as an advertisement of the dangerous or noxious characteristics of the animal. In short, the conspicuous

[20] Cott, 1940, 265–68.
[21] F. M. Jones in 1932; see Cott, 1940, 195, 272.
[22] Carpenter in 1921; see Cott, 1940, 271.

color patterns exhibited by some animals constitute a warning to their enemies and potential attackers.

As Cott states:

The fact is that over and over again, in group after group of animals belonging to widely separate families and orders, the same hues—black, red, orange, and yellow—are employed as the outward and visible sign of dangerous or distasteful properties. Indeed, in such animals, protected as they are from enemies by a variety of disagreeable attributes, the type of colour-scheme is in general the one factor common to their appearance, which in other respects varies so widely—in form, size, structure, and texture. Thus it is that the use of particular colour-schemes, such as the highly effective black-and-yellow combination, cuts right across the natural classification of the animals wearing them.[23]

Examples of unrelated black and yellow animals with poisonous or noxious characteristics are the salamanders (*Salamandra maculosa*), sea snakes (*Pelamydrus platurus*), wasps (Vespa), bees (Bombus, Nomada), various caterpillars, various beetles (Coccinellidae and others), and various bugs.[24]

The theory of warning coloration presupposes that the potential enemies of the warningly colored animals can and do learn to recognize them by forming a mental association of their unpleasant characteristics with their color patterns, and, having learned, then refrain from attacking them. Experiments carried out by Poulton and others in the nineteenth century and by Carpenter, Cott, Swynnerton, Darlington, Brower, and others in the twentieth, have established the correctness of these premises.[25] Evidence of learning and of discrimination in feeding based on visual associations has been obtained for representatives of every major group of vertebrate animals from the mammals and birds to the fishes.[26] Some exercise of discrimination in the choice of food insects is suggested also for several kinds of invertebrate animals.[27]

The monarch butterfly, *Danaus plexippus*, which occurs widely throughout North America, is conspicuously colored with bright orange wings bordered with dark brown (Plate Ie). It has been generally believed for a long time that this butterfly is unpalatable to vertebrate animals. To test this supposition, Brower captured some Florida scrub

[23] Cott, 1940, 195; quoted from *Adaptive Coloration in Animals*, 1940.
[24] Cott, 1940, 196.
[25] Cott, 1940, 251, for references.
[26] Summaries in Cott, 1940, 275 ff., 290–304.
[27] Linsley, Eisner, and Klots, 1961.

jays (*Cyanocitta coerulescens*) in the wild, caged them, fed monarch butterflies alongside palatable swallowtail butterflies (*Papilio glaucus* and *P. palamedes*) to them, and observed the reactions of the birds to each kind of butterfly. The butterflies were presented in an immobile condition and with their wings folded together, so that the reactions of the birds were based on their experiences in tasting the butterflies and not on their behavior. Whereas the swallowtail butterflies were always eaten in every trial, the monarchs were at first attacked and pecked but not eaten, and later, after a few trials, were not even touched by the

Table 3. *Number of bees eaten by a group of 18 toads on successive days of a feeding experiment*[a]

Day of experiment	1	2	3	4	5	6	7	22	23	24	25	26	27	28
Number of bees eaten	24	33	14	8	5	3	0	12	9	4	6	3	2	0

[a] Cott, 1940.

birds. Evidently the jays find the monarch butterflies inedible and learn by experience to reject them as food.[28]

Cott tested the reactions of the common British toad, *Bufo bufo*, to honeybees. At first the toads snapped up the bees eagerly. But after being stung a few times during the first two days, the toads lost their appetite for bees, ignoring or actively avoiding these insects, while continuing to eat other, innocuous insects such as mealworms. The number of bees eaten by all 34 experimental toads decreased day by day as follows: first day, 45 bees; second day, 41; third day, 18; fourth day, 10; fifth day, 8; sixth day, 3; seventh day, 0. By the seventh day, therefore, every toad, even the slowest, had learned its lesson.

A second series of trials was carried out with some of the same individual toads after an interval of two weeks, during which they had been kept without food, in order to test their memory for bees. The results are summarized in Table 3. Most of the toads remembered their lesson regarding bees for two weeks, and the few individuals that had forgotten were quick to relearn.[29]

Similar results were obtained by Brower and collaborators working with the Southern toad, *Bufo terrestris*, in Florida. Several individuals of this toad were collected in wet places and caged. When they were hungry and ready to eat insects, as shown by their quick acceptance of

[28] Brower, 1958a.
[29] Cott, 1940, 281–89.

live dragonflies, they were presented with live bumblebees (*Bombus americanorum*). When bumblebees whose stings had been removed were given to inexperienced toads, the latter readily attacked and ate the bees, showing that the bees are palatable. The toads at first attacked and tried to eat the normal unmutilated bumblebees also; but after a few such attempts they left the bees alone and exhibited a defensive manner in the presence of these insects. The toads learned to avoid the bumblebees as a consequence of the unpleasant experience of being stung, and they rejected the bees thereafter on sight.[30]

Members of the beetle family Lycidae are distinctively colored and are distasteful to insectivorous animals which usually reject these beetles. When touched they exude a fluid from their body, which probably has repellent properties. Darlington found that the iguanid lizard Anolis, though hungry, consistently refused to eat the bright yellow and blue lycid beetle Thonalmus in Cuba. Other workers have found that Lycus and other genera are rejected by such insect-eating animals as monkeys, mice, birds, spiders, scorpions, and preying mantids in Africa, Massachusetts, and Arizona.[31]

The defenses of bees and wasps by poison stings associated with warning colors do not have to be absolute in order to be effective. Bees are sometimes eaten by experimental toads, as we have just seen, as well as by various birds, lizards, and frogs in nature. But in spite of this they are protected if predators refrain from attacking them to some extent. And many if not most insectivorous animals do leave bees alone. Examination of the stomach contents of eight species of frogs by Cott showed that these frogs only rarely eat bees or wasps.[32] The stomachs of 375 cliff swallows, *Petrochelidon lunifrons*, yielded remains of 35 honeybees. These were all drones. Since the stinging worker bees far outnumber the harmless drones in nature, it is evident that the swallows must have deliberately selected the drones.[33] Starlings, *Sturnus vulgaris*, in Holland were observed to bring 799 hymenopterans to their nests to feed to the nestlings, but not one of these food animals was a bee or vespid wasp, although such types of Hymenoptera are common in the area.[34]

[30] Brower, Brower, and Westcott, 1960.
[31] Linsley, Eisner, and Klots, 1961.
[32] Cott, 1940, 302–3.
[33] Beal; Cott, 1940, 294.
[34] Kluijver; Cott, 1940, 293.

Noxious flesh is similarly a successful form of protection to its possessors even though some protected individuals are occasionally killed by inexperienced predators. Monarch butterflies are sometimes attacked and killed, in experiments as well as in the wild, but the over-all mortality of the species is reduced by the unpalatability of these visibly distinctive insects. Starlings in Holland were observed to feed 4,490 individual beetles to their young during 3 seasons. Only 2 of these 4,490 beetles were ladybirds (Coccinella), which are abundant but unpalatable and distinctively colored.[35] It follows that ladybird beetles, though occasionally captured by starlings, are not taken in proportion to their relative abundance in the feeding area of these birds, and thus receive, if not absolute immunity, at least a relatively high degree of protection from their enemies.

The sacrifice of some protected individuals is in fact a necessary part of the process of educating the predators regarding the poisonous or distasteful characteristics of the species as a whole. It is of course to the advantage of the protected species to reduce as much as possible the number of individuals sacrificed in educating predators. Clear and conspicuous color patterns, which can be easily recognized and easily remembered by predatory animals, when associated with noxious or dangerous qualities will facilitate the learning process in the predators and will therefore reduce the burden of sacrifices that must be borne by the protected species. The types of warning coloration met with in nature meet the requirements of conspicuousness and ease of recognition, as we have already seen.

Insect-eating animals must select their prey from among many hundreds of insect species. The number of distasteful or poisonous insect species which they must learn to recognize and avoid may run into scores or hundreds. Under these conditions, the memory of the predators is in danger of becoming overburdened, with the result that mistakes are likely to occur in feeding. It will be recalled that such mistakes on the part of the predators are usually fatal to some individuals belonging to the protected species. It is advantageous for the latter, therefore, to simplify as much as possible the predator's task of distinguishing and remembering the warning colorations of different protected species. The possession by different protected species of similar

[35] Kluijver; Cott, 1940, 293.

color schemes and color patterns is one means which can facilitate recognition by predators and hence reduce the number of mistaken attacks committed by them.[36]

The same types of warning coloration are in fact often found in protected species which are more or less distantly related. The occurrence of black and yellow color schemes in animals as different in other respects as salamanders, sea snakes, wasps, bees, caterpillars, and coccinellid beetles has already been mentioned. The black and yellow color scheme is common again to many species of bees belonging to different genera and families, and runs as a common theme throughout the vespid wasps.

The resemblance extends to both the color pattern and the shape of the wings in some unpalatable tropical American butterflies belonging to different genera and subfamilies of the family Nymphalidae. The same colors—black, yellow, and brown—are arranged in similar patterns on wings of similar shape in species of Lycorea (Nymphalidae subfamily Danainae), Melinaea and Mechanitis (subfamily Ithomiinae), and Heliconius and Eueides (subfamily Nymphalinae), which are all unpalatable to birds and which fly together in Brazil and other parts of the American tropics (Plate I, a–c).[37] (The subfamilies mentioned above

[36] Fr. Müller in 1879; see Wallace, 1889, ch. 9.

[37] Bates, 1862 (I am indebted to Professor F. M. Brown for loaning me a copy of this classical paper); Müller in 1879; Weismann, 1904, ch. 5.

Plate I. Warning coloration and mimicry in butterflies

The upper wings of female butterflies are shown in each case.

(a)–(d) A Brazilian mimetic assemblage consisting of three Müllerian mimics (a), (b), (c) and a Batesian mimic (d).

(e)–(g) The monarch in North America (e), its mimic the viceroy (f), and a non-mimetic relative of the viceroy (g).

(h)–(1) Mimetic polymorphism in the African *Papilio dardanus* (j), (k), (l), with two of its models (h), (i).

The names and family or subfamily relationships of these butterflies are as follows: (a) *Lycorea ceres* (= *L. halia*) (Nymphalidae-Danainae), (b) *Heliconius narcaea* (= *N. eucrate*) (Nymphalidae-Nymphalinae), (c) *Melinaea ethra* (Nymphalidae-Ithomiinae), (d) *Dismorphia astynome* (Pieridae), (e) *Danaus plexippus* (Danainae), (f) *Limenitis archippus* (Nymphalinae), (g) *Limenitis astyanax*, (h) *Danaus chrysippus* (Danainae), (i) *Amauris niavius* (Danainae), (j) *Papilio dardanus* forma *trophonius* (Papilionidae), (k) *Papilio dardanus* f. *hippocoon*, (l) *Papilio dardanus* f. *meriones*.

(a) model Lycorea

(b) co-model Heliconius

(c) co-model Melinaea

(d) mimic Dismorphia

(e) model monarch

(f) mimic viceroy

(g) non-mimetic relative of viceroy

(h) model Danaus

(j) mimetic form trophonius

(i) model Amauris

(k) mimetic form hippocoon

(l) non-mimetic form

(j), (k), (l): 3 forms of *Papilio dardanus*

are frequently treated as families; I am following the classification of Ehrlich here.[38])

A remarkable case of convergence in warning coloration between distasteful species belonging to different insect orders involves the Arizona beetle, *Lycus fernandezi* (Lycidae), and the associated moth, *Seryda constans* (Pyromorphidae), and again the beetle *Lycus loripes* and the moth *Holomelina ostenta* (Arctiidae) (Fig. 25).[39]

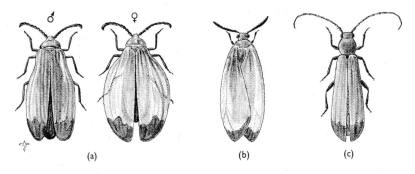

Fig. 25. Common warning coloration of a noxious beetle (a) *and a noxious moth* (b), *and mimicry of the former by an edible beetle* (c)

(a) The model, *Lycus fernandezi* (Coleoptera, Lycidae).
(b) Co-model, *Seryda constans* (Lepidoptera, Pyromorphidae).
(c) Batesian mimic, *Elytroleptus apicalis* (Coleoptera, Cerambycidae).

Redrawn from Linsley, Eisner, and Klots, *Evolution*, 1961.

Common warning coloration is denoted as Müllerian mimicry after Fr. Müller, who correctly explained this phenomenon in 1879. The different members of a Müllerian mimetic assemblage benefit alike from their mutual similarities, for by pooling their recognition marks they share the losses due to mistaken attacks by uneducated or forgetful predators, and the losses suffered by any single species are thereby reduced.

MIMICRY

Many insects and some spiders and birds which are not themselves protected, either by poisonous secretions or distasteful flesh, resemble

---

[38] Ehrlich, 1958.
[39] Linsley, Eisner, and Klots, 1961.

unrelated animals which are thus protected and which usually exhibit warning coloration. The unprotected animals, the mimics, thus gain some protection from predators on the basis of the reputation of the warningly colored models.[40] The innocuous mimics are, as Cott has aptly put it, "the sheep in wolves' clothing."[41] The resemblance of an unprotected animal to an unrelated protected species is known as mimicry (or as Batesian mimicry).

Ants are mimicked in various parts of the world by spiders, grass-hoppers, beetles, flies, and bugs. All such mimics deviate from their own related forms in appearance while assuming the disguise of ants. The ant mimics, for example, belong to orders characterized by stout waists, but they simulate the narrow-waisted outline of their hymenop-teran models. This effect is attained by different means in different groups of ant mimics. Some spiders which mimic ants have an actual constriction in the body. Grasshoppers, beetles, flies, and bugs which mimic ants produce the effect of a narrow waist by an optical illusion, wherein the antlike outline is painted in dark colors on the body of the mimic and the broad waist is left light-colored and inconspicuous.[42]

Bumblebees are well protected by their poison stings and are easily recognized by their distinctive color pattern and buzzing flight. Various kinds of harmless flies and moths which occur together with bumblebees in the same areas resemble the bees more or less closely in appearance and behavior. The asilid fly, *Mallophora bomboides* (Asilidae), for example, looks and sounds like the bumblebee species *Bombus ameri-canorum*, and occurs in the same habitats with it in Florida.[43] Bumble-bees are also mimicked by the syrphid fly *Volucella* (Syrphidae), and honeybees by drone flies (*Eristalis*, Syrphidae).[44] The bee-mimicking moth, *Hemaris*, diverges markedly from the norm of the family Sphing-idae to which it belongs in its beelike appearance and behavior. Whereas the hawk moths are typically nocturnal and cryptic and possess densely scaly wings, *Hemaris* is active by day, exhibits yellow, brown, and black colors on its body, and has transparent wings.

Lycid beetles, which are distasteful to a wide variety of insect-eating

---

[40] Bates, 1862.
[41] Cott, 1940, 435.
[42] Cott, 1940, 408–10.
[43] Brower, Brower, and Westcott, 1960.
[44] Weismann, 1904, ch. 5.

animals, both vertebrate and invertebrate, serve as models for a great many mimics in different parts of the world. According to Carpenter, "They are resembled in colour and pattern by members of every order of Insects whose bodily form yields the foundation on which a Lycoid appearance can be built." Thus there are moths, wasps, flies, bugs, and beetles which resemble the Lycidae.[45] It is interesting to note that the hymenopterous mimics of lycid beetles lack the clear wings of most Hymenoptera and instead have orange and black wings like the elytra of their models.[46] Living with Thonalmus in Cuba are a number of beetles belonging to several families—Cantharidae, Elateridae, Cerambycidae, and Oedemeridae—which have the same color pattern as the lycid beetle. Captive lizards consistently rejected both the Thonalmus and its mimics in feeding tests, while taking other non-mimicking beetles.[47] The orange, black-tipped *Lycus fernandezi* in Arizona is mimicked by the edible beetle, *Elytroleptus apicalis* (Cerambycidae), and the uniformly yellow-orange *Lycus loripes* is mimicked by another edible species of Elytroleptus (*E. ignitus*) as well as by a harmless and probably edible moth (*Eubaphe unicolor*, Geometridae) (Fig. 25).[48]

The effectiveness of mimetic resemblances in deceiving predatory animals has been demonstrated by many observations and by several controlled experiments. Drongos in Rhodesia (*Dicrurus afer* and *D. ludwigi*) have unpalatable flesh and are well marked by black coloration. A flycatcher (Bradyornis) and a cuckoo-shrike (Campephaga) resemble the drongos closely in size and dark colors. Swynnerton's cat developed a strong dislike for the drongos after trying to eat them. Swynnerton then offered his cat the mimicking flycatcher and cuckoo-shrike, whereupon the cat refused these too. But it immediately took another, non-mimicking bird, a tit (*Parus niger*).[49]

The monarch butterfly, *Danaus plexippus*, as noted in the preceding section, is distasteful to birds and is easily recognized by its distinctive color pattern. We have seen that after a few unpleasant experiences with the monarch, jays quickly learn to leave it alone and thereafter rarely attack it. The viceroy butterfly, *Limenitis archippus*, occurs with

[45] Linsley, Eisner, and Klots, 1961.
[46] Cott, 1940. 400.
[47] P. J. Darlington in 1938; Cott. 1940, 413–15.
[48] Linsley, Eisner, and Klots, 1961.
[49] Cott, 1940, 413.

the monarch in eastern North America and closely resembles it in color pattern. In its superficial visible characters, though not in its basic morphological features, the viceroy differs from other species of the genus Limenitis and converges toward a member of a different sub-family (Plate I, e–g). Since the viceroy is not itself protected from birds to any considerable extent, it has long been believed that it gains a vicarious protection by mimicking the monarch.

Brower tested this assumption in feeding experiments carried out with Florida scrub jays. Birds which had no previous experience with the monarch ate most of the specimens of viceroy that were offered to them, showing that the viceroy is acceptable though apparently not highly palatable to jays. Another group of jays were fed monarchs for a while, until they had learned about the distasteful properties of this species, and were thereupon offered viceroys in place of monarchs. The viceroys were usually left untouched and were never eaten by these jays. The birds that had experience with the monarchs rejected the viceroys to a much greater extent than did the birds which lacked any prior experience with the monarch. The mimicry of the viceroy for the monarch thus affords the viceroy a good measure of protection against jays.[50]

We have mentioned above that the bumblebee, *Bombus americanorum*, is mimicked by an asilid fly, *Mallophora bomboides*, in Florida. And we saw in the previous section that toads quickly learned to leave this species of bumblebee alone after being stung a few times. Captive toads that were presented with *Mallophora bomboides* but not with *Bombus americanorum* regularly attacked and ate the flies. Those toads, on the other hand, that were alternately given individuals of the Bombus and of the Mallophora soon acquired a habit of avoiding the mimic as well as the model. The learned reactions toward the well-defended bumblebee became transferred to its relatively defenseless mimic. Here again the mimicry has a proven effectiveness under controlled conditions.

The protection of a mimic lies in the distaste or fear acquired by predators for its model. Now the presence of innocuous mimics in the territory of unpleasant models is bound to slow down the learning process in predators, on which the protection of the mimics (as well as of the models) depends. Whereas the warning coloration of a poisonous

[50] Brower, 1958a.

or noxious insect teaches an insectivorous animal to leave that kind of insect alone, the false warning coloration of a palatable and harmless mimic teaches it the opposite lesson—to associate the particular color pattern with desirable qualities. It is advantageous for a mimic to be mistaken by a predator for an inedible model; it does not benefit the mimic to be mistaken for another mimic; nor does it benefit the mimic to resemble a model which has lost its reputation for inedibility. Consequently the effectiveness of mimicry as a means of protection of the mimic is affected by the relative abundance of mimic and model in the feeding territory of a predator. Mimicry affords the greatest protection to the mimicking species when it is present in substantially fewer numbers than its model.

This relation between the numbers of model and of mimic, which was recognized by the early students of mimicry in the nineteenth century, is well described by Cott.

If the numbers of the less-protected mimic rise in relation to those of the more protected model, the degree of protection will decline, and will eventually be converted into a disadvantage when the coloration becomes associated by predators with something palatable more often than with something repellent. There is therefore an optimum proportion between the numbers of model and mimic, and this limits the spread of the latter.[51]

Many observations made by field naturalists show that mimics are in fact frequently uncommon in comparison with their models. In 1862, Bates reported that the inedible and gaily colored Heliconia butterflies (family Nymphalidae) fly in large swarms, immune from attacks by birds, in the Amazon valley. In the swarms of Heliconias, Bates found single individuals of edible mimicking butterflies belonging to the family Pieridae (Plate Id).[52] The pipe-vine swallowtail butterfly, *Battus philenor* (Papilionidae), which occurs widely in North America, has a strong disagreeable odor and is distasteful to birds, as has been shown in feeding experiments.[53] The distinctive color pattern of *Battus philenor* is mimicked by several species of edible swallowtails, *Papilio troilus*, *P. polyxenes*, and the black form of *P. glaucus*, which fly with the Battus in parts of its range. In the Great Smoky Mountains of the eastern United States the three mimicking species combined are less numerous

[51] Cott, 1940, 424.
[52] Bates, 1862.
[53] Brower, 1958*b*.

than the model.[54] Models may not always outnumber their mimics however.

The optimal proportion between models and mimics is undoubtedly affected by various factors, such as the degree of distastefulness of the model, the learning ability of the predator, its keenness of sight, and the closeness of the mimetic resemblance. In general, a highly distasteful or highly dangerous model will protect its mimic better than will a slightly unpleasant model, and therefore a larger proportionate number of mimics could be expected to live in association with a strong model than with a weak one. A high proportion of mimics to models would be expected to interfere more seriously with the learning process in the less intelligent predatory animals, which are slower to learn under the best of conditions, than with learning by more intelligent animals. Consequently models which are preyed on by intelligent predators can probably tolerate a higher proportion of mimics than models preyed on by less intelligent animals.

It seems probable that most mimics do not rely solely on mimicry as a protection against predators, for most animals in general have several alternative means of defense against their enemies. The viceroy butterfly mimics the monarch, as we have seen, but in addition the viceroy is apparently slightly distasteful to birds in its own right.[55] A slightly distasteful mimic, by virtue of its alternative method of protection, could theoretically exist in larger numbers relative to its model than could a completely palatable mimic, other factors remaining constant. The alternative form of protection in some other mimicking species might be an exceptionally high net fecundity, which would also tend to depreciate the importance of mimicry per se in its survival. Since more losses due to predation could be supported by the highly fecund mimic than by a less fecund mimetic species, the former might be able to exist in a greater relative abundance with its model than the latter could.

Where the same model is mimicked by two or more species, the numbers of any single mimetic species will be restricted by the abundance of the competitive mimicking species. Consider for example the case of *Papilio troilus*, *P. polyxenes*, and *P. glaucus* all mimicking *Battus philenor*. It is probable that *Papilio troilus* cannot rise to the same

[54] Brower, 1958*b*.
[55] Brower, 1958*a*.

frequency in an area inhabited by *Papilio polyxenes* and/or *P. glaucus* (dark form), as it can in an area where it has the field to itself. The presence or absence of competitive mimics is thus another factor affecting the abundance of any given mimetic species.[56]

Because of the foregoing and other modifying factors, the proportion of mimics to models which affords satisfactory protection to the mimics is not a fixed quantity, but may be expected to vary from one set of conditions to the next.

In an initial attempt to quantify the numerical relations between models and mimics, Brower fed captive starlings (*Sturnus vulgaris*) with mealworms (*Tenebrio molitor*). Some of the mealworms were made distasteful by dipping them in quinine solution, and others were dipped in pure water and were quite palatable to the birds. The quinine-flavored mealworms were the models and the other mealworms were their mimics in a series of feeding experiments. The models and mimics were offered to the birds in different relative proportions. Where the numbers of models were much greater than the numbers of mimics, as nine models to one mimic, the birds rejected 80 percent or more of the mimics. Where the mimics slightly outnumbered the models, in the ratio of four models to six mimics, the mimicry was also about 80 percent effective. And, surprisingly, even where the mimics were present in much greater numbers than the models, as in the ratio of one model to nine mimics, the mimicry was somewhat effective, in that 17 percent of the mimics were rejected by the birds.[57]

The restriction on the abundance of a mimicking species, imposed by the conditions under which the mimicry is most effective, works against the natural tendency of a species to increase in numbers. Some mimicking species of butterflies make use of a device which affords a way out of this numerical restriction. A mimetic species can increase in numbers by copying two or more different models. The system by which this is accomplished is polymorphism, wherein the population of the mimetic butterfly contains two or more alternative forms.

A polymorphic population of butterflies may consist of a mimetic form and a non-mimetic form hatching out of the same batch of eggs; it may contain two or more mimetic forms copying as many different

[56] Sheppard, 1959, 162–63.
[57] Brower, 1960.

models; or it may consist of a combination of different mimetic and non-mimetic forms. In each possible case, by means of polymorphism, a favorable balance can be preserved between the numbers of a mimic and those of its model, while permitting an increase in the total numbers of the mimicking species.

A relatively simple polymorphism is found in the common swallowtail butterfly of eastern North America, *Papilio glaucus*. The females of this palatable species exist in two forms, a non-mimetic yellow form, and a black form which mimics the inedible *Battus philenor*.[58]

A more complex case was described by Trimen in 1869 in the African butterfly, *Papilio dardanus*.[59] The males of *Papilio dardanus* are non-mimetic throughout the distribution area of the species. A non-mimetic form of the female (forma *meriones*), possessing yellow wings with black markings and a tailed appendage, occurs on Madagascar. On the African mainland there are several mimetic forms of females which mimic different models. The forms *hippocoonides* and *hippocoon* mimic *Amauris niavius* (Nymphalidae subfamily Danainae); forma *cenea* mimics *Amauris albimaculata* and *A. echeria*; and forma *trophonius* mimics *Danaus chrysippus*. As shown in Plate I, h–l, the mimetic females differ strikingly from the non-mimetic female in color pattern and wing outline, and resemble just as strikingly their models. The very different mimetic and non-mimetic forms of *Papilio dardanus* can occur as sister individuals in the same brood.

It has been shown recently that two blocks of genetic factors, more specifically two supergenes, control most of the visible differences between the alternative polymorphic forms of this butterfly. One of the supergenes controls the color pattern of the wing, while the other determines the presence or absence of tailed appendages. The alternative forms or alleles of these genes act as switches to determine the development of clear-cut alternative wing types in the females of *Papilio dardanus*.[60]

COORDINATED COMPLEXES OF CHARACTERS

The adaptations of organisms which enable them to live successfully in their normal environments generally involve the coordination of

[58] Brower, 1958*b*.
[59] Trimen, 1869. A recent brief review is given by Sheppard, 1959, 163–65.
[60] Clarke and Sheppard, 1959–60, 1960*a*, 1960*b*.

different characteristics. The successful functioning of one part of the body depends on the way in which some other part functions. An organism is more than a bundle of separate adaptations; it is a co-ordinated complex of adaptations. The adaptation of a hawk for making a living by hunting small animals involves the combination of several features: soaring flight, telescopic vision, sharp grasping talons, strong body, and hooked tearing beak.

Table 4. Character combinations found in two types of black mustard (Brassica nigra) *in western Asia*[a]

Plant part	Race occurring in open fields	Race occurring in cultivated fields of yellow mustard
Leaf rosette in young plant	Well developed	Absent
Branching	Few branches in upper part of plant	Numerous branches from base
Stem	Thick and tall	Slender and short
Lower leaves	Long with 5–7 pairs of lobes	Short with 2–3 pairs of lobes
Pubescence	Abundant and rough	Sparse and soft, or lacking
Time of flowering	Late	Early
Inflorescence	Compact	Open
Flowers	Small	Large
Pods	Short and narrow, lying close to stem	Long and broad, standing out from stem
Dehiscence of pods	Pods opening easily and seeds falling out spontaneously	Pods opening with difficulty under pressure, seeds not shedding spontaneously
Seeds	Small	Large

[a] Sinskaja, 1931.

The black mustard, *Brassica nigra*, is a coarse weed in various parts of the world. In western Asia where the yellow mustard, *Brassica campestris*, is cultivated as a crop plant, the black mustard is represented by another more slender form specialized to survive as a weed in the yellow mustard fields. The populations of black mustard which grow with yellow mustard differ from the common weedy forms of black mustard found in open fields and waste ground in 11 separate characters (Table 4). The different characters—mode of branching, stem size, flower size, shape and dehiscence of seed pods, seed size, and time of flowering—work together to help the specialized race of black mustard grow and ripen simultaneously with yellow mustard and to produce seeds which are threshed and dispersed with those of the crop plant whose fields it inhabits.[61] Its adaptation as a successful weed in yellow mustard fields is based not on any one single character but on a combination of characters.

CONCLUSIONS

The main body of this chapter has been devoted to a consideration of the chief types of coloration which help to protect animals from their enemies. The topics considered have a close connection with the development of evolution theory. Bates discovered mimicry in Amazonian butterflies in 1862; he and A. R. Wallace suggested warning coloration for butterflies and for caterpillars in 1862 and 1866; Darwin took a keen interest in both subjects[62] and in concealing coloration. These topics were considered again by other nineteenth-century evolutionists, like Poulton and Weismann, by such early twentieth-century evolutionists as R. A. Fisher and Carpenter, and by various modern workers down to the present time. Partly because of their historical connection with evolution theory, but more particularly because of their great intrinsic interest, the forms of adaptive coloration in animals comprise a worthwhile subject to focus attention on in this book.

In singling out for detailed consideration one group of adaptive phenomena, the danger exists that we will create the impression that adaptations are confined to special structures of organisms performing special functions. This is of course not the case. The most general characters of organisms are adaptations. The wings of birds in general are as much an adaptation for flying as the particular type of bill and clinging feet of a woodpecker are an adaptation for a specialized method of food-getting. Nor are adaptations exemplified only by morphological structures; many of the physiological and biochemical processes carried out by animals, plants, and microorganisms are clearly adaptive.

Many of the morphological and physiological characteristics of living organisms are thus adaptive. Are *all* the characteristics of organisms adaptive? It would be impossible to prove a universal generalization of this kind even if it were true. Moreover, there are reasons for believing

[61] Sinskaja, 1931, 61–66.
[62] See Darwin, 1871, ch. 11; Wallace, 1889, ch. 9.

that some characters of organisms may be non-adaptive, as we shall see in Chapter 11. Some biologists have made the mistake, however, of assuming that if the adaptive value of a character is not immediately evident to a human observer, it does not exist. Few observers would guess from mere inspection that the difference between yellow onions and white onions is adaptive. Yet it has been found that yellow onions are resistant to a fungus disease, a smudge, and that white onions are susceptible to the same disease.[63] The color of the onions is not adaptive in itself; it is, however, the visible manifestation of biochemical processes which play a vital role in the fitness of onions for life in a smudge-infected environment.

Adaptation is thus a general feature of the relationships between organisms and their environments. How has this universal condition of adaptedness come about? It is, as Pierre Lamarck proposed in 1809 and as Charles Darwin demonstrated 50 years later by a thorough marshalling of the evidence, a result of evolutionary changes which have taken place in the history of each kind of organism living in its particular environment. In short, evolution is guided by the environment, as Lamarck and Darwin recognized, and its main goal is adaptation. In order to explain the origin of adaptations, therefore, we must analyze the mechanism which brings about evolutionary changes in general.

[63] H. A. Jones and others; see Dobzhansky, 1951a, 99–100.

SIX

Adaptability and Heredity

> Any evolution theory which disregards the established genetic principles is faulty at its source. TH. DOBZHANSKY

ORGANISMS UNDERGO VARIOUS TYPES OF CHANGES in response to their environment. Changes in the hereditary characteristics of organisms, as between ancestral and descendant generations, constitute evolution, and the latter is consequently a process of change, not in individuals, but in groups of individuals or populations. Individual organisms also make responses of their own to the environment. The adjustments of individuals to their environment, while not constituting evolutionary changes as such, do play a role in evolution, which we will consider in this chapter.

In order to understand the role of adaptability in evolution, we must first understand how the individual organism develops and what it passes on to its descendants in reproduction. Consequently this chapter also includes a brief discussion of development and heredity.

DEVELOPMENT OF THE INDIVIDUAL

One of the links in the chain of events resulting in evolution is the growth and development of the individual organism. Among multicellular plants and animals each individual develops from a single cell. There is to begin with the unfertilized egg cell produced by the female parent. This egg cell is already a highly differentiated organic structure. The nucleus of the egg contains a set of genes and chromosomes contributed by the mother individual. The cytoplasm surrounding the nucleus consists of materials organized in a definite way during the formation of the egg in the mother's body.

When a sperm fuses with the egg at fertilization a set of chromosomes and genes derived from the male parent is added to the maternal set.

118

The fertilized egg—the zygote—now contains the materials which will guide the future course of development of the new individual: it contains the genes of both parents in its nucleus, and the cytoplasm produced largely or entirely by the maternal parent. These genic and cytoplasmic materials will control future developmental processes.

The zygote undergoes an orderly series of cell divisions leading first to the development of an embryo, then to the formation of an immature individual, and finally to the appearance of a new adult plant or animal. The pattern of development from zygote to mature offspring is channelized and guided within certain limits by the specific genes and cytoplasm contained within the fertilized egg. The materials in the zygote of a cat develop into a cat; the fertilized egg of a corn plant becomes finally a mature corn plant.[1]

The sum total of the genes—the genotype—determines a certain general course of development of the organism. However, this course of development is influenced by the environmental conditions to which the developing organism is exposed. The genetic constitution of the zygote predetermines the general pattern of development; it does not predetermine the exact final outcome of that developmental process, which is affected by other modifying influences.

The morphological and physiological characteristics of the organism—the phenotype—are a result of the action of a particular genotype as modified by particular environmental conditions. The individual organism is a product of two sets of factors: its complement of genes and the environment in which it develops. The genetic determinants and the environmental factors act jointly in the control of development.

PHENOTYPIC MODIFICATIONS

One and the same genotype can, in the course of development in different environments, give rise to a variety of phenotypes. The various phenotypic expressions induced by the action of different environmental conditions on any given genotype are known as phenotypic modifications.

The genotype of an organism usually leads to the production of phenotypic characteristics which are adaptive in the environments

[1] For an introductory account of development see Simpson, Pittendrigh, and Tiffany, 1957, ch. 14.

normally inhabited by the organism, as we have seen in the previous chapter. And, as noted above, each genotype also has the capacity to react phenotypically to its environment in a multitude of ways. Some of these reactions are such as to bring about further and finer adjustments between the individual organism and its environment. In other words, many phenotypic modifications are also adaptive.

Potentilla glandulosa, a perennial herb belonging to the family Rosaceae, has long-lived roots which can be divided and propagated. Clausen, Keck, and Hiesey dug up a single individual of this plant from a natural colony in the California Coast Ranges and transplanted it to their experimental garden at Stanford. Here they divided the root into four parts and propagated each division under a different set of environmental conditions. One plant was raised in a plot of dry soil and full sun; a second, in equally dry soil but under shade; a third, in moist soil and full sun; and a fourth in a combination of moisture and shade. The same original genotype reacted to these four environments by producing four strikingly different phenotypes. The shade plants were taller with longer stems and broader leaves than the sun plants; the moist plants were more vigorous than the dry plants (Fig. 26).[2]

The phenotypic reactions of *Potentilla glandulosa* to normal changes in its environment evidently help to adjust the plants to their immediate circumstances. In order to manufacture the food it requires for growth, the green leaves of this plant require a certain amount of light. The plants growing in the shade plots compensate for the lower intensity of light by forming broad, expanded leaf surfaces. The sun plants have sufficient light but are exposed to the danger of excessive loss of water in the drying sun. Their smaller and more compact form and their smaller and tougher leaves enable these plants to reduce the amount of water they use and give off into the air.

When a human is in the hot sun, he sweats; in the cold, he shivers. These changes help the body to adjust to differences in temperature. Perspiring helps by cooling the skin through evaporation, and shivering by increasing the flow of warm blood to the body surface. When a human uses his muscles, they develop and become strong; when those muscles are not exercised, they shrink and fall into disuse.

Chameleon lizards possess the ability to turn green on a green

[2] Clausen, Keck, and Hiesey, 1940, 53 ff.

background and brown on a brown background. It is advantageous for the lizard to blend into its background so that it is inconspicuous to the insects it catches for food and to its enemies. Since the habitat of the chameleon includes green leaves and brown bark, the range of its phenotypic responses corresponds to its normal range of environmental conditions. The lizard does not often find itself on a red or blue

DRY SUN   DRY SHADE   MOIST SUN   MOIST SHADE

Fig. 26. Different phenotypes produced by the growth of one genotype of
Potentilla glandulosa *in four different environments*

From Clausen, Keck, and Hiesey, *Carnegie Inst. Wash. Publ.* No. 520, 1940; photograph by courtesy of Dr. Hiesey.

background, however, and it lacks the ability to turn red or blue. There are limits to the modifiability of the phenotype produced by a given genotype.

The plasticity of the phenotype is relatively great in some types of organisms and slight in others. In general, herbaceous plants are capable of producing a wide range of phenotypic modifications. The case of *Potentilla glandulosa* described above is typical of perennial herbs. Annual herbs are if anything even more plastic in the sense that they can respond to varying environmental conditions by producing very different phenotypes. Botanists have frequently observed that the same strain of annual plants when raised under conditions of ample moisture will produce a luxuriant growth with tall stems, large leaves, numerous flowers, and great quantities of seeds, but when grown in dry ground will send up a low stem only a few centimeters high bearing a single flower and only a few seeds.[3] The opposite extreme is exemplified by the insects which in general are rather inflexible phenotypically. Individuals of Drosophila raised in different environments, for example, are likely to be morphologically identical; this is in marked contrast to the variety of phenotypes obtained when herbaceous plants are exposed to a comparable range of conditions.

Perhaps these differences in the modifiability of the body are related to the different methods of growth and development in the two groups. The adult body of the insect develops within a hard external skeleton which is formed in a preadult growth stage. The main external features of the insect body are laid down at a stage when they do not function, and have become rigid and unchanging by the time they can be used.[4] The plant body, by contrast, develops from growing points which are exposed to and strongly influenced by the environmental factors prevailing while the new parts are forming, so that drought-resistant leaves appear during seasons of drought, succulent leaves in times of abundant moisture, and so on.

PHYSIOLOGICAL HOMEOSTASIS

An individual organism may adjust to variable environmental conditions by varying phenotypically. Or it may adjust by remaining constant and stable in the face of environmental changes. The human body reacts to cold by shivering and to heat by sweating; the ultimate consequence of these bodily changes is the preservation of a constant body temperature at different air temperatures. This property of organic stability under variable environmental conditions is known as physiological homeostasis.[5]

Physiological homeostasis is a type of phenotypic reaction wherein the organism, instead of conforming to changes in the external environment, resists those changes and by means of self-regulating physiological processes maintains the internal environment constant in spite

[3] See for example Khoshoo, 1958.
[4] Warburton, 1956.
[5] Cannon in 1932; Lerner, 1954, 1–2.

of fluctuating external conditions. Homeostasis is obviously an advantageous property of organisms, since it permits the body to develop in constant ways or to function with constancy in a changeable environment.

Homeostatic properties are not developed to the same degree in all organisms. Plants, which have a great amount of phenotypic plasticity, are relatively slightly homeostatic; whereas higher animals, which tend to maintain a steady state in a changing environment, exhibit a relatively high development of physiological homeostasis.

HABITAT SELECTION

Organisms possessing means of locomotion, as most animals and many single-celled forms do, can become adjusted to the environment by making use of their motility. An animal sweltering under the desert sun does not have to stay in the desert; it can move to some other region where the conditions are more favorable.

The environment is heterogeneous and normally affords numerous different habitats and niches within the radius of dispersal of an animal or microorganism. The organism can find and occupy those habitats or niches in its heterogeneous territory which are most suitable for its life. Within limits the organism, by habitat selection, can adjust the environment to its particular tolerances, in addition to adjusting itself phenotypically to the environment.

An active selection by animals of those habitats and niches in which they can live most successfully can be observed on every hand. Grazing animals range up the mountain slope with the advance of spring and summer and work down slope during fall and winter. Plains animals wallow in the rivers and water holes when the heat of summer is on the land.

In these cases all the individuals of a species tend to respond in a uniform manner to the same environmental conditions. This response follows from the fact that all the genotypes belonging to one species possess many features in common. To the extent that the various genotypes within a species produce phenotypes with different tolerances, we would expect preferences for different environmental conditions to be exhibited by the various individuals.

The females of the sulphur butterfly, *Colias eurytheme*, occur in different color phases determined by different genotypes. Hovanitz observed that the white females are most active during the early morning hours and just before sunset, becoming relatively inactive at midday, while the orange females reach their peak of activity in the middle of the day.[6]

Waddington and his co-workers have tested experimentally the environmental preferences of various genotypes of the fruit fly, *Drosophila melanogaster*.[7] They constructed an apparatus consisting of a central compartment connected by tunnels with eight peripheral chambers. Each of the chambers was provided with a different set of environmental factors. The conditions that varied were light, moisture, and temperature. Known numbers of normal and mutant flies were introduced into the central compartment at the beginning of the test and left to themselves for five or six hours, after which their numerical distribution among the various chambers was recorded.

Flies like other insects are generally attracted toward light, and in this experiment 74 percent of the wild-type individuals of Drosophila moved into the light chambers and 26 percent into the dark chambers. Some of the mutant forms (namely, "purple," "apricot," and "black") shunned the dark to an even greater extent than the wild-type flies, however, as shown by the concentration of between 90 and 95 percent of the individuals in the light chambers, as compared with 10 to 5 percent in the dark chambers. Flies of the mutant type "rough" moved into the light dry cold environment twice as frequently as non-mutant flies. Similarly, mutant individuals of the type "forked" were apparently attracted more strongly to dark wet cold conditions than were mutants of the types "black," "apricot," "purple," or "aristaless."[8]

THE NON-INHERITANCE OF PHENOTYPIC REACTIONS

The normal phenotypic reactions of an organism are adaptive insofar as they enable the organism to make the adjustments necessary in order to live in a changeable environment. Among the adaptations which comprise an organism's heritage from past periods of evolution is the ability of its genotype to respond with the appropriate phenotypic

[6] Hovanitz, 1953; Waddington, Woolf, and Perry, 1953.
[7] Waddington, Woolf, and Perry, 1953.
[8] Waddington, Woolf, and Perry, 1953.

modifications and homeostatic adjustments to a certain range of environmental conditions. In fact, the term genotype was originally defined by Johannsen as "the reaction norm."[9]

However, it does not follow that phenotypic modifications are themselves the initial stages of evolutionary change. Evolution consists of changes in the heredity of organisms. We must find out whether environmentally induced reactions of the phenotype are inherited or not before we can decide whether they enter into evolutionary changes.

Offspring usually possess phenotypic characteristics similar to those of their parents. The simplest and most obvious theory of heredity holds that phenotypic traits are transmitted from generation to generation. Since it is readily observable that phenotypes can be modified by direct influences of the environment and by the effects of use or disuse of organs, it was frequently assumed by earlier biologists that these modifications are also inherited. This view is so simple and obvious that it has been held almost universally until recent times. It forms an important part of the doctrine of Lamarckism, the part known as the theory of the inheritance of acquired characters. But the eighteenth and nineteenth century French evolutionist, Pierre Lamarck, only stated a generally held idea, as did Hippocrates before him, Darwin after him, and Lysenko in modern times.[10]

A hypothetical mechanism was invented to account for the alleged inheritance of phenotypic traits. Darwin suggested in 1868 that hereditary particles are constantly passing from all parts of the body to the reproductive organs.[11] Modifications induced by environmental conditions in the body cells consequently enter the hereditary stream via these particles. This hypothesis of Darwin's was again only a restatement in more exact terms of an age-old belief. The idea that semen is derived from all parts of the body can be traced through numerous Renaissance and medieval writers to Hippocrates in the fourth century B.C. The Father of Medicine wrote that "the seed [semen] comes from all parts of the body, healthy seed from healthy parts, diseased seed from diseased parts."[12]

[9] Johannsen, 1911.
[10] Zirkle, 1946.
[11] Darwin, 1868.
[12] Zirkle, 1946; Grant, 1956d.

As often happens with self-evident truths, the direct theory of heredity turned out to be wrong. In the first place, a growing body of embryological knowledge in the latter nineteenth century failed to reveal any process corresponding to the hypothetical migration of hereditary units from the body cells to the gametes. August Weismann pointed out in 1892 that the line of cells which leads ultimately to the gametes is set apart from the rest of the body at an early stage of development. On this fact he based his important concept of the independence of the germplasm from the so-called soma or the rest of the body.[13]

The hypothesis of the continuity of the germ-plasm depends on the assumption of a contrast between the *somatic* and the *reproductive* cells, such as can be observed, in fact, in all multicellular plants and animals, from the most highly differentiated forms to the lowest heteroplastids amongst the colonial Algae.

Since heredity consists of the transmission of the germplasm, and not of the somatic cells of the body, to future generations, it follows that "acquired characters," or somatogenic variations as Weismann called them, or phenotypic reactions as we call them today, cannot be transmitted from one generation to the next. In the words of Weismann again[14]:

By *acquired* characters I mean those which are not preformed in the germ, but which arise only through special influences affecting the body or individual parts of it. They are due to the reactions of these parts to any external influences. . . . It is an inevitable consequence of the theory of the germ-plasm, and of its present elaboration and extension so as to include the doctrine of determinants, that somatogenic variations are not transmissible, and that consequently every permanent variation proceeds from the germ, in which it must be represented by a modification of the primary constituents. . . .

It is impossible to assume transmission of somatogenic variations in any theory which accepts the nuclear substance of the germ-cells as germ-plasm or "hereditary substance". . . .

All permanent—i.e., hereditary—variations of the body proceed from primary modifications of the primary constituents of the germ. . . . Neither injuries, functional hypertrophy and atrophy, structural variations due to the effect of temperature or nutrition, nor any other influence of environment on the body, can be communicated to the germ-cells, and so become transmissible.

The morphological separation of the germplasm from the rest of the body is most clear-cut in many invertebrate animals, such as

[13] Weismann, 1892, 183.
[14] Weismann, 1892, 392, 395.

roundworms and insects, but is somewhat less clear in the vertebrate animals. The independence of the germinal line of cells from the somatic tissues is still less evident in plants, where both the vegetative and reproductive organs arise from a common growing point. On considering the special case of plants, Weismann concluded that although the germ cells may bud off from somatic tissues, they still preserve their independence from vegetative organs which have budded off separately.[15] Furthermore, Weismann's concept of a germplasm independent of somatic modifications was found to fit the experimental evidence generally, in plants and vertebrate animals as well as in invertebrate animals. Weismann notes to begin with that[16]:

There are no observations which prove the transmission of functional hypertrophy or atrophy, and it is hardly to be expected that we shall obtain such proofs in the future, for the cases are not of a kind which lend themselves to an experimental investigation. The hypothesis that acquired characters can be transmitted is therefore only directly supported by [various] instances of the transmission of mutilations. For this reason, the defenders of the Lamarckian principle. . . have endeavoured to show that these observations are conclusive, and therefore of the highest importance. . . .

It can hardly be doubted that mutilations are acquired characters: they do not arise from any tendency contained in the germ, but are merely the reaction of the body under external influences. They are, as I have recently expressed it, purely somatogenic characters. . . .

If mutilations must necessarily be transmitted, or even if they might occasionally be transmitted, a powerful support would be given to the Lamarckian principle, and the transmission of functional hypertrophy or atrophy would thus become highly probable. For this reason it is absolutely necessary that we should try to come to a definite conclusion as to whether mutilations can or cannot be transmitted.

Weismann then proceeds to examine a number of cases of mutilations, such as the loss of the tail in cats, or slit ear lobes in humans, that have reportedly been passed on to the offspring. He finds that in many cases the account is based on hearsay without proper documentary evidence. "In a great number of such cases every guarantee for the trustworthiness of the statements is entirely wanting, and . . . they are of no greater value as evidence than the merest tales."[17] Other

---

[15] Weismann, 1892, 197.
[16] Weismann, 1889–92, I, 436–37.
[17] Weismann, 1889–92, I, 436.

cases cannot be accepted as evidence of inherited mutilations for various other reasons, which it is not worthwhile to consider here.

On the other hand, many forms of mutilations have been practiced consistently over numerous generations without showing any tendency to become hereditary. Thus the amputation of tails in dogs and sheep has not led to tailless breeds of these animals.[18]

Furthermore, the mutilations of certain parts of the human body, as practised by different nations from times immemorial, have, in not a single instance, led to the malformation or reduction of the parts in question. Such hereditary effects have been produced neither by circumcision, nor the removal of the front teeth, nor the boring of holes in the lips or nose, nor the extraordinary artificial crushing and crippling of the feet of Chinese women. No child among any of the nations referred to possesses the slightest trace of these mutilations when born: they have to be acquired anew in every generation.

It is recorded that the Jews have practiced circumcision since the time of Abraham who lived about 5,500 years ago. If we make the conservative assumption that the time of Abraham and the beginning of circumcision in the Jews was only 5,000 years ago, and that the average length of a generation in man is 25 years, we can conclude that this mutilation has been kept up consistently for at least 200 generations, without its effects having become hereditary.

Weismann undertook to investigate the heritability of mutilations by means of a simple experiment, commenting wryly that "such a course might, perhaps, have been more natural to those who maintain the transmission of mutilations," but who contented themselves instead with evidence of an anecdotal nature.[19] He amputated the tails of white mice during five generations from 1887 to 1889, and raised 901 individuals from artificially mutilated parents. The results are reported and discussed in the following words[20]:

There was not a single example of a rudimentary tail or of any other abnormity in this organ. Exact measurement proved that there was not even a slight diminution in length. . . .

What do these experiments prove? Do they disprove once for all the opinion that mutilations cannot be transmitted? Certainly not, when taken alone. If this conclusion were drawn from these experiments alone and without considering other

[18] Weismann, 1889–92, I, 446–47.
[19] Weismann, 1889–92, I, 444.
[20] Weismann, 1889–92, I, 445–46.

facts, it might be rightly objected that the number of generations had been far too small. It might be urged that it was probable that the hereditary effects of mutilations would only appear after a greater number of generations had elapsed. . . .

Hence the experiments on mice, when taken alone, do not constitute a complete disproof of such a supposition: they would have to be continued to infinity before we could maintain with certainty that hereditary transmission cannot take place. But it must be remembered that all the so-called proofs which have hitherto been brought forward in favour of the transmission of mutilations assert the transmission of a single mutilation which at once became visible in the following generation. Furthermore, the mutilation was only inflicted upon one of the parents, not upon both, as in my experiments with mice.

To cite a case from an invertebrate animal, the fruit fly Drosophila was reared and bred for 69 generations in the dark where the eyes were unable to function. Neither the morphology of the eyes nor the vision of the flies was affected by this prolonged period of disuse.[21]

The nineteenth-century botanist Anton Kerner transplanted scores of plant species, such as the field violet (*Viola arvensis*), common groundsel (*Senecio vulgaris*), veronica (*Veronica polita*), parnassia (*Parnassia palustris*), campion (*Lychnis viscaria*), and others, from the lowland valleys of Austria to an experimental garden at 7,200 feet elevation in the Tyrolean Alps. The lowland plants grown in the alpine environment produced shorter stems, smaller leaves, smaller and fewer flowers standing closer to the ground, and a more brilliant coloration of both leaves and flowers than parallel lots of the same species grown in the lowlands. The plants grown in the alpine garden gave rise to seedling progeny exhibiting the same modifications as their parents as long as they were grown in the same alpine environment. But as Kerner noted[22]:

As soon as the seeds formed in the Alpine region were again sown in the beds of the Innsbruck or Vienna Botanic Gardens the plants raised from them immediately resumed the form and colour usual to that position. The modifications of form and colour produced by change of soil and climate are therefore not retained in the descendants. . . . In no instance was any permanent or hereditary modification in form or colour observed.

Johannsen showed in 1903 that the differences in size and weight of bean seeds which were dependent solely on nutritional conditions were not inherited. Johannsen then suggested the fundamental distinction

[21] Payne in 1911; Huxley, 1943, 459.
[22] Kerner, 1894–95, II, 514; order of sentences changed in quotation.

of genotype and phenotype, referring to the sum total of the genes and the sum total of their character expressions, respectively. This distinction has come to be accepted generally as a refinement over the more strictly morphological distinction between germplasm and body tissues.

MODERN VERSUS TRADITIONAL THEORIES OF HEREDITY

It may be useful at this point to contrast in summary form the modern genetical conception of heredity with the views held by the earlier biologists. Figure 27 is an attempt to show diagrammatically the three theories of heredity.

1. The theory of the inheritance of acquired characters, as held by Lamarck, Darwin, and other traditional authors, postulated that hereditary particles migrated from the various body parts to the sex organs at the time of reproduction (Fig. 27a). In this way environmental influences affecting the body parts were supposed to be transmitted to future generations. This theory is invalidated by both the embryological and the genetic evidence.

2. The theory of the germplasm, proposed by Weismann in the latter nineteenth century, held that the hereditary material or germplasm is set aside from the rest of the body at an early stage. As a result of this morphological isolation of the germplasm from the body or somatic tissues, environmental influences on the body cannot be transmitted in inheritance (Fig. 27b). This theory fits the genetic data in all organisms and is generally consistent with the facts of embryology. However, it postulates a difference in the genetic constitution of the reproductive and somatic cells which is not in fact there.

3. The chromosome theory of heredity was formulated by Sutton, Boveri, Wilson, Johannsen, and other geneticists around the turn of the century. It maintains that the essential distinction is not between a special tissue, the germplasm, and the rest of the body, but rather between the set of genes and the morphological and physiological traits they determine (Fig. 27c). This theory is in agreement with all the available embryological, genetic, and cytological information.

Most of the genes were found to be located on the chromosomes in the nucleus of each cell in the body. The body grows by a process of

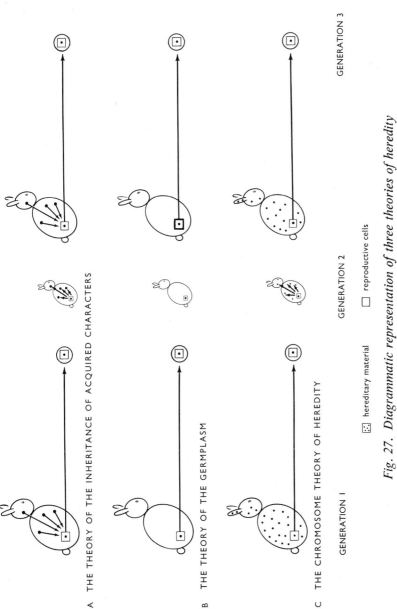

Fig. 27. Diagrammatic representation of three theories of heredity

The hereditary material which determines the development of similar bodily characters in successive generations is indicated by black dots. The reproductive cells which pass this hereditary material on from one generation to the next are shown by a hollow square in the animal's lower body.

Based on a diagram by Darlington, 1953.

cellular division involving the distribution of equal daughter chromo-somes to each daughter nucleus. About 24 cell divisions take place in some cell lines from the fertilized egg to the mature tissue in a Drosophila or a man.[23] The different cells of the body—the muscle, nerve, and blood cells in an animal or the leaf, root, and fruit cells in a plant—therefore possess identical copies of the same chromosomes and the same genes. (There are exceptions to this generalization which do not need to concern us here.) The physical basis of the genotype, the chromosome complement, is thus scattered throughout the body, and is not pre-served intact only in the reproductive parts as Weismann had believed. The cells composing the non-reproductive organs and those composing the reproductive organs have the same genetical constitution, but are specialized for performing the different functions of their respective body parts.

The special function of the reproductive organs is to pass a set of the organism's genes on to the next generation. What these organs transmit in inheritance is not phenotypic characters but the genes which deter-mine phenotypic characters. Furthermore, the reproductive cells transmit only the genes that they carry and not those found in other cells of the body. Therefore the only changes that are hereditary are those occurring in the genes of the reproductive cells.

THE THEORY OF THE INHERITANCE OF ACQUIRED CHARACTERS IN RELATION TO EVOLUTION

As we have already seen, the traditional view of heredity held that an organism passes the phenotypic characteristics acquired during its lifetime, the modifications of its body produced by environmental influences and by the use or disuse of organs, on to its descendants. This view was stated by numerous medieval and Renaissance writers.[24] In 1809 Lamarck took up the theory of the inheritance of acquired characters, without attempting to test its validity by experiments, and used it as a means of explaining evolutionary change. Observing that organisms are now fitted for their particular environments and habits of life, and postulating that these organisms have developed their present characters by evolutionary transformations from their ancestors, he

[23] Haldane, 1954, 101.
[24] Zirkle, 1946.

argued that their fitness for their conditions is due to direct molding by the environment and the transmission of the environmental influences to future generations.[25] In a word, adaptability is supposed to become converted directly into adaptation.

Before the theory of adaptive evolution by the inheritance of acquired characters can be discussed profitably, it is necessary to clarify two of the concepts involved. First, what is it that is transmitted in inheritance? The modern view on this question was summarized in the preceding section. Secondly, what exactly do we mean by acquired characters?

Hereditary diseases are both acquired and inherited. Mutations acquired during the lifetime of an individual will be passed on to its progeny if they occur in the line of cells which produces the gametes. An individual can pick up such inherited changes as a result of exposure to particular environmental influences, such as disease organisms or x-rays. But since hereditary diseases and mutations usually produce deleterious effects on their carriers, their acquisition leads more to degeneration than to adaptation.

What the Lamarckian theory of evolution really requires therefore is the inheritance of acquired adaptive characters. The most common kind of adaptive change in an individual is its phenotypic modifications. But, as shown in an earlier section, experiments with different plants and animals have repeatedly demonstrated that, contrary to the Lamarckian theory, phenotypic modifications are not inherited. What is inherited, indeed, is not phenotypes at all, still less their particular responses to particular stimuli, but rather a constellation of genes which can determine the formation of phenotypes. Adaptive changes, to be inherited, must begin by being changes in the genic material of the reproductive cells; and this conclusion is fatal to the old idea that the reactions of the non-reproductive parts of the body to environmental influences can enter the stream of heredity and evolution.

The foregoing considerations greatly restrict the range of possibilities for the operation of a Lamarckian mechanism of externally directed hereditary changes. This mechanism is restricted in theory to the special case in which an environmental influence can modify the

[25] Lamarck, 1809, 1815–22. Lamarck suggested other causes of evolution too, which are irrelevant to this discussion, but are mentioned briefly in Chapter 18.

phenotype in a given direction, and at the same time change the heredity so that the altered phenotype is reproduced in subsequent generations. Does such a situation ever arise, and does such a process ever take place? We will examine two possible situations.

It will be recalled that the genic material is normally located on the chromosomes within the nucleus of the cell. Some self-reproducing particles, however, reside outside the nucleus in the cytoplasm. These cytoplasmic genes include plastids of various sorts. Now the cytoplasm of the cell, and the self-reproducing particles contained in it, are a good deal more susceptible to environmental influences than is the nucleus.

The unicellular green flagellates possess chlorophyll-bearing plastids in their cytoplasm with which they perform photosynthesis in sunlight. The same organisms can be grown artificially in darkness, where their chloroplasts are functionless, providing sugar is fed to them in a nutrient solution. *Euglena gracilis* and *Euglena mesnili* both respond to darkness by gradually losing their chlorophyll. In the former species, however, the plastids continue to reproduce in cells growing in the dark, and the capacity to produce chlorophyll in the presence of light is retained indefinitely. When *Euglena mesnili* is grown in the dark, on the other hand, the chloroplasts decrease in number and eventually become lost altogether in some cell lines, which are then incapable of becoming green and photosynthetic again on reexposure to light.[26] As Haldane points out, the case of *Euglena mesnili* (but not that of *E. gracilis*) illustrates the hereditary loss of an organ and function through disuse.[27]

In *Euglena mesnili* an externally directed hereditary alteration is possible because the same organ that responds phenotypically to a certain environmental stimulus, the chloroplasts in the cytoplasm, also forms a constituent of the set of genes in the reproductive cells. In other words, the phenotype coincides with the genotype in this case.

In higher multicellular organisms the cytoplasmic genes appear to control various general physiological processes, which are not yet thoroughly understood.[28] It is hypothetically possible, though not demonstrated, that the self-reproducing cytoplasmic determinants of

[26] Lwoff in 1944; Haldane, 1954, 86.
[27] Haldane, 1954, 86; Michaelis, 1954, 389.
[28] See Michaelis, 1954.

some general process, like growth rate, might respond directly in concentration and intensity of action to external environmental conditions. Such responses, if carried out in the reproductive as well as the somatic cells, as might easily happen in plants with their production of flowers or cones on vegetative shoots, could conceivably be transmitted in inheritance.[29]

The second possibility to be considered is that the environmental influence affecting some organism might itself be the genetic material of another organism, in which case, of course, it could produce a directed hereditary change in the former. Hereditary transformation is known to take place in the bacteria. The pneumonia bacterium, Pneumococcus, exists in various true-breeding forms, some of which are characterized by a smooth cell surface and others by a rough surface. In one experiment the rough form was grown in a sterile and lifeless extract containing the genetic material of the smooth form. In this medium the rough bacterial cells produced a smooth coat. The change moreover was permanent and was transmitted to subsequent bacterial generations. The heritable properties of the donor strain were conferred on the receptor strain by a sterile growth medium.[30]

In the example just described, the transformation of one bacterium by the incorporation of the genetic material of another bacterium was brought about artificially by man. A question naturally arises as to whether bacteria are ever exposed under natural conditions to foreign genetic substance in a way which can produce similar hereditary transformations. Bacterial viruses infecting different bacterial cells are now known to transfer bits of genetic material from one bacterium to another. This method of transfer of genetic substance by infective bacterial viruses, known as transduction, leads to the acquisition of new hereditary characteristics by the recipient bacterium.[31]

The processes of transformation and transduction in bacteria have no known counterparts in higher organisms. To be sure, hereditary disease organisms may introduce foreign genetic material into the body of a host, with results that are perpetuated in later generations; but otherwise it is not clear how foreign genes could be implanted in the

[29] Crosby, 1956.
[30] Avery, Macleod, and McCarty, 1944.
[31] See Zinder, 1958.

highly organized cellular apparatus of plants or animals in such a way as to become permanent components of the genotype. Occasional claims to have accomplished this result experimentally, as by permanently transforming breeds of ducks by injections of germplasm, require confirmation.

With regard to higher organisms, and many lower ones as well, a mechanism does exist in nature which regularly brings together dissimilar genetic material. That mechanism is sex. Sexual reproduction plays an extremely important role in the normal process of evolution in higher organisms, and is promoted accordingly by numerous devices in most species. The hereditary changes produced by the sexual mechanism are sorted out by environmental selection and under certain conditions by genetic drift. This method of organic evolution is more than a mere hypothetical possibility.

As regards the higher organisms (which we are concerned with primarily in this book), the available evidence justifies the following statement of Weismann[32].

If the transmission of acquired characters is truly impossible our theory of evolution must undergo material changes. We must completely abandon the Lamarckian principle, while the principle of Darwin and Wallace, viz. natural selection, will gain an immensely increased importance.

THE BALDWIN EFFECT

Although the phenotypic reactions of organisms are not inherited, and hence do not lead directly to the evolution of new adaptive characteristics, the ability of an organism to adjust phenotypically to its environment does have evolutionary consequences. The evolutionary role of phenotypic reactions is indirect, but is not less real on that account.

Each genotype engenders a certain range of phenotypic reactions to varying environmental conditions. There are limits to the modifiability and adjustability of the phenotype produced by any given genotype. The heat-regulating mechanism of the human body, for example, can function only within a certain temperature range, beyond which the body succumbs to freezing or heat stroke.

[32] Weismann, 1889–92, I, 435.

Within the tolerance limits, moreover, there is a smaller range of conditions to which the genotype and its normal phenotypic expression are best adapted. Although the genotype can respond to extreme environmental conditions by making temporary adjustments, it is specialized for producing a "normal" phenotype under the conditions usually encountered by the population, and the extreme modifications represent an expediency. The human body may be able to endure the midday summer temperatures in hot deserts, but it is not comfortable and cannot perform well under those conditions. The genotype, placed in an environment close to its tolerance limits, may barely survive in a state of poor adaptedness.

But survival is obviously a necessary first step in the evolution of new adaptive properties. The next step is the appearance of heritable gene changes, or mutations, which cause a phenotype to develop which is better adapted than the old forms in the new environment. Heritable gene changes, as we shall see later, do occur with a low but definite frequency in all organisms, and while most of these mutations have deleterious effects, a small fraction of them are beneficial. The final step in the evolution process is the establishment of the rare favorable mutations in a population of organisms.

In the over-all process of evolution, phenotypic adjustability, by enabling a genotype to hang on in a marginal environment where it is poorly adapted, may give it time in which to mutate to new forms that are better adapted to these conditions. A population of animals capable of adjusting phenotypically, albeit with difficulty, to hot desert temperatures might be able to survive long enough in the desert to accumulate gene mutations for greater heat tolerance. At least the population, if it survives, has a chance of becoming better adapted by hereditary changes. If the animals cannot adjust phenotypically to the new conditions, they have lost their first and perhaps their only chance of evolving to a state of better adaptedness.

The role of phenotypic adjustments in evolution is thus an indirect one.[33] Phenotypic reactions permit a population to exist in an environment to which it is not well adapted and give it time to acquire, by a more or less random process of gene mutations, genotypes which are

---

[33] See Huxley, 1943, 304.

adapted to the new environment. This process was called organic selection by Baldwin in 1896, and has been renamed the Baldwin effect in recent years.[34]

SUMMARY

It is clear that living organisms meet the challenges of the environment actively in various ways. An individual can adjust itself to a range of conditions by means of phenotypic modifications, physiological homeostasis, and habitat selection. A population, a group of individuals existing through successive generations, can evolve.

[34] Baldwin, 1896; Simpson, 1953.

Part 3 *Evolution within Populations*

SEVEN

The Statics and Dynamics
of Populations

REDUCED TO its barest essentials, evolution is a change in the frequency of gene alleles. The field of space and time in which such changes take place is the population. A population containing two alternative forms or alleles of a gene will not undergo changes in the proportions of these alleles unless certain forces are at work. The population can be considered to have an equilibrium condition when the alleles remain at constant frequencies. That equilibrium can then be changed by the action of known evolutionary forces.

In this chapter we shall consider first the equilibrium condition of a population at rest, and then the forces which can produce evolutionary changes in the population. These aspects can be thought of respectively as the statics and dynamics of the evolution process. Since the evolutionary forces will be discussed in more detail in the chapters following this one, the account given here is limited to an introductory preview.

THE BREEDING POPULATION

Under natural conditions animals, plants, and microorganisms exist in populations consisting of scores, hundreds, thousands, or millions of interbreeding individuals. One can think of a herd of buffalos, a school of fish, a stand of pine trees, or a field of wild flowers. The breeding population is a unit of reproduction held together by bonds of mating and parenthood. It has a continuity of its own in space and time.[1]

Wherever such breeding populations have been studied, they have been found to be variable, or polymorphic. A population of the butterfly *Papilio dardanus* may contain forms distinguished by different wing

---

[1] Dobzhansky, 1950.

colors and wing shapes (Plate I, h-k). A population of humans is likely to consist of brown-eyed and blue-eyed individuals, tall and short persons, individuals with varying degrees of resistance to the common cold, and so on. Any given population is, moreover, characterized by certain frequencies of the different polymorphic types. Thus 73 percent of the individuals of one human population may be blue-eyed and 27 percent brown-eyed; in another population the respective frequencies of eye colors may be 12 percent and 88 percent.

Underlying much of the observable diversity in a natural population is a diversity in the allelic forms of one or more genes. The polymorphism in bodily characteristics is based in part on a genetic polymorphism. In terms of a hypothetical gene A present in two allelic forms (a_1 and a_2), individuals with three different genetic constitutions may exist together in the population (thus a_1a_1, a_1a_2, a_2a_2).

Polymorphic genes usually exist in more than two allelic forms. A series of several or many alleles is found within the species for the gene in question; besides a_1 and a_2, there are a_3, a_4, a_5, and so on. The number of alleles in a multiple allelic series may be rather large. It is estimated that there are 160 alleles of the gene B in cattle,[2] and over 200 alleles of the gene S in red clover.[3] (The B gene in cattle determines the immunological type of blood. The S gene in red clover controls self-incompatibility or the inability of an individual plant to set seeds with its own pollen.)

The number of possible combinations of alleles in pairs increases exponentially as the allelic series is extended. With two alleles there are two kinds of gametes which can be combined to form three genetically different kinds of individuals. Three alleles (a_1, a_2, a_3) can combine in pairs to form six genotypes (a_1a_1, a_1a_2, a_1a_3, a_2a_2, a_2a_3, a_3a_3). With four alleles there are ten genotypes, and with five alleles 15 kinds of individuals. In general, where r represents the number of alleles of any gene, the number of different genotypes (g) is given by the formula[4] $g = r(r + 1)/2$.

Each individual animal or plant is derived from the fusion of two gametes and hence carries one pair of alleles for each gene. If, therefore,

[2] Wright, 1960.
[3] Crow, 1960, 57.
[4] Also Stern, 1960, 176.

any gene is present in three or more allelic forms, no single individual can carry all the genetic variants found within the population. As the series of multiple alleles increases in size, each individual organism possesses a proportionately smaller fraction of the total genetic variation. The maintenance and storage of genetic variation under these conditions is a function of the population as a whole.

A polymorphic population is a storehouse of genetic variation. In general, the larger the population, the greater the number of genetic variants that can be kept in storage.

GENE INERTIA IN POPULATIONS

The genetically different individuals in the population commonly mate without regard to their genetic constitutions. This would be the case, for example, if brown-eyed women accepted blue-eyed men as readily as brown-eyed men. The offspring can then be regarded as the products of random combinations of the various kinds of gametes produced by the parental individuals.

It is possible, in other words, to treat both the ancestral generation and the descendant generation collectively and as a whole; and in order to do so it is necessary to treat the connecting link between them, namely, the gametes produced by the parental generation, as a collective whole. We can consider the sum total of the gametes produced by the parental generation as a pool of gametes. Under conditions of random mating the individuals comprising the descendant generation then represent the products of different pairs of gametes drawn at random from this gamete pool.

The frequency of any type of gamete in the gamete pool (i.e., an a_5 gamete) will depend on the initial frequency of the parental individuals carrying the a_5 allele. It so happens that in the next generation the individuals derived from the a_5 gametes will preserve the same allele frequency. Unless certain forces to be described later are at work in the population, the frequency of any given allele will remain approximately the same from generation to generation. This generalization, known as the Hardy-Weinberg law,[5] can be illustrated by a series of hypothetical examples.

[5] Hardy and Weinberg in 1908; see Srb and Owen, 1952, ch. 20; and Stern, 1960, ch. 10.

For the sake of simplicity in our first example let us imagine a population polymorphic for a single gene, A, present in two allelic forms, a_1 and a_2. These two alleles are distributed among three genotypes (a_1a_1, a_1a_2, and a_2a_2). Assume arbitrarily that the population consists of 20 individuals. We will suppose that 5 of these individuals have the constitution a_1a_1, that 10 individuals are a_1a_2, and 5 are a_2a_2. Our first step is to determine the allele frequencies in the original population. The 20 individuals carry between them a total of 40 alleles. There are 20 a_1 alleles contained in the population (10 a_1 alleles in the 5 a_1a_1 individuals and 10 a_1 alleles in the 10 a_1a_2 individuals). The a_1 allele thus has a frequency of 20/40, or 50 percent. Likewise the a_2a_2 and a_1a_2 types collectively carry 20 a_2 alleles, and the allele frequency of a_2 is also 20/40, or 50 percent.

In order to determine the allele frequencies in the next generation we must consider separately the two steps of gamete formation and fertilization. It is assumed that each individual in the original population produces equal numbers of gametes, and that the heterozygotes produce equal numbers of a_1 and a_2 gametes. The frequencies of the two types of gametes will then correspond exactly to the frequencies of the two alleles. In other words, the gamete pool will consist of 50 percent a_1 gametes and 50 percent a_2 gametes.

We are assuming that random mating prevails within the population. This means that a given female gamete has an equal chance of being fertilized by any male gamete in the gamete pool. Where, as in our present example, two classes of gametes are present in equal frequencies, an a_1 female gamete will be fertilized as often by an a_2 male gamete as by an a_1 male gamete.

The products of the random combinations of the gametes can be shown by constructing a checkerboard with the female a_1 and a_2 gametes arranged on the left-hand side and the male a_1 and a_2 gametes across the top (see Fig. 28, left). The percentage frequency of each class of gametes in the gamete pool is also indicated. The individuals expected to be derived from this array of gametes can now be written in the appropriate squares (Fig. 28, right). The union of an a_1 female gamete with an a_1 male gamete will produce an a_1a_1 individual. The frequency with which an a_1a_1 individual will be formed in the new generation, moreover, is the product of 50 percent a_1 female gametes $\times$

50 percent a_1 male gametes, or 25 percent. We expect the next generation to consist of 25 percent a_1a_1 individuals, 50 percent a_1a_2 individuals,
and 25 percent a_2a_2 individuals.

Our final task is to compute the frequencies of alleles in the derived
generation. Proceeding by the same method that we used for the
parental population, we discover that the a_1 and a_2 alleles both have a
frequency of 50 percent in the new generation. The allele frequencies
are thus unchanged.

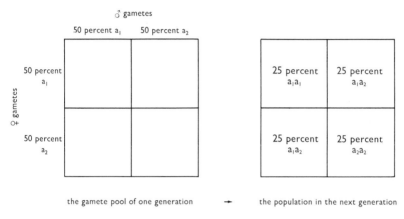

*Fig. 28. The composition of the gamete pool produced in one generation
(left), and the composition of the population derived from the random
union of these gametes in the next generation (right)*

Suppose now that the original breeding population consists of
15 a_1a_1, 8 a_1a_2, and 3 a_2a_2 individuals. Thirty-eight of the 52 alleles
present are a_1, and this allele thus has an initial frequency of 73.1
percent. The allele frequency of a_2 is 26.9 percent. The gametes
carrying these alleles combine in proportion to their relative frequencies.
It can be shown by the same type of checkerboard as was used above
that the population will quickly reach an equilibrium in which 53.4
percent of the individuals are a_1a_1, 39.4 percent a_1a_2, and 7.2 percent
a_2a_2 (see Fig. 29). A summation of the alleles in the derived generation
reveals that their proportions are $(.534 + .534 + .197 + .197)$ a_1:
$(.197 + .197 + .072 + .072)$ a_2. This turns out to be 73.1 percent a_1
and 26.9 percent a_2, the same frequencies as in the original generation.

♂ gametes

	.731 a₁	.269 a₂
.731 a₁	.534 a₁a₁	.197 a₁a₂
.269 a₂	.197 a₁a₂	.072 a₂a₂

♀ gametes

Fig. 29. *The random union of two classes of gametes, a₁ and a₂, present in the relative frequencies .731 and .269, will lead to the production of different genotypes in the frequencies shown*

♂ gametes

	.6 a₁	.2 a₂	.1 a₃	.1 a₄
.6 a₁	.36 a₁a₁	.12 a₁a₂	.06 a₁a₃	.06 a₁a₄
.2 a₂	.12 a₁a₂	.04 a₂a₂	.02 a₂a₃	.02 a₂a₄
.1 a₃	.06 a₁a₃	.02 a₂a₃	.01 a₃a₃	.01 a₃a₄
.1 a₄	.06 a₁a₄	.02 a₂a₄	.01 a₃a₄	.01 a₄a₄

♀ gametes

Fig. 30. *The frequencies of the genotypes derived from the random union of four classes of gametes present in given frequencies*

The same result is obtained when the breeding population is polymorphic for a series of three or more alleles. Let the initial frequencies of four alleles and of the gametes carrying them be 60 percent a_1, 20 percent a_2, 10 percent a_3, and 10 percent a_4. The gametes combine with one another in the proportions shown in the checkerboard in Fig. 30. The genetic constitutions of the individuals comprising the new generation are given in the boxes of the checkerboard, along with the relative frequency of each genotype. Distributed among the boxes is a total of 100 individuals, each of which carries one pair of alleles. Collecting the a_1 alleles distributed among the different boxes, we find that this allele is represented 120 times; its percentage frequency is therefore 120/200 or 60 percent. The a_2 allele has a collective frequency in the new generation of 40/200 or 20 percent. Both the a_3 and a_4 alleles are present in the proportions 20/200 or 10 percent. The allele frequencies are thus the same in the new generation as in the preceding one.

The process of gene reproduction thus operates to conserve indefinitely an array of variants in a large breeding population. Not only is the genetic variation conserved, but it tends to be conserved in unchanging proportions by the reproductive process. The net result of gene replication, gamete formation, and random fertilization is that the allele frequencies remain constant in successive generations. The allele frequencies in a large breeding population do not change by themselves. The system of polymorphic genes in a population can be said to have inertia.

EQUILIBRIUM FREQUENCIES OF GENOTYPES
IN A POPULATION

Implicit in the foregoing discussion is the notion that the relative frequencies of alternative genotypes reach an equilibrium and then remain constant from generation to generation in a large random mating population. The Hardy-Weinberg law describes this equilibrium condition of genotype frequencies in the following terms.

Assume that two alleles a_1 and a_2 of the gene **A** are present in the population, and let p represent the frequency of a_1, and q the frequency of a_2 (p + q = 1, so q = 1 − p). The zygotes produced by random combinations of the a_1 and a_2 gametes in pairs will then be expected to occur in the proportions given by expansion of the binomial square $(p + q)^2$. The expected proportions of genotypes are thus

$$p^2(a_1a_1): 2pq \ (a_1a_2): q^2(a_2a_2).^6$$

If the frequency of a_1 gametes in the gamete pool is 50 percent, and that of a_2 gametes is likewise 50 percent, as in the first hypothetical case considered in the preceding section, so that $p = q = .50$, the distribution of genotypes at equilibrium under conditions of random mating will be

$$.25 \ a_1a_1 + .50 \ a_1a_2 + .25 \ a_2a_2.$$

Or if, as in the second case considered earlier, the frequencies of a_1 and a_2 in the gamete pool are $p = .73$ and $q = .27$, the substitution of these values in the Hardy-Weinberg formula gives us

$$p^2 = .534 : 2pq = .394 : q^2 = .072.$$

It will be noted that these are the same frequencies for the genotypes a_1a_1, a_1a_2, and a_2a_2, respectively, as we derived by the checkerboard method previously (see Fig. 29).

In the case of a population polymorphic for a series of multiple alleles, the frequencies of the different alleles can be represented by three or more terms, the values of which add up to unity. The frequencies of the various genotypes at equilibrium can then be calculated from the polynomial equation containing these same terms. Thus if the population contains a mixture of four alleles of **A** and hence four classes of gametes (a_1, a_2, a_3, a_4), the frequencies of which are represented by p, q, r, and s, so that $p + q + r + s = 1$, the frequencies of the different genotypes derived from the random pairing of the four gametes are given by the expansion of the polynomial equation $(p + q + r + s)^2$.

Let us reconsider, as an example, the third case in the preceding section, where a_1, a_2, a_3, and a_4 gametes occur in the gamete pool in the frequencies of 60, 20, 10, and 10 percent, respectively. The four alleles can be represented as follows:

$$a_1 \quad p = .60 \qquad a_2 \quad q = .20 \qquad a_3 \quad r = .10 \qquad a_4 \quad s = .10$$

By expanding the polynomial $(p + q + r + s)^2$, we derive the following

[6] See Srb and Owen, 1952, ch. 20; Stern, 1960, ch. 10.

expected frequencies of the various genotypes resulting from random combination of these gametes:

$$p^2 = .36 \quad a_1a_1 \qquad 2pq = .24 \quad a_1a_2 \qquad 2qr = .04 \quad a_2a_3$$
$$q^2 = .04 \quad a_2a_2 \qquad 2pr = .12 \quad a_1a_3 \qquad 2qs = .04 \quad a_2a_4$$
$$r^2 = .01 \quad a_3a_3 \qquad 2ps = .12 \quad a_1a_4 \qquad 2rs = .02 \quad a_3a_4$$
$$s^2 = .01 \quad a_4a_4$$

The calculated frequencies of the ten genotypic classes are the same as those shown in the checkerboard in Fig. 30.

Thus not only alleles but also the various classes of genotypes, once they reach their equilibrium frequency in a population, tend to remain indefinitely in constant proportions under conditions of random mating.

THE PRIMARY EVOLUTIONARY FORCES

It is an observable fact that breeding populations do undergo changes in the frequencies of different alleles during a succession of generations. Such changes may be detected directly by sampling the gamete pool at intervals and finding a systematic increase or decrease in the frequency of a given allele; or the allele frequency changes may be inferred from an observed change in the relative abundance of a morphologically recognizable type of individual. In either case the changes in allele frequencies are by definition evolution.

What causal factors can bring about evolutionary changes in populations? Four such causal factors or forces are known. They are mutation, gene flow, natural selection, and genetic drift. The primary evolutionary forces will be introduced briefly here and will be discussed in more detail in the following chapters.

Mutations are discrete heritable changes in the genes and chromosomes. In our hypothetical example, an a_1 allele may occasionally change or mutate to the new form a_2, thereby increasing the frequency of the latter. Or individuals carrying the allele a_2 may migrate into the population with the same net effect (gene flow).

Individuals with the different genotypes a_1a_1, a_1a_2, and a_2a_2 will possess different characteristics, which may affect their ability to survive and reproduce. If the mutant individuals a_2a_2 are superior to their siblings a_1a_1 and a_1a_2 in ways affecting their ability to reproduce, the allele a_2 will increase in frequency. The change of allele frequencies in this case is due to the effect of natural selection. Finally, the allele a_2 may undergo a change in frequency due to chance alone (genetic drift).

Each one of the four primary evolutionary forces has been measured quantitatively in experimental populations. The frequency with which a gene mutates per generation is conventionally represented by the letter u. The gene **A**, for example, might be perfectly stable, or unstable, or might mutate to new allelic forms at a certain low rate; thus u would equal 0, 1, or some intermediate value. The rate of gene flow is symbolized by m.

The average increase in frequency of one allele relative to another in each generation as a result of natural selection is indicated by the selection coefficient (s), which varies from 0 to 1. If $s = 0$, no selection is taking place, that is, the carriers of the alleles a_1 and a_2 have an equal ability to survive and reproduce. The value $s = 1$ indicates a complete replacement of one allele by another in a single generation, and intermediate values of s measure intermediate rates of gene replacement.

The operation of drift, finally, is indicated by a certain relationship between the selection coefficient and the number of breeding individuals in the population (N); drift may take place when $N \leq 1/2s$, assuming that m and u have low values.

CONCLUSIONS

The first two forces, mutation and gene flow, produce variability. The second two forces, selection and drift, sort out this variability and establish the variant types in new frequencies in a population.

The variation-producing forces which form the starting point of the evolution process are not oriented in the direction of adaptation, as we shall see later. Mutational changes in the genes and chromosomes occur more or less at random with respect to the adaptive requirements of the organism. Mutations, alone or in cooperation with gene flow and drift, could not bring about the non-random combinations of genes embodied in the genotypes of organisms. The agency of selection, with or without drift, is required in order to carry the evolution process through to the stage of adaptation.

"The unit of variation," as Darlington has put it, "is not the unit of selection."[7] Indeed the variation-producing forces and the selective forces operate on different levels of organization. For as Darlington has said[8]:

Changes in the chromosomes are determined by conditions of molecular stability. They are biologically at random. The combinations of these changes together with the selection of environments [are] what take us from the chemical level of mutation to the biological level of adaptation.

Evolutionary changes in populations thus come about as a resultant of the opposing forces that produce and sort out heritable variations. Evolution is due, not to the operation of any single factor, but to the interplay between a limited number of primary forces.

[7] Darlington, 1939, 127.
[8] Darlington, 1939, 127, quoted from *The Evolution of Genetic Systems*.

EIGHT

Hereditary Variation

Any variation which is not inherited is unimportant for us. CH. DARWIN[1]

THE EVOLUTION PROCESS begins with the origin of variation. An understanding of variation, its nature and causes, is therefore necessary for an understanding of the dynamics of evolutionary change. The classical theories of evolution, and some recent accounts as well, have gone on the rocks in one way or another because of incorrect notions concerning variation. As Darwin and Weismann pointed out, only the hereditary variations enter directly into evolutionary changes.

THE CAUSES OF GENETIC VARIABILITY

A certain amount of hereditary variation is generally found in breeding populations of plants and animals. Among sexual organisms, in fact, no two individuals are likely to have the same genetic constitution, with rare exceptions such as identical twins. Genetic variability, the occurrence of individuals with different genetic constitutions, or, in other words, the presence of genotypically different individuals, is a universal feature of breeding populations.

The genotypic variation between individuals is the starting point of evolutionary changes. It is appropriate, therefore, to consider here the kinds of genetic differences that we find between individuals of a population, and the causes of these differences.

It is convenient for the purposes of our discussion to classify genetic variation into two general categories: single-gene variation and multiple-gene variation. We can consider the differences between individuals of a population with respect to a single gene, **A**, or the individual variation for two or more genes, **A** + **B** + **C**, etc.

The variation for a single gene in the population arises from two

[1] Darwin, 1859, ch. 1.

152

sources: mutation and gene flow. Multiple-gene variation is caused by mutation, gene flow, and an additional process, recombination.

In the following sections of this chapter we will consider each of the causes of heritable individual variation in turn.

GENE MUTATION

It is generally believed that each gene allele has a specific and characteristic molecular configuration which determines its biochemical actions and its phenotypic effects. The idea that gene structure predetermines gene action and hence hereditary traits, stated in general terms by Weismann, Wilson, and other late nineteenth-century biologists, and restated in terms of protein chemistry by many geneticists in the early twentieth-century, is expressed today as a correlation between a particular sequence of base-pairs on a segment of the DNA chain and the specific metabolic actions initiated by that base-pair sequence.

The structural configuration of any given allelic form of a gene is copied identically during successive cell divisions. The allele a_1 of the gene **A** is replicated in all the cells of the body of one individual. And since the production of daughter individuals can be regarded, in a sense, as an extension of the continuous process of cell division, the same allele a_1 is perpetuated from generation to generation. The hereditary resemblance of offspring to parents is due to the self-replicating properties of genes.

Nevertheless, the power of an allele to replicate itself unchanged during cell divisions is not unlimited, and in the process of forming a new copy, alterations can arise in the molecular structure. This is important, for if the genes always produced identical copies of themselves in reproduction, and if no new forms of the hereditary units ever arose, there could be no evolution.

To take a hypothetical example, let us assume that five of the many nucleotides within the limits of the gene **A** carry the bases CCTTA. This sequence might reproduce itself faithfully for a long time but might give rise on one occasion to an altered copy CCTTT. The new sequence of bases will then reproduce its molecular structure, and the alteration or mutation is thus permanent. The standard allele a_1 containing the pattern CCTTA can be said to have mutated to a mutant allele a_2, distinguished by the pattern CCTTT. Since the allele a_2 has different properties than a_1, it is likely to alter in some way the course of development and produce different phenotypic effects.

Mutations, then, are changes in the genes, and presumably in the molecular structure of the genes, which bring about permanent and hereditary changes in the phenotype.

The phenotypic expression and the heritability of mutations, as of genetic factors generally, are subject to numerous conditions. Mutations may occur in any cell of the body. Mutant alleles arising outside the germ track in the somatic tissues die with the body of the individual organism in the case of animals or must be perpetuated by vegetative reproduction in plants. If and only if a mutation arises in the gametes or the cell line leading to the gametes can it be passed on to the next generation.

Most though not all new mutations in diploid organisms are recessive and hence do not manifest themselves phenotypically in heterozygotes, where the mutant allele can be covered up by the dominant standard allele. But individuals heterozygous for the mutations can, by inter-crossing, produce progeny homozygous for the new allele, and such homozygous recessive individuals then appear as mutant types visibly different from their normal parents and siblings.

In a wide variety of organisms which have been studied genetically, mutant phenotypes have been observed to arise with a rare but definite frequency in the populations and have been traced back to changes at particular gene loci. It will be sufficient for our purposes here to merely allude briefly to the fruit fly, *Drosophila melanogaster*, and other species of this genus, which because they can be reared easily and quickly in the laboratory and have giant chromosomes in the salivary glands of the larvae, have been very thoroughly studied from the standpoint of gene mutations.[2]

Most of the time the flies breed true to type. Occasionally, however, one or a few mutant individuals appear in the progeny of normal flies. Among the mutants described are flies with yellow bodies, reduced wings, vermillion instead of the normal red eyes, or bodies lacking the normal covering of bristles. A few mutant types in Drosophila are portrayed in Fig. 31; see also the mutant frogs and tomatoes in

[2] Good reviews of mutations in relation to evolution are given by Timofeeff-Ressovsky, 1940; and Dobzhansky, 1951a, chs. 2–3.

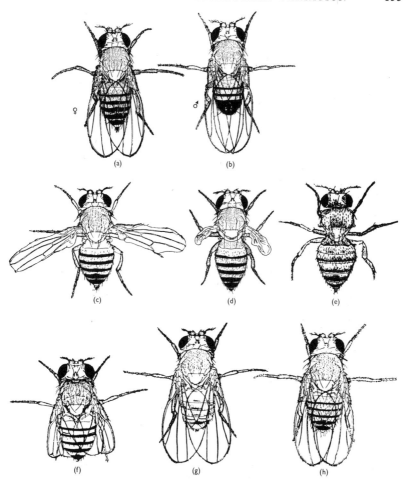

Fig. 31. Normal and several mutant types of the fruit fly, Drosophila melanogaster

The mutants are females

(a) Normal female   (d) Vestigial   (g) Bobbed
(b) Normal male   (e) Apterous   (h) Frizzled
(c) Fringed   (f) Dumpy

Morgan, Bridges, and Sturtevant, 1925; Bridges and Brehme, 1944.

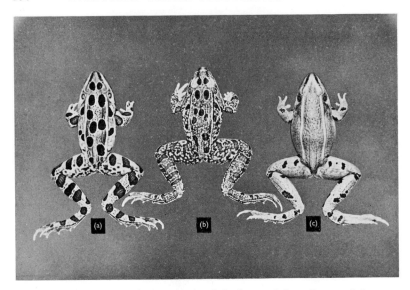

Fig. 32. Mutant color patterns of the leopard frog, Rana pipiens

(a) Normal form.
(b) Mottled, due to a dominant mutant allele.
(c) Burnsi, a dominant mutant at a different gene locus.

From E. P. Volpe, *J. Heredity*, 1960, with permission of the American Genetic Association. Photograph by courtesy of Dr. Volpe.

Figs. 32 and 33. A given mutant reproduces its altered phenotype in subsequent generations and therefore represents a permanent change in the genetic constitution of the pedigree. By appropriate cytogenetic techniques the hereditary change can be located in a particular region on a chromosome. By breeding tests the mutation can be shown to be homologous with some standard allele and hence is identified as a new allelic form of a particular gene.

The frequency with which mutations arise is expressed as the average number of mutations per generation. The mutation rate, the proportion of mutant alleles arising in a generation, varies in different organisms

Fig. 33. Mutant leaf forms of the wild tomato,
Lycopersicon pimpinellifolium

(a) Normal form	(c) Bipinnata	(e) Carinata	(g) Bullosa
(b) Accumbens	(d) Diminuta	(f) Dwarf	

Rearranged and redrawn from Stubbe, 1960.

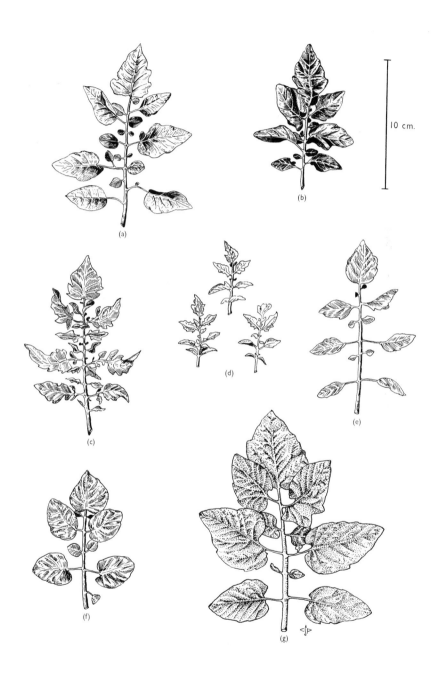

and for different genes in the same organism. Each gene which has been studied has been found to exhibit a characteristic mutation rate under normal environmental conditions. In corn, *Zea mays*, one out of 10,000 gametes can be expected to carry a mutant form of the gene **I**, a color inhibitor, whereas the gene **Sh** affecting seed shape mutates in one out of 1,000,000 gametes. The mutation rate of the gene **I** is thus .0001 and that of the gene **Sh** is .000001. Some genes in corn have mutation rates between these two values, others are more stable than **Sh**, and still others are more mutable than **I**.[3] In *Drosophila melanogaster* the average mutation rate per gene for a class of genes capable of giving rise to lethal mutations is 1 per 100,000 gametes per generation, or .00001.[4] Various disease-producing loci in man show mutation rates per gene ranging from .0001 to .00001.[5]

Since the genotype of a higher plant or animal consists of thousands of genes which are individually subject to mutation, the mutability of the organism as a whole is naturally much higher than that of any single gene. It is estimated that between 2 and 10 percent of the sperms and eggs of Drosophila carry some new mutation in each generation.[6] If we make the arbitrary but reasonable assumptions that the average mutation rate per gene in Drosophila is .00001, and the number of genes is 5,000, the average over-all mutation rate for the genotype would be 5 percent.

ADAPTIVE PROPERTIES OF GENE MUTATIONS

Many biologists during the early period of modern genetics failed to see any evolutionary significance in the mutation studies. They pointed out that the mutant flies or plants were freaks and monstrosities which could have no place in evolution. It was true that the mutants which the geneticists were studying in the laboratory and experimental garden were less viable than the wild-type organisms and represented alterations having no counterpart in nature. As the science of genetics developed, however, more and more gene mutations were discovered which affect vital processes. One group of plant geneticists found a gene controlling the vitamin content in a higher plant; genes were

[3] Stadler; see Dobzhansky, 1951*a*, 59.
[4] Wallace and Dobzhansky, 1959, 41–42.
[5] Wallace and Dobzhansky, 1959, 37.
[6] Timofeeff-Ressovsky, H. J. Muller; see Dobzhansky, 1951*a*, 56.

found for enzyme activity in both lower and higher plants; genes were discovered in corn and tobacco which determine the formation of chlorophyll, the sugar-manufacturing material of green leaves; genes became known which in different allelic states bring about fertility or sterility. These facts gradually convinced most biologists that many of the changes established in the evolution of a group of organisms could be traced ultimately to gene mutations.

The first mutations to be studied in Drosophila and plants were those producing conspicuous phenotypic effects. Many mutations were later found which have small phenotypic effects, such as changing slightly the number of bristles on the hind part of a fly.

It has been found in Drosophila that minor mutations arise far more frequently than mutations with conspicuous effects. Timofeeff-Ressovsky obtained a quantitative expression of the viability of a series of mutations induced by x-rays in *Drosophila melanogaster*. Over half of all the mutations in the sample studied were found to bring about only slight changes in viability. Thus 45 percent of the mutations lowered viability from the normal 100 percent to 95 percent, and another 10 percent of the mutations raised the viability from normal to 105 percent. Mutations with drastic effects on viability, that is, lethal and semilethal mutations, comprised only about 20 percent of the sample.[7]

Many geneticists now believe that the minor mutations are in general more important for evolution than the large conspicuous mutations. One reason for believing this is that an organism represents a complex assemblage of processes, which to be effective must work together harmoniously. A gross mutation of major effect is apt to throw the delicate machinery of the body out of kilter at some crucial point, and is therefore likely to be eliminated by natural selection. Mutations which alter the workings of the organic system only slightly, on the other hand, have a better chance of fitting in with the other genes and hence of being incorporated as a permanent element of the genotype.[8] Furthermore, small mutations occur in greater numbers than major mutations, so that a larger store of the former is available for natural selection or drift to establish as regular components in the population.[9]

[7] Timofeeff-Ressovsky, 1940, 84.
[8] Timofeeff-Ressovsky, 1940; Dobzhansky, 1951a, ch. 2.
[9] Timofeeff-Ressovsky, 1940; Dobzhansky, 1951a, ch. 2.

Direct evidence confirming the importance of minor mutations in evolution is provided by experimental studies on the inheritance of the characteristics which differentiate related races or species. Such studies show that races and species differ not in just a few genes of major effect but in numerous genes which individually produce minor phenotypic effects.[10]

Mutant forms are usually less viable than the normal forms of an organism.[11] Thus 90 percent of a series of mutations induced by x-rays in *Drosophila melanogaster* were found to lower the viability of their carriers to some extent.[12] If this were always the case, however, mutations could hardly form the raw materials for evolution.

A number of examples are known, actually, of mutations which improve the viability of their carriers. Ten percent of the sample of mutants of *Drosophila melanogaster* mentioned above had an enhanced viability. In barley the mutation "erectoides I" causes an increased strength of the straw and higher yield. The yield of the mutant form was found to be about 2 percent higher than that of the mother strain of barley at three stations in Sweden over a period of years.[13] Mutant flies of the type known as "eversae" in *Drosophila funebris* have an average viability 4 percent higher than that of wild-type flies belonging to the same stock when reared at 25°C.[14]

The reason why most mutations are deleterious is not far to seek. Each kind of organism has passed through many generations of mutation and selection. Most of the mutations capable of improving its fitness for its conditions have already occurred in the organism's past history and have become a part of its standard genotypic equipment of the present day. An organism at any given moment of time represents the product of a long period of accumulation of the favorable mutations. Nevertheless, adaptation is never perfect. Among the many new mutations that arise in the organism's present phase are a small minority which can lead to further improvements in adaptedness.[15]

[10] Timofeeff-Ressovsky, 1940; Dobzhansky, 1951a; *inter alia.*
[11] Timofeeff-Ressovsky, 1940; Dobzhansky 1951a; *inter alia.*
[12] Timofeeff-Ressovsky, 1940, 84.
[13] Gustafsson, 1951.
[14] Timofeeff-Ressovsky; see Dobzhansky, 1951a, 84.
[15] Timofeeff-Ressovsky, 1940; Dobzhansky 1951a; *inter alia.*

A mutant may be inferior to the wild type under one set of environmental conditions but superior under other conditions. The average viability of the mutant "eversae" in Drosophila funebris is 2 percent lower than that of non-mutant flies at 16° and 29°C but is 4 percent higher than the controls at 25°C.[16]

In the snapdragon, Antirrhinum majus, the standard type produces a better growth than any one of six mutant types when cultivated under normal greenhouse conditions; but under abnormal conditions of continuous light, high humidity, and high temperature the six mutants grow better than the normal non-mutant snapdragon (see Table 5). In the normal environment the standard strain grows well and the mutant "matura" grows poorly; the relative positions of the two genotypes are reversed under a different set of conditions (Environment C in Table 5) where "matura" grows well and "standard" is inviable. The mutants "olive-green" and "delicate," which do not grow well under the usual greenhouse conditions (Environment A in Table 5), prove to be of normal or nearly normal vigor in Environments D and E.[17] The relative vigor of the mutant "delicate" is compared graphically with that of the non-mutant strain in all five environments in Fig. 34.

The relative viability of a mutation thus depends not only on the particular gene change itself but also on the environment in which the mutant gene expresses itself. Translated into evolutionary terms, these results mean that if the environment were to remain static, most mutations would be valueless to a population. But the environment does not remain static. Mutations which are relatively inviable in the contemporary environment may produce a superior viability in some future environment.[18]

The adaptive value of a mutant allele is affected by the other genes in the complement as well as by environmental conditions. Timofeeff-Ressovsky found that the mutant alleles of two genes in Drosophila funebris, "miniature" and "bobbed," lower the viability of flies carrying either mutant alone, yet the same alleles when combined in one genotype give rise to flies with approximately normal viability.[19]

[16] Timofeeff-Ressovsky; see Dobzhansky, 1951a, 84.
[17] Brücher, 1943; Gustafsson, 1951.
[18] Dobzhansky, 1951a, 82 ff.
[19] Timofeeff-Ressovsky; see Dobzhansky, 1951a, 84.

Table 5. *The relative vigor of a standard strain and six mutants of Antirrhinum majus in five different environments*[a]

Conditions	A	B	C	D	E
Light	Light by day, dark at night	Continuous light	Continuous light	Continuous light	Continuous light
Temperature	Warm days, cool nights	5°C for one month, then 12°C	Constant 25°C	Constant 12°C	Constant 5°C
Humidity	Moderate	Low (67%)	High (98%)	High (90–95%)	High (100%)
Vigor					
1 (Normal growth)	Standard	Argentea	Matura	Olive-green	Delicate
2	Argentea	Delicate	Olive-green	Delicate	Argentea
3	Small 2	Small 2	Delicate		Olive-green
4	Delicate	Standard	Small 2		Small 2
5	Olive-green	Matura	Salicifolia	Small 2	
6	Salicifolia		Standard	Standard	Standard
7 (Poor growth)	Matura			Argentea	

[a] Modified from Brücher, 1943, Gustafsson, 1951.

The mutation process produces a varied array of mutant types. It is sometimes stated that the types of mutations arising in a population of organisms are at random. The randomness of mutations is not un-restricted however. Any genotype at any moment possesses an archi-tecture which probably imposes physical limitations on the forms of

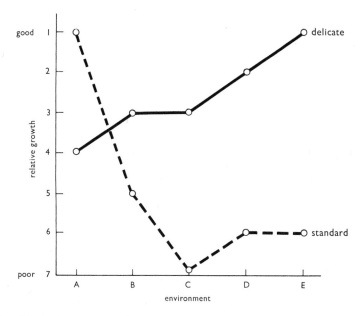

Fig. 34. *Relative vigor of a normal and a mutant strain of snapdragon,* Antirrhinum majus, *under five environmental conditions*

Growth is rated on a numerical scale where 1 is normal growth and 7 is very poor growth. The different sets of environmental conditions are symbolized by letters, A being a normal greenhouse environment and B to E being various abnormal conditions as specified in Table 5.

Drawn from data of Brüchner as given in Table 5.

gene alterations that can arise. The mutations in a gene are no doubt channelized along certain lines predetermined by the existing gene structure.[20] Among the array of mutant genes that are physically possible, however, there is no known tendency for the mutation process to produce preferentially those which fit the adaptive requirements of

[20] Blum, 1955, ch. 12; Waddington, 1957, 188.

the organism. In fact the majority of the known mutations are detrimental in some degree to their carriers. The mutation process, though non-random in a physical sense, is unoriented with reference to any standard of adaptedness, and hence is random in a biological sense.

One of the older theories of evolutionary mechanisms, the theory of orthogenesis, postulates that the guiding force in evolution lies within the organism. Evolutionary changes are assumed to proceed along a certain line, usually toward a more perfect state of adaptation, but sometimes toward overspecialization and extinction, because the mutations are oriented in this direction. Geneticists studying mutations have not, however, found that gene alterations are directed along any particular course of evolution. Most mutations are deleterious and only a minority are superior to the standard type in either a standard or an abnormal environment. Evolution takes place when the opportunistic processes of selection and drift preserve and fix in a population the few advantageous gene changes that come out of a biologically random mutation process. The present evidence indicates that among the higher organisms there is an internally controlled mutation rate, as we shall see next, but not an internally directed mutation process.

ON THE CAUSES OF SPONTANEOUS GENE MUTATIONS

Under normal conditions genes mutate at a certain low frequency, which is known as the spontaneous mutation rate. The mutation rate is greatly increased when the organism is exposed to x-rays, ultraviolet light, mustard gas, some other chemicals, and some ionizing radiations.[21]

Various authors have suggested that cosmic rays and other forms of natural radiation might be the cause of the spontaneous mutations of organisms. Measurements of the intensity of natural radiation in the world, and calculations based on these measurements and on the known mutation-inducing or mutagenic effects of artificial radiation, have shown, however, that the amount of natural radiation is sufficient to cause only about 0.1 percent of the observed spontaneous mutations in Drosophila.[22] The amounts of natural radiation in recent geological periods of earth history have been too small to account for more than

---

[21] The classical work on radiation and mutation was reviewed by Timofeeff-Ressovsky, 1934; and Lea, 1955, ch. 5; *inter alia*.

[22] H. J. Muller, Timofeeff-Ressovsky, etc.; see Lea, 1955, 180–81.

a minor fraction of the spontaneous mutations of plants and animals.

It is doubtful, furthermore, whether mustard gas or other mutagenic chemical compounds play a direct role in the evolution of higher organisms either, because most plants and animals never come into contact with these chemicals.

Bearing these considerations in mind, geneticists have often been inclined to regard spontaneous mutations as chance failures of the genes to form perfect copies of themselves in reproduction. It is inevitable that changes should occasionally be made during gene replication. Spontaneous mutations, according to this view, are the infrequent but inevitable mistakes in the copying process of the genes.

That this is not the only possible interpretation is indicated by the fact that the mutability can be increased by certain genes. The gene **Dt** in corn, for example, increases the mutation rate of the gene **A** which is located on a different chromosome.[23] (**A** is a gene controlling the formation of color pigments.) A mutator gene, **Hi**, in *Drosophila melanogaster*, when present in heterozygous condition, increases the mutation rate of many other genes two to seven times, and when present homozygously increases the mutation rate of other genes ten times.[24]

Since various chemicals are known to cause mutations, and since genes are known to control the production of enzymes and other chemicals, it is logical to suppose that mutator genes bring about their effects by elaborating mutagenic chemical compounds within the cell.[25] Higher organisms, while they do not normally encounter mustard gas and other mutagenic chemicals in their natural external environment, do perhaps come into contact with chemical mutagens of their own making in the internal milieu of the cell.

If, as seems plausible, mutator genes are widespread in the organic world, spontaneous mutations are not necessarily merely imperfect copies of the genes, but could be events caused by the specific activity of mutator genes.[26] The mutation rate of a species, then, is one of its

[23] Rhoades, 1941.
[24] Ives, 1950.
[25] Ives, 1950.
[26] Ives, 1950.

gene-controlled attributes affecting its chances for success in evolution. It follows that the mutability of a species is subject to control by natural selection.

The genotypic control of mutability would account for some of the characteristic differences which have been found in the spontaneous mutation rates of different kinds of organisms. Thus the genes of bacteria are in general more stable than the genes of Drosophila and corn.[27] Some strains of *Drosophila melanogaster* are more mutable than others.[28]

Related species of Drosophila sometimes differ in mutability. *Drosophila willistoni* and *D. prosaltans* both occur in forests in the American tropics, where the former species is widespread and the latter is rare. For the accurately measurable class of lethal mutations one particular chromosome (chromosome II) has a collective mutation rate of .000022 in *Drosophila willistoni* and .000011 in *D. prosaltans*. The common and widespread *Drosophila willistoni* is twice as mutable in chromosome II and one-half again more mutable in chromosome III than its rare relative *D. prosaltans*.[29]

Dobzhansky, Spassky, and Spassky argue that a numerically abundant species like *Drosophila willistoni* can tolerate a relatively high production of mutants, most of which are inviable. The large store of variability can then enable this species to evolve forms fitting into a greater number of habitats within the tropical forest, increasing still further its abundance and hence its ability to support a high mutation rate. A rare species like *Drosophila prosaltans*, on the other hand, cannot afford to produce a high proportion of mutant individuals; it is in equilibrium with its conditions when it has a relatively low mutation rate. But its low mutation rate and consequently its small store of variability may be one of the reasons why it has remained restricted to a narrow range of habitats. Its rarity and its low mutability work together in a vicious circle.[30]

If the mutability of a species is simply the sum total of the chance imperfections in gene replication, it is difficult to understand why two

[27] Dobzhansky, 1951a, 59.
[28] Dobzhansky, 1951a, 57, 60.
[29] Dobzhansky, Spassky, and Spassky, 1952.
[30] Dobzhansky, Spassky, and Spassky, 1952.

closely related species, *Drosophila willistoni* and *D. prosaltans*, should exhibit very different mutation rates. The differences in mutability between the two kinds of Drosophila can, however, be explained on the alternative view that the mutation rate is a heritable trait which has been adjusted by natural selection in each species according to its particular circumstances.[31]

An as yet limited body of facts thus supports the hypothesis that mutation rates are due partly to selection rather than solely to chance. More information concerning the comparative mutability in a greater number of organisms is obviously desirable.

We should note in conclusion that the hypothesis that spontaneous mutations are chance alterations in the copying of a molecular structure is not necessarily contradicted, but may be supplemented and extended, by the idea that mutations are often induced by mutator genes, for the latter may produce mutations which are in addition to those originating from copying mistakes. In other words, there is perhaps a large measure of truth in both prevailing explanations of the causes of gene mutations.

CHROMOSOMAL MUTATIONS

Mutations in the most general sense of the term are changes in the structure and functioning of the genetic material. This material is organized into a hierarchy of functional units ranging from single nucleotides and subgenes at the lower level to whole chromosomes and chromosome sets at the upper level. Changes in the genetic material at any one of these levels are by definition mutations. We have so far considered intragenic changes, or gene mutations. Next we must mention—all too briefly—mutations in the chromosomes and in chromosome sets.

Chromosomal mutations are structural changes in the chromosomes which lead to rearrangements in the order of the genes. Chromosomal rearrangements are initiated by breaks and completed by reunion of the broken ends in new ways. The mode of reunion determines the type of rearrangement, as shown in Table 6. One break near the end of one chromosome can give rise to a terminal deficiency. With two breaks in one chromosome, three kinds of rearrangements are possible, namely,

[31] Dobzhansky, Spassky, and Spassky, 1952.

Table 6. *Types of chromosomal rearrangement*

Rearrangement	Number of breaks	Number of chromosomes involved	Gene order		
			Chromosome I	Chromosome II	Chromosome III
Standard arrangement			ABCDEFGHI	MNOPQRST	UVWXYZ
Deficiency, terminal	1	1	ABCDEFGH–	"	"
Deficiency, interstitial	2	1	A–DEFGHI	"	"
Duplication	2	1	ABCBCDEFGHI	"	"
Inversion	2	1	ADCBEFGHI	"	"
Translocation, single	2	2	ABCDEQRST	MNOPFGHI	"
Translocation, successive	3 or more	3 or more	ABCDEQRST	MNOPXYZ	UVWFGHI

interstitial deficiencies, duplications, and inversions (Table 6). When the broken ends of two or more chromosomes rejoin, translocations are produced (Table 6).

A discussion of the origin, cytological behavior, and genetic effects of the various kinds of chromosomal rearrangements would fall outside the scope of this book.[32] Suffice it to say here that chromosomal aberrations of various kinds are known to occur as polymorphic variants in natural populations of plants and animals.[33] These chromosomal mutations evidently become established in a uniform or monomorphic condition in the evolution of some of the populations, since in many groups of plants and animals the species, races, and even the local breeding populations differ from one another with respect to the structural arrangement of the chromosomes.[34]

The phenotypic effects and adaptive properties of chromosomal rearrangements, which could account for their role in evolution, are varied. Changes in the gene order lead to changes in the action and expression of the genes involved, or position effects, in some cases. Inversions and translocations bring about new linkage relations between genes.[35] Duplications, by adding gene loci to the complement, could be the starting point of a divergence in function and division of labor between neighboring genes controlling the same process.

Whole chromosomes can be the unit of mutation, as where a particular chromosome of the complement in a diploid organism is represented singly, or in a threefold or fourfold manner, rather than in the normal duplex condition. This condition is referred to as aneuploidy. Such aneuploid changes have been incorporated into the genetic endowment of many evolving plant and animal groups.[36]

Finally, the chromosome set as a whole is subject to mutation. A diploid organism can mutate to the haploid condition, or to a polyploid condition characterized by three, four, or more chromosome

[32] For a concise classification of chromosomal rearrangements see Dobzhansky, 1951a, 28–29. Introductory treatments of the subject are given in genetics texts, i.e., Srb and Owen, 1952, ch. 10. For more advanced treatments see, *inter alia*, Stebbins, 1950; White, 1954; and Swanson, 1957.

[33] Dobzhansky, 1951a, ch. 5; Stebbins, 1950, 76–85; White, 1954, ch. 7–8; da Cunha, 1960.

[34] Grant, 1956a.

[35] Grant, 1956a, for review and references.

[36] Reviews in Stebbins, 1950, ch. 12; and White, 1954.

sets. Polyploidy is a common and important aspect of evolutionary changes in plants.[37]

In the remainder of this chapter and in the immediately following chapters we shall describe evolutionary changes in terms of shifting frequencies of genes rather than of chromosomes. The operation of the evolutionary forces is essentially the same for mutational units at any functional level from subgene to chromosome set. The forces which control the relative frequencies of alternative forms of a chromosome in a population are the same as those controlling the relative frequencies of the alleles of a gene. It is convenient to be able to use a single term to designate the units of hereditary variation, and the word gene, used in its more general sense, is adequate for this purpose. Furthermore, the gene in the strict sense of the word, that is, the crossover unit on the chromosome, is a basically important unit of heredity, variation, and evolutionary change. This statement does not, of course, imply that chromosome segments, whole chromosomes, and chromosome sets are not also important units.

GENE FLOW

A population composed of a_1a_1 individuals may acquire a mutant allele, a_2, either by a mutation $a_1 \rightarrow a_2$, occurring in some individual of the same population, or by the immigration of an individual or gamete carrying the allele a_2 from some other population. This latter process is gene flow.

It is evident that the dispersal of individuals or gametes is a factor affecting genetic variability only if the foreign individuals are genetically different from the population they enter. An immigrant carrying an a_1 allele into a population composed of a_1 alleles does not contribute to the variability of the latter. An immigrant carrying an a_2 allele into this same population does change the genetic composition of the recipient population.

An individual carrying a new allele into a population must have obtained that new allele as a result of some prior event of mutation. In the history of the species either the allele a_2 arose as a mutant form of a_1, or a_1 arose from a_2, or both a_1 and a_2 diverged by separate mutations from some common ancestral form of the gene **A**. Gene flow as a

---

[37] Review in Stebbins, 1950, chs. 8–9.

source of genetic variations thus depends on the previous existence of mutations. Gene flow may be regarded as a delayed effect of the mutation process (Fig. 35).

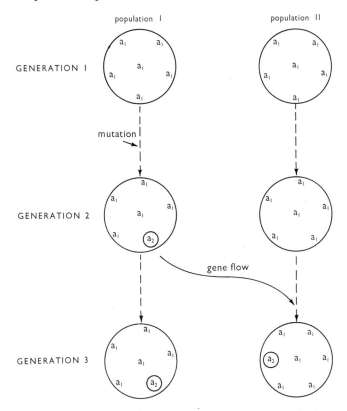

Fig. 35. The sources of single-gene variation in populations

The gamete pools of the two populations diagrammed here are originally homogeneous for the allele a_1. Population I becomes polymorphic for the alleles a_1 and a_2 of the gene **A** as a result of a mutation. Mutations do not occur in Population II. This population eventually becomes polymorphic for the gene **A** too, however, as a result of the immigration of an individual or gamete carrying the mutant allele a_2 from Population I.

Nevertheless, the direct and immediate source of some variations in a population at any given moment may be the immigration of carriers of different alleles from other populations. A population of plants might consist entirely of blue-flowered individuals. A mutant type with white flowers might never appear in this population. The mutation for white flowers could, however, arise in some other population of the species, and seeds carrying the new allele could be dispersed from the variable blue and white population to the pure blue population. If these seeds grow successfully in the new site, a population which was formerly invariable with respect to flower color will come to consist of a mixture of blue- and white-flowered individuals. The *ultimate* source of the variability in the first population may be a gene mutation occurring in another population at some past time; but the *effective* cause of the variability in the first population at the present time is gene flow.

The land snail, *Cepaea nemoralis,* exists in a series of semi-isolated colonies in France. The most common type of individual has brown bands on the shell; however, some individuals in the same colonies lack these bands. These differences are due to a gene **B** present in different allelic forms, the dominant allele b+ determining bandless shells and the allele b producing banded shells. The polymorphism of the colonies for the presence or absence of bands on the shell depends on the coexistence of the different alleles in the populations. The genic variability is due in turn to both mutation and gene flow.[38] The mutation rate of b → b+ is .0001, and the reverse mutation b+ → b occurs at a rate of .0005.[39] Mutation thus leads to some variability in banding pattern within any moderate-sized population.

In the province of Brittany where the snail colonies are widely spaced and little migration occurs between them, the variability is due primarily to the factor of mutation. This is not to say, however, that gene flow contributes nothing to the variability of the populations in Brittany. In the district of Aquitaine, by contrast, the colonies are less isolated from one another than in Brittany, and migrant individuals can be found fairly frequently in the spaces between neighboring populations. The amount of gene flow in this area is m = .003 to .004. This value of m is higher than the mutation rate (u = .0001 to .0005). It follows that the variability in banding pattern in the colonies of Aquitaine is due primarily to gene flow and secondarily to mutation. The two sources of single gene variation, mutation and gene flow, are active in both

[38] Lamotte, 1951.
[39] Lamotte, 1951.

Brittany and Aquitaine, but have a different order of importance in the two areas.[40]

The vehicles of gene flow are individuals or gametes migrating from one population or segment of a population to another. The amount of gene flow in any species thus depends on the size and structure of its breeding populations and on the dispersal range of its individuals or gametes. These factors vary widely among different kinds of organisms. In general, if individuals interbreed freely over wide distances, the genes carried by any one individual will be able to spread quickly throughout the population, whereas if the interbreeding is confined predominantly to small local neighborhoods or colonies the gene dispersal will proceed at a much slower rate.

In an extensive pine forest, the pollen grains of which are carried by the wind, the genes present in one part of the forest can be dispersed to some distant part in the course of successive generations of cross-pollination by wind. The range of dispersal of pine pollen under calm weather conditions has been measured by Colwell. This worker released 1-liter lots of pollen from inverted jars located at a height of 12 feet, and caught the grains in petri dishes and traps placed at regular intervals along various radii from the point of release. On a mild day with gentle breezes the main bulk of the pollen was dispersed to a distance of 10 to 30 feet downwind from the pollen source. Beyond this zone of maximum pollen concentration the amount of pollen fell off rapidly, as Fig. 36 shows. A small proportion of the pollen grains reached distances of 150 feet or more. These results suggest that a female cone in a pine forest will be swamped with pollen from neighboring trees and will receive only small amounts of pollen from trees standing a few hundred feet away.[41]

Many kinds of flowering plants in mountainous regions occur only in special habitats, such as outcrops of limestone or serpentine soils, near seeps or springs, or on bare summits. If the favorable habitats are scattered and widely separated, as is often the case, so are the populations of the plant species. The species then exists in a series of spatially separated colonies. The individuals forming one colony may cross more or less freely with one another. Only occasionally, however,

[40] Lamotte, 1951.
[41] Colwell, 1951.

will a seed or pollen grain be carried by wind or animals from one colony to another. Gene flow is reduced under these circumstances. It is even more reduced in colonial plants which are self-fertilizing.

The higher animals, the insects, and the vertebrates, possess efficient means of locomotion and are capable of ranging widely in search of food, mates, and nesting territories. However, the fact that these

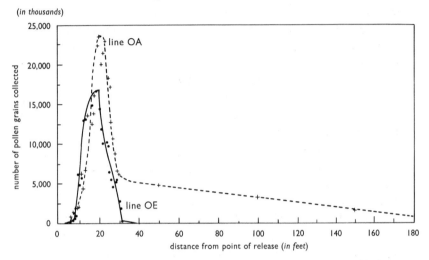

Fig. 36. The amount of pine pollen carried by breezes on a mild day to various distances from the point of release

Line OA is a radius directly downwind from the pollen source, and line OE is a radius 45° from downwind.

Colwell, 1951.

animals can swim, walk, or fly long distances does not entitle us to assume that they regularly do at the time of mating. Many birds and rodents, in fact, have a strong attachment to a home territory, and most individuals in any new generation settle and breed in or close to the area in which they were born.[42] It has been shown by marking, releasing, and recapturing butterflies of the species *Euphydryas editha* that most individuals restrict their activities to a very limited area, in some cases to a space 100 feet in diameter.[43]

[42] Mayr, 1942, 240, 1947; Blair, 1953.
[43] Ehrlich, 1961.

In a study designed to measure the rate of dispersion of a gene through a population of *Drosophila pseudoobscura,* Dobzhansky and Wright released mutant flies at a specific locality in the Sierra Nevada of California and set traps at various distances along lines radiating from the point of release to recover the carriers of the mutant allele. Analysis of the recoveries at various distances from the point of release and at given intervals of time from the date of release leads to the following estimates of the rate of dispersal of the mutant allele. In one year and hence in several generations 50 percent of the progeny of the original mutant flies will occur within a radius of .85 kilometer, 95 percent of the progeny within a radius of 1.76 kilometers, and 99 percent within a radius of 2.2 kilometers from the point of release.[44]

The picture of gene dispersion which is shaping up on the basis of studies of the distribution of pollen grains by wind, bees, birds, and other pollinating agents, and of the dispersal of individual flies, butterflies, rodents, and other animals is that of a combination of much sedentariness with some long-range dispersal.[45] Most pollen grains of a plant and most individuals of a fly or butterfly remain within a localized neighborhood, but a few gametes are carried far and wide. As Bateman puts it, and as Fig. 36 shows, the curve representing the pattern of gene dispersion is leptokurtotic or peaked with a high proportion of short-range and a significant number of long-range migrations.[46] Consequently, although close inbreeding between neighboring individuals may prevail in a population, a certain, usually small, amount of gene flow over larger distances is also expected to occur.

RECOMBINATION

We have considered the causes of variation for a single gene **A**. Mutations are likely to occur in other genes of the complement besides **A**. The gene **B** may also occur in different allelic forms. Assume that one individual with the genotype $a_1a_1 b_1b_1$ mutates in the **A** gene and becomes $a_1a_2 b_1b_1$, and that another $a_1a_1 b_1b_1$ individual has a mutation in the **B** gene and becomes $a_1a_1 b_1b_2$. The population now contains

[44] Dobzhansky and Wright, 1947.
[45] Mayr, 1947; Dobzhansky and Wright, 1947; Bateman, 1949, 1950; Grant, 1950*a*, 381–82, 1958, 344; Blair, 1953; Andrewartha and Birch, 1954, 88–104; Ehrlich, 1961.
[46] Bateman, 1950.

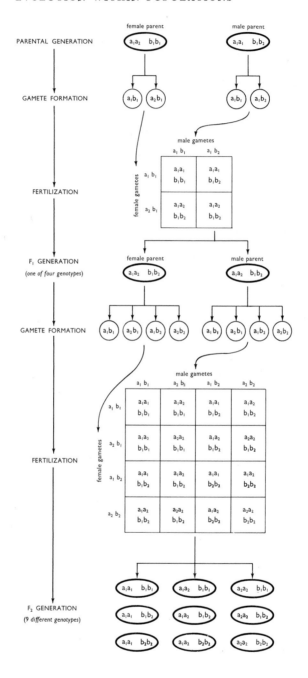

three genotypes, the two mutant types and the original form, as a result of mutations occurring independently in different genes. One or both of the mutant alleles, a_2 and/or b_2, could be introduced into the population by gene flow with the same consequences for individual variation.

A much greater amount of individual variation can now be generated by the mechanism of sexual reproduction. The a_1a_2 b_1b_1 individual produces gametes of two types, a_1b_1 and a_2b_1. The a_1a_1 b_1b_2 individual yields the two gametic classes, a_1b_1 and a_1b_2 (see Fig. 37). Different combinations of two classes of male gametes with two classes of female gametes give rise to four genotypes in the next generation. As shown in Fig. 37, these are a_1a_1 b_1b_1, a_1a_2 b_1b_1, a_1a_1 b_1b_2, a_1a_2 b_1b_2.

Sexual reproduction in individuals heterozygous for two or more genes borne on separate chromosomes has interesting consequences. During sex-cell formation the separate chromosomes, and the genes carried by these chromosomes, are assorted independently of one another to the different gametes. The two alleles of one pair (a_1 and a_2) are shuffled into different gametes; the two alleles of another gene borne on another chromosome (b_1 and b_2) are also separated and distributed to different gametes; and the shuffling process goes on independently for the two pairs of alleles **A** and **B**. As a result, the double heterozygote (a_1a_2 b_1b_2) produces four types of gametes (a_1b_1, a_1b_2, a_2b_1, a_2b_2).

The random union of four kinds of male gametes with four kinds of female gametes leads to the formation of nine different genotypes (see Fig. 37). In other words, two individuals of the doubly heterozygous constitution a_1a_2 b_1b_2 can cross to produce nine genetically different types of individuals in the next generation. Among the individuals thus produced are such previously unknown genotypes as a_1a_1 b_2b_2, a_2a_2 b_1b_1, a_2a_2 b_2b_2, and others. The new types possessing novel gene combinations not found in either parent are called recombination types.

Fig. 37. Crossing between two individuals, one of which carries a mutant allele of the gene **A** *and the other a mutant allele of* **B**, *leads to the formation of four kinds of genotypes in the next generation, and nine genotypes in the second generation*

Gametes are outlined by thin lines and diploid individuals by thick lines.

We can translate these postulations into real examples of observable characters and their recombination in successive sexual generations. Mendel crossed a homozygous pea plant with yellow, round seeds to another homozygous pea plant with green, wrinkled seeds. The color and surface conformation of the pea seeds are determined by separate and unlinked genes. In the F_2 generation progeny of the cross, Mendel obtained some plants like one or the other parental type and in addition some individuals unlike either parent. One of the recombination types was characterized by yellow, wrinkled seeds; another by green, round seeds.

We have considered recombinations between pairs of alleles of two genes, **A** and **B**. Other genes, **C**, **D**, and so on, may also be present in a heterozygous condition. An individual carrying two allelic forms for each of ten unlinked genes will produce 1,024 classes of gametes, and by crossing with another like individual can yield 59,049 different genotypes in the next generation. The general relation is as follows. Let n represent the number of unlinked genes present in two allelic forms; then the number of possible gametic classes is 2^n, and the number of different genotypes that can be produced by various combinations of these gametes is 3^n.[47]

A single individual carries only two alleles of each heterozygous gene. In a population of individuals, however, there are likely to exist several or many alleles of each gene. This would be the case if, in addition to a_1 and a_2, the alleles a_3, a_4, and a_5 are present in the population; if the **B** gene is also represented by a series of multiple alleles (b_1, b_2, b_3, b_4, and b_5); and if **C**, **D**, and other genes are polymorphic to a similar extent. The gamete pool of the population will then consist of a very large number of types of gametes, which can combine to form an even larger number of genotypes.

The number of diploid genotypes (g) that can be assembled from any number of alleles (r) of the gene **A** is given by the formula: $g_A = r(r + 1)/2$ (as was shown in Chapter 7). The number of possible combinations in pairs of any number of alleles (r) of the gene **B** is: $g_B = r(r + 1)/2$. If the two genes, **A** and **B**, are borne on separate chromosomes and can be recombined freely during gamete formation,

----

[47] Mendel, 1866.

the total number of possible genotypes for the two genes considered together is the product of $g_A \times g_B$.

If two alleles of **A** are present, $g_A = 3$; with two alleles of **B**, $g_B = 3$ also; and the number of genotypes considering both **A** and **B** is $g_A \times g_B = 3 \times 3 = 9$. This is the number of genotypes discussed previously as originating from the intercrossing of two double hetero- zygotes ($a_1 a_2\, b_1 b_2 \times a_1 a_2\, b_1 b_2$) (see also Fig. 37).

Table 7. The number of genotypes which can be produced by recom- bination between various numbers of unlinked genes each of which possesses various numbers of alleles

Number of alleles of each gene	Number of independent genes				
	2	3	4	5	n
2	9	27	81	243	3^n
3	36	216	1,296	7,776	
4	100	1,000	10,000	100,000	
5	225	3,375	50,625	759,375	
6	441	9,261	194,481	4,084,101	
7	784	21,952	614,656	17,210,368	
8	1,296	46,656	1,679,616	60,466,176	
9	2,025	91,125	4,100,625	184,528,125	
10	3,025	166,375	9,150,625	503,284,375	
r	$\left[\dfrac{r(r+1)}{2}\right]^2$	$\left[\dfrac{r(r+1)}{2}\right]^3$	$\left[\dfrac{r(r+1)}{2}\right]^4$	$\left[\dfrac{r(r+1)}{2}\right]^5$	$\left[\dfrac{r(r+1)}{2}\right]^n$

Two unlinked genes each with three alleles can be combined into $6 \times 6 = 36$ genotypes. With four alleles in each of the separate genes, 100 recombination types are possible; and with five alleles for each gene, 225 recombination types can be formed.

Several unlinked genes, **A**, **B**, **C**, etc., each with several alleles, can be recombined in a number of ways equivalent to the product of $g_A \times g_B \times g_C \ldots$. Thus three independent genes with three alleles each can form $6 \times 6 \times 6 = 216$ genotypes. Three genes with four alleles each can form 1,000 genotypes. Three genes with three, four, and five alleles, respectively, can give rise to $6 \times 10 \times 15$ or 900 geno- types. And with five alleles recombining freely in three separate genes, the possible number of genetically different individuals is 3,375 (see Table 7).

Some indication of the tremendous power of the sexual mechanism for generating individual variation by recombination is given by the numbers of possible genotypes in Table 7 and by the curve drawn from

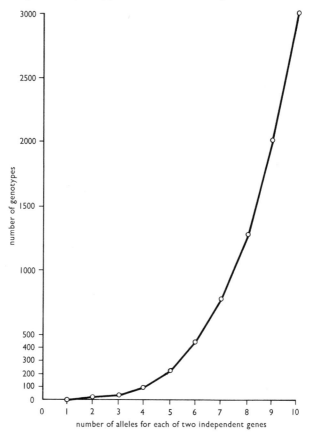

Fig. 38. The rise in individual variability due to recombination as the numbers of alleles in two unlinked genes become larger (see Table 7)

some of those data in Fig. 38. If the number of independently assorting genes present in two allelic forms is 19 or more, the number of different genotypes that can be generated by recombination must be reckoned in the billions. Or if a series of ten or more alleles exists for each of six independently segregating genes, the number of recombination types is again in the billions.

Genes do not, of course, have to be borne on separate chromosomes in order to undergo recombination. Widely separated genes on the same chromosome recombine more or less freely, and even closely linked genes become separated from one another and recombined to some extent.

With even moderately small numbers of polymorphic genes, the number of genotypes that can be produced by recombination is likely to be greater than the existing number of individuals in the species.[48] Most natural populations are in fact normally polymorphic for many genes. Recombination is therefore a virtually limitless source of individual variation.

The capacity of the sexual mechanism for producing recombinations between a limited given number of genetic variations accounts for the fact that, in most higher animals and many plants, no two individuals developing from different zygotes are exactly alike genotypically, and offspring are not just like their parents.

ADAPTIVE PROPERTIES OF GENE COMBINATIONS

Most of the characteristics of organisms on which their adaptations depend, and all of the complex adaptive characters, are determined by combinations of genes. A mechanism for putting mutant alleles together in new combinations thus plays a vital role in the evolution of complex adaptations. Sexual reproduction, as we have just seen, is such a mechanism.

Let us consider the difficulty of assembling a new combination of genes in an organism which reproduces by asexual means exclusively. Assume that this organism is haploid and has the constitution $a_1b_1c_1$, and that a more favorable adaptation would be determined by the triple mutant condition $a_2b_2c_2$. One asexually reproducing lineage may in the course of time acquire the mutation a_2, another line the mutation b_2, and still another the mutant allele c_2. Since fitness depends on the gene combination $a_2b_2c_2$, however, the separate lineages are no better adapted than the original stock. The new favorable genotype can be assembled in the absence of a sexual process only by multiple mutations within one lineage. If, however, mutation of any single gene is a rare

[48] Fisher, 1930, 96; Dobzhansky, 1955a, 34.

event, the chances of simultaneous or nearly simultaneous mutations in three genes are infinitesimally slight. And, if the organism is diploid instead of haploid, the change from one homozygous gene combination $(a_1a_1b_1b_1c_1c_1)$ to another $(a_2a_2b_2b_2c_2c_2)$ by the mutation process alone might not occur even once during the entire history of the species.

But if the various individuals carrying different mutant alleles can cross sexually, they can pool their separate mutations and produce the new gene combination directly. Intercrossing between diploid individuals with the genotypes $a_1a_2b_1b_1c_1c_1$, $a_1a_1b_1b_2c_1c_1$, and $a_1a_1b_1b_1c_1c_2$, for instance, can lead in just three generations to the formation of the homozygous recombination type $a_2a_2b_2b_2c_2c_2$. The gene combination which could not be assembled once in millions of generations by simultaneous multiple mutation is readily brought together in a few generations by the sexual mechanism.

HYBRIDIZATION AND VARIATION

A certain amount of genetic variation which can segregate and recombine is produced by mutation within a population. This variation is increased by the flow of genes from neighboring populations. It can be enhanced still further by interbreeding between races, which differ genotypically to a much greater extent than do the populations of one race, and again by hybridization between species, which are even more strongly differentiated genetically.

Hybridization between individuals belonging to different species is usually prevented by various barriers known collectively as isolating mechanisms (Chapter 13). These barriers are not absolute in many groups of plants and animals, however, and may break down occasionally or locally so as to permit an exchange of the genes of distinct species. The occasional formation of interspecific hybrids and the occasional reproduction by these hybrids may raise the level of genetic variation in the populations affected by the hybridization far above the normal level for unhybridized populations.

The pokeweed, Phytolacca, forms highly variable populations in Brazil. This variation is correlated with the presence in that area of several coexisting species which hybridize with one another. In most of North America, by contrast, there is only one species of Phytolacca, namely, *P. americana*, and this species is highly uniform throughout a vast area from the Gulf of Mexico to Canada and from the Great Plains to the Atlantic Coast.[49]

The Leafy-Stemmed Gilias comprise a group of species of annual plants in the foothills and valleys of California. A high level of genetic variation is found in five of these species, namely, *Gilia capitata*, *G. achilleaefolia*, and others. The variation between populations of *G. achilleaefolia* for two morphological characters is mentioned and illustrated graphically in Chapter 11. It is known from experimental crossings and population studies in the wild that the variable species of Leafy-Stemmed Gilia can and do hybridize.[50]

Gilia tricolor, belonging to this same group of species and growing in the same area, is relatively uniform by comparison with the other Leafy-Stemmed Gilias. A search for evidences of natural hybridization between *Gilia tricolor* and other species failed to reveal any traces of such hybridization. All attempts made during a four-year period to produce artificial hybrids in the experimental garden between *Gilia tricolor* and its congeners, as represented by 23 strains belonging to 11 other species, failed completely. Hybridization between *Gilia tricolor* and other species is thus blocked, both in nature and in the breeding plot, by a very strong if not absolute barrier of incompatibility.[51] The conclusion is inescapable that the variability of *Gilia tricolor* is limited by its genetic isolation, while the extraordinary variability of the related species is due to hybridization.[52]

Anderson, on the basis of a critical examination of the variation patterns in 30 genera of flowering plants in the eastern United States, has reached the following conclusion[53]:

For each of these genera, all the readily detectable variation can be ascribed to introgression [a form of hybridization] . . . . For critical genera such as *Cercis* and *Phytolacca* which have no close relatives over much of the area, there is plant-to-plant and place-to-place ecological and morphological uniformity except in those sectors of their distributions which abut on the area of a related species. Variation, in other words, is strictly proportional to the opportunity for introgression. From a consideration of all the evidence we are forced to the conclusion that introgressive

[49] Sauer, 1951.
[50] Grant, 1953*b*.
[51] Grant, 1952*a*.
[52] Grant, 1953*b*.
[53] Anderson, 1953, 300, quoted from *Biological Reviews*.

hybridization is certainly more important than all other factors combined in providing raw material for natural selection to work upon. It must be one of the chief immediate causes of variability if not the chief one.

Interspecific hybridization is a potential source of genetic variation in animals, protista, and fungi, as well as in plants. In any sexual organism the variability of the species can be increased by natural hybridization. Whether in fact the natural hybridization does or does not occur on a scale sufficient to raise the level of variation is another question. In many plant groups it does. In many though not all groups of animals it does not. The protista, fungi, and large parts of the animal kingdom have not yet been surveyed by methods which would reveal whether the variability in those organisms has been enhanced significantly by hybridization.

THE PROBLEM OF DISTINGUISHING GENE MUTATIONS FROM RARE RECOMBINATIONS AND MINNTE CHROMOSOMAL REARRANGEMENTS

Linked genes are separated by crossing-over and recombined by the sexual mechanism. The recombination types then differ genotypically, and may differ phenotypically as well, from their parents. If the linked genes are very close, so that crossing-over between them occurs only rarely, the recombination types will also appear rarely in the progeny of heterozygous individuals, which ordinarily breed true to type.

Various examples have been analyzed of rare crossing-over between closely linked genes or subgenes in corn, Drosophila, and other organisms.[54] Crossover values as low as .0002 or possibly even .00005 have been found in Drosophila. Such crossover values approach the mutation rates of genes. The possibility exists, therefore, that any assortment of unanalyzed mutant types may include a proportion of rare recombinants, in addition to true gene mutations, considering the latter as alterations in the molecular structure of the crossover units themselves.[55]

In the identification of various mutations in corn and Drosophila as recombinations, history is repeating itself. The concept of mutations as sudden changes in the hereditary material was introduced into modern

[54] Pontecorvo, 1958, 57–58, for summary.
[55] Sax, 1931; Anderson, 1953; Mangelsdorf, 1958b.

genetics by DeVries in 1901 on the basis of his studies of the evening primrose, Oenothera. This plant is true-breeding for the most part, but regularly gives rise to a small proportion of sports or mutant forms. The "mutations" of Oenothera were later shown to be recombination types produced occasionally by highly heterozygous plants. Now we see that some of the mutations in Drosophila and corn and other organisms, on which the contemporary mutation theory is based, are likewise recombinations.

Although some mutations have to be reclassified after further study as recombinations, there is no reason to believe that *all* mutations are recombinations in disguise, whereas good reasons do exist for postulating that *some* mutant types are due to true gene mutations.

In the first place, an organism must be heterozygous in two or more homologous parts of its genetic material before it can give rise to new recombination types. The effectiveness of recombination as a source of genetic variability presupposes the existence of some process by which allelic differences can arise in different genes or subgenes, and that process is gene mutation. Gene recombination is a mechanism for extracting the maximum variability from gene mutations. In other words, the mutation process is necessary in order to provide the original genetic variations which can be worked over by the recombination process.

Assuming, furthermore, that each gene possesses a specific molecular structure, it is inconceivable that that structure could be reproduced forever without the occasional appearance of alterations in some of the copies. If, as is widely believed, the specific physical-chemical organization of the gene resides in its particular sequence of base-pairs on the DNA chain, a change in one or more of the bases is bound to occur from time to time during gene replication. We have every reason to expect the genes to undergo true gene mutations of this sort, and the wonder is not that the genes should mutate, but rather that they are as stable as they are.

Chemical and genetical methods for mapping the pattern of bases in a genic segment of DNA, for describing mutational changes therein, and hence for discriminating directly between mutations and recombinations, are, as Pontecorvo has pointed out, likely to be developed in the bacterial viruses, bacteria, and other microorganisms, where "the resolving power of genetic analysis" is greater than in higher organisms, and the genetic material can be analyzed in fine detail.[56]

Gene mutations, or intragenic changes in the molecular structure and presumably in the sequence of base-pairs, are also indistinguishable for all practical purposes from chromosomal rearrangements that are too small to be detected cytologically, particularly if the latter produce a position effect. Some geneticists, in carrying this difficulty to what they consider to be its logical conclusion, argue that a valid distinction cannot be drawn, even in theory, between gene mutations and small rearrangements. Their argument is that since chromosomal rearrangements are known to occur in the genetic material, whereas direct evidence regarding intragenic changes on the molecular level is lacking, we should equate the unknown with the known and reduce all mutations to some form of rearrangement.

Goldschmidt concludes a critical discussion of this problem by stating[57]:

To me it seems that the facts favor the assumption that there is no difference in principle between the processes resulting in so-called gene mutations and those resulting in rearrangements; the differences observed are only in the size of the effect, that is, invisible versus visible rearrangement.

The contrary view that gene mutations do differ in kind from small rearrangements is expressed by Muller and other geneticists.[58]

In at least one case in *Drosophila melanogaster*, involving the gene "forked" which is concerned with the body bristles, and probably in some other genes as well, it has been possible to show that reverse mutations from the mutant to the normal type are not associated with small rearrangements, but involve wholly intragenic changes.[59] It must be granted, however, that the possibility that gene mutations are in reality small rearrangements has not been rigorously excluded in the vast majority of cases. But neither has it been demonstrated by adherents of the opposite viewpoint that gene mutations are, in all or even most cases, changes of arrangement. Nor is it clear where the burden of proof lies. In the present state of our knowledge, certain presumptions

[56] Pontecorvo, 1958, 42–48, and ch. 1.
[57] Goldschmidt, 1955, 119, quoted from *Theoretical Genetics*.
[58] H. J. Muller in 1950; see Goldschmidt, 1955, 117–18; also Green, 1957.
[59] Green, 1957.

are implicit in either alternative hypothesis. Pending enlightenment from future research, our best course is to adopt the assumptions that seem most plausible.

Let us try to sum up the problem with the aid of a metaphorical example. Assume that the sequence of base-pairs on a small segment of DNA inside a gene spells in coded form the phrase RAT ON FLOOR. To be sure, a short inversion involving the first word (subgene) would change the sense to TAR ON FLOOR. But the sense could be changed equally by an alteration in one or two of the base-pairs; thus one gene mutation might spell out the thought RAT ON DOOR; while another gene mutation could be CAT ON FLOOR. There is no reason why the recognition of the mutational possibilities inherent in rearrangements should require us to reject the possibilities of direct biochemical changes within the genes, that is, true gene mutations.

To continue with this hypothetical example, an individual heterozygous at both mutational sites could give rise to other new forms by recombination. If the original gene spells RAT ON FLOOR, and if a new rearrangement arises spelling TAR ON FLOOR, and a gene mutation occurs to RAT ON DOOR, the double heterozygote could produce the new recombination type TAR ON DOOR.

The change from RAT ON FLOOR to TAR ON DOOR has involved a small rearrangement, a gene mutation, and a recombination. To assume that only one or two of these three modes of change in the genetic material is realized in nature is unjustified and doctrinaire. The practical difficulty of distinguishing between the various forms of genic change is, however, very great, but perhaps biochemical studies of the genes and gene products will provide the answers to this problem.

A SERIES OF PROPOSITIONS REGARDING VARIATION

1. Evolution is a change in the genetic composition of a population, and, in its most elementary form, consists of a change in allele frequencies.

2. Reactions of the individual organism to environmental conditions (phenotypic reactions) are not inherited, and hence do not constitute or initiate evolutionary changes.

3. The starting point for evolutionary change is the formation of individuals with different genotypes (genetic variability).

4. The ultimate source of all genetic variability is the mutation process.

5. Mutations may be changes in the molecular structure of the genes (gene mutations), or in the linear sequence of the chromosome segments (chromosomal rearrangements), or in the number of chromosomes. Gene mutations and minute chromosomal rearrangements cannot normally be distinguished in practice.

6. The immediate source of single-gene variation in any population may be: mutations occurring in this population, or mutations occurring in a second population and introduced into the first population by the immigration of individuals or gametes carrying the mutant allele (gene flow).

7. The immediate source of multiple-gene variation in a population is: independent mutations in two or more genes, the immigration of carriers of different mutant genes (gene flow), or crossing between individuals carrying mutations in different genes followed by the formation of new gene combinations in their progeny (recombination).

8. Once a few mutations are present in a population, they can be recombined in numerous ways by the sexual process. Gene recombination is thus a second-order effect dependent on the prior existence of mutations in two or more genes.

9. Interspecific hybridization, by permitting recombination to occur between the genes of distinct species, represents a vast extension of the potentialities of sexual reproduction. In many plant groups, at least, these potentialities are exploited, and the level of variability in the species is raised substantially by hybridization.

10. The distinction between mutation and recombination, though sound in principle, is difficult to make in many actual cases where crossing-over and recombination between closely linked genes or subgenes occur at low frequencies comparable to mutation rates. An assortment of mutations at a locus is likely to include a proportion of changes due to rare crossovers in addition to alterations in the molecular structure of the genes themselves.

11. Heritable individual variation is a necessary preliminary condition for evolutionary change, but does not constitute evolution in itself. Evolution can be said to occur when the relative frequencies of different genotypes undergo changes. Such changes are brought about

by the primary evolutionary forces: mutation, gene flow, selection, and drift.

12. Some of the causes of genetic variability, namely, mutation and gene flow, may directly alter the allele frequencies in a population, and insofar as they do they are also causes of evolutionary changes.

13. Gene recombination assembles an existing array of allelic forms of different genes into a variety of combinations, but does not alter the frequencies of these alleles; the recombination process is therefore not an evolutionary force.

14. Although mutation is the ultimate source of genetic variation, recombination is by far the chief effective source of individual genetic variability in sexually reproducing organisms. Recombination generates most of the genotypic differences between individuals in a population, and thus provides the major proportion of the variations which are worked over by the forces of selection and drift.

NINE

Natural Selection

Natural selection is a mechanism for generating a high degree of improbability. R. A. FISHER[1]

DARWIN'S THEORY OF SELECTION

IN HIS BOOK, *The Origin of Species*, first published in 1859, Charles Darwin gave reasons for believing that the main guiding force in evolution is natural selection. The full title of Darwin's book—*On the Origin of Species by Means of Natural Selection or the Preservation of Favoured Races in the Struggle for Life*—is a brief statement of the theory in itself. The theory of natural selection was proposed independently at the same time by A. R. Wallace.

It was Darwin's thesis that the individuals comprising any species of plant or animal differ by small variations, that these varying individuals compete with one another for the means of life, and that some of the variants might prove better adapted to survive and reproduce than others. The reproduction preferentially, generation after generation, of the individuals possessing the more favorable hereditary variations would result in gradual evolutionary changes toward a better adaptedness within each species. Darwin summarized the theory in these words[2]:

As many more individuals of each species are born than can possibly survive; and as, consequently, there is a frequently recurring struggle for existence, it follows that any being, if it vary however slightly in any manner profitable to itself, under the complex and sometimes varying conditions of life, will have a better chance of surviving and thus be *naturally selected*. From the strong principle of inheritance, any selected variety will tend to propagate its new and modified form.

[1] Quoted by Huxley, 1943, 474.
[2] Darwin, 1859, Introduction.

190

In order to show how natural selection acts to improve the adaptations of a species, Darwin gave a number of illustrations. We will quote him on the subject of the giraffe.[3]

The giraffe, by its lofty stature, much elongated neck, forelegs, head and tongue, has its whole frame beautifully adapted for browsing on the higher branches of trees. It can thus obtain food beyond the reach of the other Ungulata or hoofed animals inhabiting the same country; and this must be a great advantage to it during dearths.... So under nature with the nascent giraffe the individuals which were the highest browsers, and were able during dearths to reach even an inch or two above the others, will often have been preserved; for they will have roamed over the whole country in search of food. That the individuals of the same species often differ slightly in the relative lengths of all their parts may be seen in many works of natural history, in which careful measurements are given. These slight proportional differences, due to the laws of growth and variation, are not of the slightest use or importance to most species. But it will have been otherwise with the nascent giraffe, considering its probable habits of life; for those individuals which had some one part or several parts of their bodies rather more elongated than usual, would generally have survived. These will have intercrossed and left offspring, either inheriting the same bodily peculiarities, or with a tendency to vary again in the same manner; whilst the individuals, less favoured in the same respects, will have been the most liable to perish.

Darwin and his followers emphasized the life and death value which various characteristics might afford to their possessors in the struggle for existence. Natural selection, in their view, was an aspect of differential mortality. As Darwin put it[4]:

Can it ... be thought improbable [that] variations useful in some way to each being in the great and complex battle of life should sometimes occur in the course of thousands of generations? If such do occur, can we doubt (remembering that many more individuals are born than can possible survive) that individuals having any advantage, however slight, over others, would have the best chance of surviving and of procreating their kind? On the other hand, we may feel sure that any variation in the least degree injurious would be rigidly destroyed. This preservation of favourable variations and the rejection of injurious variations I call Natural Selection.

In Darwin's time the causes and nature of variation in populations were poorly understood; the laws of heredity remained to be worked out (in 1865) and rediscovered (in 1900); and the genes were unknown. The rise of modern genetics in the first decades of the present century

[3] Darwin, 1872, ch. 7.
[4] Darwin, 1859, ch. 4.

made it possible for R. A. Fisher (1930)[5] and others to restate the Darwinian theory of selection in more precise and more general terms.

THE GENETIC THEORY OF SELECTION

In a population composed of individuals with the constitution a_1a_1, mutations in the **A** gene will lead to an occasional a_1a_2 individual. The initial frequency of the a_1a_2 individuals and hence of the a_2 allele will be low. The Hardy-Weinberg formula tells us that the mutant gene a_2 will not become more frequent in a large breeding population by the process of gene reproduction. In fact, if a mutation having no selective

Table 8. Probability of survival of a selectively neutral mutation appearing in a single individual[a]

Generations	Probability of survival
1	.6321
3	.3741
31	.0589
63	.0302
127	.0153
1,000	approx. .0010
10,000	approx. .0001
40,000	approx. .000025

[a] From Ronald A. Fisher, *The Genetical Theory of Natural Selection* (New York, Dover Publications Inc., 1958, 2d rev. ed.), ch. 4.

advantage or disadvantage appears in a single individual, the mutant gene can easily become extinct just by chance. Table 8 records the prediction that a single selectively neutral mutation has 6,321 chances in 10,000 of surviving for one generation, but only about 6 chances in 100 of surviving for 31 generations, and 1.5 chances in 100 of still being present after 127 generations. The probability that a unique mutation will remain in a population at the end of a period of many generations is real but slight.

Recurrent mutations from a_1 to a_2 may in time increase the relative numbers of a_2 alleles slightly, but this source of additional a_2 genes is eventually counterbalanced by reverse mutations from a_2 to a_1. From our knowledge of the lack of orientation of the mutation process, we can predict that the allele a_2 will usually not become frequent by the process of gene mutation.

[5] Fisher, 1930.

If, however, the mutant allele improves even slightly the ability of the individual carrying it to live and reproduce, it will increase in frequency. The increase in this case is due to the factor of selection, which is the most important means by which allele frequencies are changed.

Natural selection can be regarded as the differential and non-random reproduction of different alleles. Or, more generally, since these alleles are components of whole genotypes, selection is the differential and non-random reproduction of different genotypes. What observable properties of one allele or genotype (a_2) can bring about its regular increase in frequency relative to another allele or genotype (a_1)?

Whatever selective advantage a_2 possesses over a_1 is manifested in the kinds of phenotypes these alleles or genotypes, respectively, determine. If in a population the individuals carrying a_2 are, on the average, more viable, or longer-lived, or more successful in mating, or more fecund than the carriers of a_1, the former will leave more offspring than the latter, and the frequency of a_2 will consequently increase relative to a_1 in the next generation. The differential mortality or reproductive potential of the a_2 and a_1 genotypes could rest, in turn, on any one of innumerable effects of gene action. Thus a_2 could stand for a gene allele which enhances the resistance to some disease, or leads to a better protective coloration in an animal which has many enemies, or increases the number of eggs which can successfully develop into young.

Whereas natural selection meant differential mortality to Darwin, it means differential reproduction to the modern evolutionist. And the differential and non-random reproduction of genetically different individuals can be the result of various components: differential viability, differential longevity, differential emigration, differential success in mating, and/or differential fecundity. Only the first two of these components are synonymous with differential mortality and hence with the selection theory of Darwin's time.

Selection has both a positive and a negative aspect. The preferential survival and reproduction of some genotypes go hand in hand with the preferential elimination of other genotypes. The substitution of one allele for another entails the non-reproduction of a higher proportion of the carriers of the old allele than of the selectively favored allele.

The reduced contribution of some individuals to the hereditary pool of subsequent generations was considered in Darwin's time to be due to the premature death of the unfavored individuals. The modern evolutionist visualizes this preferential elimination of certain genotypes in a more general way, and calls it genetic death. The carriers of a relatively unsuccessful allele may be eliminated from the population due to lack of normal physical vigor in one case; in another case the unfavored types may be perfectly vigorous physically but are sterile; or they may lack normal sex drive; or again the unsuccessful genotypes may seem fully vigorous as adults but have a high mortality rate in embryonic or juvenile stages when the deaths are not noticed.

The line of genotypes can become extinct just as surely as a result of one factor that interferes with normal reproduction as from another. The unfavored allele will eventually die genetically, whether or not it leaves a series of visible corpses behind. Natural selection can and frequently does take the form of differential mortality, as Darwin correctly supposed, but it can also take alternative forms.

It will be noted that, since differential mortality does exist in nature, Darwin's selection theory is extended rather than replaced by the genetic theory of selection. It seems to be necessary to stress this point since some modern theorists have proposed to discard the nineteenth-century terms "struggle for existence" and "survival of the fittest" on account of their alleged crudity, overlooking the fact that nature *is* crude. On the realities of life in nature the nineteenth-century naturalists were well informed. The great majority of plants die before they reach maturity; the lives of nearly all wild animals, as Seton pointed out, have tragic endings. The struggle for existence is real enough. But the genetical theory of selection properly emphasizes the fact that natural selection can and does take many forms other than differential mortality.

The differential and non-random reproduction of different genotypes will be most pronounced when competition exists among the carriers of these genotypes for the raw materials and factors necessary for life. In other words, the process of selection comes into play most effectively in environments teeming with life. The real environments in nature are usually, but not invariably, teeming, and competition for the means of life is normally, but not always, the lot of living organisms.

THE SINGLE-GENE MODEL OF SELECTION

The selective advantage of one allele over another does not need to be, in fact usually is not, an all-or-none relationship. Suppose that gene A controls the speed of swimming in a small fish which is preyed upon by larger carnivorous fishes and which relies mainly on its speed to escape from its predators. Suppose further that the old allele a_1 enables its carriers to swim rapidly and that the mutant allele a_2 determines an ability to escape with a *slightly* greater rapidity. In a typical population, individuals of the type a_1 and individuals of the type a_2 will be randomly distributed at any moment. A wide-mouthed carnivorous fish suddenly passes through the population, engulfing those individuals that lie directly in its path, while the remaining individuals dart for cover. Among the individuals devoured will probably be some carrying the allele a_2 as well as some with the allele a_1. Similarly, the individuals which successfully escape will probably include both types a_1 and a_2. The chance position of an individual at the moment of entry of the carnivorous fish on the scene is a most important factor in determining whether that individual will be caught or not. There will, however, be a few borderline situations where both a_1 and a_2 individuals are in the path of the carnivorous fish and equally distant from it; then the individuals carrying the allele a_2 will usually escape owing to their slightly greater rapidity, whereas individuals of the genotype a_1 are usually caught under the same conditions. In this particular case the genotype of the individual, and not chance, is the critical factor.

For the population as a whole there is a preferential elimination of the a_1 genotypes in addition to the random elimination of many individuals irrespective of their genotype. The net result of repeated attacks by the carnivorous fish is not only the death of many small fish but also a statistical change in the relative numbers of a_1 individuals and a_2 individuals which survive to contribute their alleles to the next generation. The allele a_2 gradually increases in frequency relative to the allele a_1. This non-random component in the elimination of genetically different individuals from the population is selection.

Only if the individuals carrying the a_1 allele and those with the a_2 allele escaped from their predator with statistically equal frequency could we say that no selection at all was taking place. In that case, for every 100 a_2 alleles passed on to the next generation, 100 a_1 alleles would also be passed on. Let us call this Case i for future reference.

If, on the other hand, the fish captured by the predator were all of the genotype a_1 and none of the genotype a_2, so that all a_2 alleles but no a_1 alleles were passed on in reproduction, a_2 would replace a_1 in a single generation (Case ii). This is a possible but unlikely situation. The implication where a mutant allele completely replaces the normal allele is that the old genotypes were very poorly adapted indeed. The fate of a population composed of poorly adapted genotypes is apt to be extinction rather than evolution.

In most actual cases of evolutionary change one allele has only a slight selective advantage over another. But how slight is the selective advantage of one allele over another? In Case iii the fish carrying the mutant allele a_2 may be able to escape from their predators a little more often than fish carrying the allele a_1, so that for every 100 a_2 alleles passed on to the next generation there are 99 a_1 alleles. Or, in another case (iv), the relative rate of reproduction of the two genotypes might be in the ratio of 100 a_2 to 95 a_1. In still another case (v) the two alleles might change in frequency each generation in the ratio of 1,000 a_2 to 999 a_1.

Even if one allele possesses only a slight selective advantage over another, the favored allele will tend to increase and unfavored allele to decrease in frequency. Over a long enough period of time, through many generations of selection, the favored allele may become fixed in the population and the unfavored allele may become extinct or rare. The time required for the complete or nearly complete replacement of one allele by another may be very long, however, if their respective selective values are only slightly different.

THE RATE OF CHANGE IN ALLELE FREQUENCIES

When competing alleles possess slightly different selective values, the changes in allele frequencies are gradual and are spread out over many generations. A quantitative measure of the relative rate of reproduction of different alleles of the same gene is given by the selection coefficient (s). We may express the reproduction of one allele as a decimal fraction of the reproduction of another allele. The coefficient of selection is then the complement of this proportion; that is, s equals unity minus the decimal fraction.

Let us reconsider in terms of the selection coefficient the several hypothetical examples mentioned above. In Case i we assumed that the rate of reproduction of allele a_1 was equal to that of a_2, with 100 a_1 alleles being contributed to the next generation for every 100 a_2 alleles. Therefore $s = 1 - (100\ a_1/100\ a_2) = 0$.

In the remaining four cases a_1 is assumed to be less successful in reproduction than a_2. This selective disadvantage of a_1 is greatest in Case ii, where 0 a_1 alleles are contributed to the next generation for every 100 a_2 alleles. Here $s = 1 - (0/100) = 1$. In Case iii the relative rate of reproduction of the two alleles is 99 a_1 to 100 a_2; therefore $s = 1 - .99 = .01$. Or, as in Case iv, if this ratio is 95 a_1 to 100 a_2, $s = 1 - .95 = .05$. Finally, in Case v, the ratio is 999 a_1 to 1,000 a_2, and so $s = 1 - .999 = .001$.

We have assumed in Cases ii–v that the mutant allele a_2 is more successful in reproduction than the normal allele a_1. Under natural conditions most new mutations have a lower selective value than the normal alleles. The method of expressing the rate of gene replacement is the same whether a_1 is replacing a_2 or vice versa. For example, in Case iv, the reproduction of a_1 is 95 percent of that of a_2, and $s = .05$. If, contrariwise (Case vi), the proportion of gene reproduction is 100 a_1 to 95 a_2, the selection coefficient will be the same, i.e., $s = 1 - (95/100) = .05$. The only difference is that selection is favoring a_2 in Case iv and a_1 in Case vi. In other words, a_2 possesses a 5 percent selective advantage in Case iv and a 5 percent selective disadvantage in Case vi.

The selection coefficient thus gives a measure of the rate at which evolutionary changes occur in the population. Although the favored allele will gradually replace the less successful allele in each of the several cases considered, this replacement will come five times faster where $s = .05$ than where $s = .01$, and, conversely, the change in allele frequencies is much slower if $s = .001$ than if $s = .01$.

An allele which is recessive (as most mutant alleles are) and has a selective advantage of 1 in 100 ($s = .01$) will require nearly a million generations to increase in frequency from .01 percent to 1.0 percent in a population. Subsequent increases up to 50 percent take place more rapidly, and the gene can be raised from a 50 percent to a 99 percent frequency in a relatively short time (see Table 9). The table and the curve drawn from similar data (Fig. 39) show that selection is most effective in increasing the frequency of a recessive gene after it has reached a frequency of 1 percent or more. At very low and very high gene frequencies selection is slow and relatively ineffective for a recessive allele.

If the mutation is dominant, its increase in frequency by selection is

Table 9. The number of generations necessary for a given change in allele frequency where s = .01[a]

RECESSIVE ALLELE		DOMINANT ALLELE	
Percent change in frequency	*Number of generations*	*Percent change in frequency*	*Number of generations*
.01–.1	900,230	.01–.1	230
.1–1.0	90,231	.1–1.0	232
1–2	5,070	1–50	559
2–50	5,189	50–98	5,189
50–99	559	98–99	5,070
99.0–99.9	232	99.0–99.9	90,231
99.90–99.99	230	99.90–99.99	900,231

[a] After Pätau, 1939.

a much more rapid process (Table 9, Fig. 39). For dominant mutations, selection is most effective at low and intermediate gene frequencies, and least effective at high frequencies.

The rate of change in allele frequencies is thus not constant for any given value of s, but varies with the absolute frequency of the allele in the population. Whether dominant or recessive, the allele increases most rapidly when present in medium frequencies, but increases slowly when at high frequencies. If the allele is recessive, it also increases slowly when present in low frequencies.[6]

PLEIOTROPY IN RELATION TO SELECTION

The hypothetical alleles a_1 and a_2 can be expected to have various pleiotropic effects in the individual organisms. Some of the phenotypic characteristics determined by a given allele may be advantageous to the

[6] Haldane, 1932.

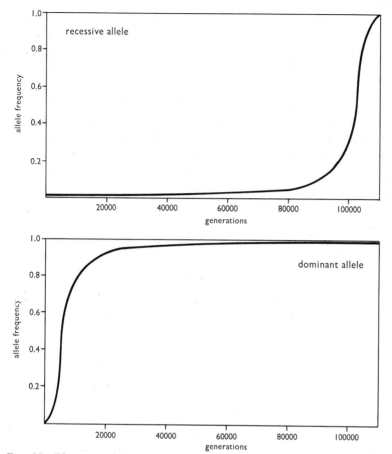

Fig. 39. The rate of change in allele frequency per generation when the allele has a selective advantage of 1 in 1,000 (s = .001) and is present at different frequencies from nearly 0 to 1.0

The allele is recessive in the upper curve and dominant in the lower curve.

From C. Stern, *Principles of Human Genetics* (W. H. Freeman and Company, Publishers, 1960).

individual while other effects of the same allele may be disadvantageous. An individual carrying the allele a_1 might have a selective advantage over an individual carrying a_2 as regards one character, but might be inferior to a_2 genotypes in some other character. Under these conditions the relative selective value of genotypes carrying, respectively, the a_1

and a_2 alleles depends on their over-all net performance. If individuals carrying a_1 have a net superiority over individuals carrying a_2, the former allele will increase in frequency in the population in spite of its deleterious side effects.[7]

THE EQUILIBRIUM BETWEEN SELECTION AND MUTATION

The initial increase in frequency of a recessive mutation by selection alone in a large population is very slow, as we have just seen. If the mutant allele (a_2) has a selective advantage of 1 in 1,000 (s = .001), it will require many thousands of generations to double in frequency from .000001 to .000002. During this long time recurrent mutations $a_1 \rightarrow a_2$ are likely to occur. If the mutation rate of $a_1 \rightarrow a_2$ is u = .000001, the doubling of the allele frequency, which requires thousands of generations without recurrent mutations, can be accomplished in a single generation.[8] At an early stage of the evolution process, therefore, mutation pressure can reinforce and accelerate the effects of selection.

At a later stage, when the favored allele is present in intermediate frequencies, the changes in allele frequencies are controlled so effectively by selection as to render the mutation process unimportant by comparison.

As the favored allele approaches fixation, mutations become a factor to be reckoned with again. Selection may be working in the direction of the complete replacement of a_1 by a_2. Reverse mutations of $a_2 \rightarrow a_1$ may prevent this process from going to completion. The population thus remains polymorphic forever.

Hemophilia is a condition in man wherein the blood fails to clot normally. Individuals suffering from this disease may bleed to death from even small cuts, and usually die before reaching sexual maturity. The condition is due to a single recessive allele, hm. Since this allele is lethal and is strongly selected against, it might be expected to have become extinct in the human population. The hm allele is in fact rare but extant, its estimated frequency in the gamete pool being in the range .0002 to .00004.[9] The fact that the hemophilia allele continues to exist at all is attributed to recurrent mutations from Hm $\rightarrow$ hm,

[7] Wright, 1960, 437.
[8] Dobzhansky, 1951a, 81–82.
[9] Haldane; see Beadle, 1957, 22.

occurring at the estimated rate of 1 per 32 million gametes.[10] The continual formation of new hm alleles counterbalances the selective elimination of these alleles and prevents the normal allele Hm from reaching a frequency of 100 percent.

THE EQUILIBRIUM BETWEEN SELECTION AND MIGRATION

Let us suppose that a population polymorphic for two alleles a_1 and a_2 of a gene **A** is undergoing selection in favor of a_1, so that a_1 is increasing and a_2 is decreasing in frequency. The immigration into this population of carriers of a_1 will obviously speed up the replacement of a_2 by a_1, whereas the immigration of carriers of a_2 will work against the effects of selection. The point of equilibrium between selection and gene flow in the case where they act in opposite directions will depend on the relative magnitudes of s and m.

It has been shown experimentally that very strong selective pressures are capable of overcoming the effects of considerable gene flow.[11] Artificial selection was carried out during eight generations in experimental populations of *Drosophila melanogaster* for an increasing number of sternopleural chaetae. The intensity of selection was such that only the 10 percent of the population with the greatest number of chaetae were allowed to reproduce in each generation. One line of flies was kept completely isolated. It showed as expected a significant rise in chaeta number as compared with an isolated and unselected line.[12]

In another line a random sample of flies was periodically introduced from outside, so that the parents of each new generation consisted of selected native individuals and unselected immigrants in the relative proportions of four to one. The number of chaetae rose from generation to generation at about the same rate in this semi-isolated population as in the completely isolated population. In other words, the effects of 20 percent gene flow did not alter significantly the response of the fly population to a very strong selection.[13]

However, in still another line, where the selected native flies and the unselected immigrants were equal in numbers in each generation, so

[10] Haldane; Neel and Falls; see Srb and Owen, 1952, 237.
[11] Thoday and Boam, 1959; Streams and Pimentel, 1961.
[12] Streams and Pimentel, 1961.
[13] Streams and Pimentel, 1961.

that the gene flow amounted to 50 percent, the selective pressure remaining the same as before, the effects of the selection were very slight.[14] In this case, evidently, a massive immigration swamped out the effects of strong selection.

In a parallel series of experiments the selective pressure was made less intense. In this series the 40 percent of the flies with the greatest number of chaetae (instead of the highest 10 percent as in the first experiments) were allowed to reproduce in each generation. The average number of chaetae rose in response to this degree of selection in a completely isolated population. In populations receiving even relatively few unselected immigrant individuals in each generation, however, the result was otherwise. Where the number of immigrants amounted to only 6 percent of the breeding stock in each generation, the number of chaetae did not rise during eight generations of selection.[15] The same selective pressure that led to marked changes in the chaeta number in a completely isolated population had only negligible effects in populations exposed to a relatively slight amount of gene flow.

The case of moderate or weak selection is probably more typical in nature than that of very intense selection. In natural populations subjected to normal selective pressures, therefore, the effects of selection may easily be swamped out by moderate or even small amounts of gene flow.

SELECTION FOR GENE COMBINATIONS

The single-gene model of selection can be extended to encompass combinations of two or more genes. The separate genes **A**, **B**, and **C** may work in combination to produce some complex character, like the length of a giraffe's neck, or some combination of characters, like number of kernels and presence or absence of chaffy bracts on a corn ear. Assume that the triple mutant genotype $a_2a_2\ b_2b_2\ c_2c_2$ determines a form of this character or character complex possessing adaptive properties which are superior to those determined by the preexisting genotype $a_1a_1\ b_1b_1\ c_1c_1$.

The several mutant alleles will at first probably arise in different individuals: $a_1a_2\ b_1b_1\ c_1c_1$, $a_1a_1\ b_1b_2\ c_1c_1$, and $a_1a_1\ b_1b_1\ c_1c_2$. Crossing

---

[14] Streams and Pimentel, 1961.
[15] Streams and Pimentel, 1961.

between these individuals followed by the recombination of genes in their progeny will next lead to the formation of some individuals of the genetic constitution a_2a_2 b_2b_2 c_2c_2. These individuals will however intercross with other genotypes in the population with the result that the gene combination which was assembled by sexual reproduction is broken up by the same process.

But if the new genotype a_2a_2 b_2b_2 c_2c_2, once formed, proves to have superior adaptive properties, selection will cause the new alleles to gradually replace the old alleles for each gene. When the several mutant alleles a_2, b_2, and c_2 reach high frequencies in the population, intercrossing between individuals will produce mainly a_2a_2 b_2b_2 c_2c_2 genotypes. A new combination of genes will have become established in the population by selection.

Selection for a single gene or single character is a burden in itself to a population, but selection simultaneously for two or more independently inherited characters causes the number of genetic deaths to rise exponentially. As will be described in more detail later in this chapter, the moth *Biston betularia* underwent strong selection for the type of coloration in industrial England in the latter nineteenth century, a dark-colored form being favored over the ancestral gray form. The two color forms differ in the alleles of a single gene. The intensity of the selection was such that the frequency of gray individuals was sometimes reduced by one-half in a day. If selection were equally intense for 10 independently inherited genes, only $(1/2)^{10}$ or 1 in 1,024 individuals of the ancestral type would survive.[16]

Selection for combinations of genes is thus very costly to a population in terms of genetic deaths. A mortality as intense as that hypothesized above, namely, 1,023 out of 1,024 individuals of the ancestral type, would probably lead to the extinction of the population. Yet most adaptations are based not on single genes but on gene combinations, and selection is required for their establishment. In other words, a population, to be successful, must somehow develop complex adaptations while maintaining its numbers. Perhaps the populations which are surviving under natural selection are those in which the establishment of gene combinations is spread out over many generations, with

---

[16] Haldane, 1957.

only a moderate number of alleles progressing toward fixation at any one time.[17]

Another and more efficient method of establishing new gene combinations in populations, as we shall see later, is by inbreeding and drift combined with selection in small daughter colonies derived from the polymorphic ancestral population (Chapters 11 and 16).

THE GENOTYPE AS A UNIT OF SELECTION

The single-gene model of selection, which considers the relative rates of reproduction of alternative alleles, for example a_1 and a_2 of a gene **A**, is clearly an oversimplification, because natural selection does not ordinarily work on single genes. The extension of this model to encompass small combinations of genes—**A**, **B**, **C**—is also an oversimplification. The primary form of selection in nature—Darwinian selection as it is sometimes called—takes place not between the alleles of one or a few genes but between individuals in a population.[18] These individuals are the carriers of genotypes in which the particular alleles occur as subordinate elements.

Since the alleles a_1 and a_2 are components of whole genotypes, consisting of hundreds or thousands of genes, opportunities exist for diverse interactions between genes which will further complicate the action of selection. The phenotypic effects of any given allele may be modified by other genes in the complement. Consequently the same allele (a_1) may be advantageous in some genetic backgrounds but not in others. Some genotypes containing a_1 will then increase in frequency at the same time that other a_1 genotypes are decreasing in relative abundance. The direction of change in the allele frequency of a_1 will then be a resultant of the opposite tendencies.

Or the composition of the genetic background itself may be changed by selection. If a_1 is a favorable allele of a major gene **A**, selection may take place for the alleles of a modifier gene system which enhance the expression of a_1. Likewise, in the same situation, selection can build up a system of alleles of the modifier complex acting to suppress the effects of a deleterious major allele a_2.

---

[17] Haldane, 1957.
[18] As emphasized by Wright (i.e., 1931, 101) and many other evolutionary geneticists.

The majority of new mutations are deleterious. Most mutant alleles arising in diploid organisms are recessive to the standard form of the gene and consequently do not produce their deleterious phenotypic effects in heterozygotes. The diploid individual carrying the new mutation in a single dose thus possesses a normal phenotype. Now the dominance of the standard allele over mutant alleles is due in part at least to the action of other genes in the complement, and more particularly to dominance modifiers. Putting these facts together, Fisher proposed that the genotypes of organisms have been selected for modifier complexes that suppress the unfavorable phenotypic effects of most new mutations in the heterozygotes in which these mutations arise. The dominance of standard over mutant alleles, in other words, is not an inevitable result of the mutation process, but is instead an adaptation to protect the diploid individual from the ill effects of mutations. This adaptation has been built into the genotype by selection of dominance modifiers.[19]

Under constant environmental conditions, a given allele (a_1) may produce phenotypic effects in some individuals carrying it, but produce no detectable effects in other carriers; or the degree of phenotypic expression of the allele may vary from slight to strong in different individuals having any expression at all. The allele has variable penetrance or variable expressivity, respectively. These differences in the manifestation of the same allele in different individuals which are genetically identical for the allele in question, and which live in the same environment, are attributable to the action of expressivity and penetrance modifiers.

If the allele a_1 brings about phenotypic effects which are favorable in a certain environment, but these effects are manifested in only a small proportion of the carriers of a_1, selection will favor the reproduction of the genotypes permitting the expression of the trait. The result of selection will be to increase the frequency of genotypes containing modifiers for high penetrance of the allele a_1 at the expense of genotypes containing low-penetrance modifiers. Similarly, if the major allele a_1 producing an adaptively valuable character is strongly expressed in some individuals and weakly expressed in others, owing to

---

[19] Fisher's theory of the evolution of dominance; Fisher, 1930, ch. 3.

allelic differences in the modifier genes, selection can increase the frequency of the genotypes which have strong expressivity modifiers.

Selection for different sets of modifier genes under different environmental conditions could also have unexpected consequences, as we shall next attempt to show.

SHIFTS IN EXPRESSIVITY

It is known that environmental shocks applied to early developmental stages of *Drosophila melanogaster* can bring about various aberrant adult phenotypes like those caused by gene mutations. Thus high temperatures during the pupal stage can lead to adult flies with wings lacking a posterior crossvein, and treatment of the eggs with ether is followed by a modification of the thorax in some adult individuals to a condition known as "bithorax." There are, in addition to these purely phenotypic modifications induced by the environment, also known genetic mutants with the same crossveinless wings or bithorax condition of the body.

In a series of experiments, Waddington exposed developing flies to specific environmental shocks and then selected for the corresponding phenotypic modification in the adult stage. The environmental treatment applied in one experiment was high temperatures and the phenotypic response selected for was crossveinless wings. In another selection experiment the eggs were treated with ether, and bithorax adults were chosen as parents of the next generation. The final result of selection at the end of the experiment, in each case, was a population of flies containing some individuals which exhibited the aberrant phenotypic trait, not only under the stimulus of heat shocks or ether treatment, but also spontaneously in a normal environment.

The foundation population with which the selection experiment for the bithorax condition was started consisted of flies with a normal thorax. Their ether-treated eggs developed into some bithorax adults as well as some normal adults. A sample of the individuals with the bithorax phenotypic modification was chosen as the parents of the next generation. Some of the eggs in the second generation were again treated with ether, and some of the bithorax adults resulting from these eggs were again chosen to continue the line. The ether treatment of the eggs and the selection of the bithorax adult phenotypes were kept up until the 29th generation.[20]

In a separate but parallel selection line, the unmodified normal phenotypes were taken as parents throughout the 29 successive generations.

In each generation of both selection lines, some eggs were collected and allowed to develop without the ether treatment. In the line selected for non-response to the ether stimulus, only normal adults emerged from the untreated eggs, as expected. Likewise, in the early generations of the line selected for the bithorax response to ether treatment, only normal adults emerged from untreated eggs. But in the 8th generation of this same line, one fly from an untreated egg showed a slight tendency toward the bithorax condition; in the 9th generation 10 such uninduced bithoraxlike types appeared; and in the 29th generation several other bithorax types arose from untreated eggs. Progeny tests showed that the bithorax character was genetically determined in these individuals. In some of them, especially in the 29th generation mutants, the bithorax phenotype was determined by a major allele and modifier genes.[21]

The selection experiment making use of heat shocks began with a population of flies with normal wings. Their pupae were exposed to high temperatures. The adult flies developing from these pupae consisted of both normal and crossveinless individuals. Flies with crossveinless wings were selected as parents to continue the line, and their offspring in the pupal stage were again treated with high temperatures. The combined environmental treatment and selection of the modified phenotypes were kept up each generation for 24 generations.[22]

As mentioned, some individuals in each generation did not respond to the heat shocks by the development of crossveinless wings. The flies with unmodified wings were selected as parents in a separate and parallel selection line.

At the end of this experiment, genetic differences were found between the descendant and ancestral generations of each selection line, and again between the two selection lines. In the line selected for readiness to make the crossveinless phenotypic response to heat, the proportion

[20] Waddington, 1956.
[21] Waddington, 1956.
[22] Waddington, 1953.

of individual flies producing this modification was 34 percent in the ancestral generation and 97 percent in the 24th generation. The high percentage of individuals producing crossveinless wings in response to the stimulus of heat in the 24th generation of the line selected for readiness to react in this fashion stands in contrast with a low percentage of modified individuals in the 24th generation of the line selected for phenotypic stability in relation to heat shocks, this frequency being 97 and 14 percent, respectively. Furthermore, some of the flies in the line selected for the crossveinless modification produced crossveinless wings even when reared under normal temperatures. They were now genotypically crossveinless. The heritable condition of crossveinlessness appears to be determined by a major gene and modifiers. In the line selected for the development of normal wings at high temperatures, on the other hand, crossveinless phenotypes did not appear whenever individuals belonging to that line were reared at normal temperatures.[23]

Some of the changes observed in these experiments can be explained easily enough. The flies respond to abnormal environmental conditions by undergoing phenotypic modifications to the bithorax or crossveinless condition. In the early generations of the experiment such phenotypic modifications are apparently the only type of change involved. But during the course of the experiment new gene mutations or recombinations producing a bithorax or crossveinless phenotype could arise and could provide genetic differences for selection to operate on. Selection of the bithorax or crossveinless individuals then increases the frequency of the genotypes that yield altered phenotypes in the abnormal environment.

This is only part of the story, however, for genotypes have arisen which produce the aberrant phenotype not only in the abnormal environment, in which the selection was carried out, but also in a normal environment, for which the aberrant types were not selected.

The latter result could be explained by assuming that the mutations arising in the environmentally treated flies are somehow channelized along pathways having favorable phenotypic effects in the new environments.[24] This assumption is contradicted, however, by much independent evidence indicating that the mutation process is, in fact, not oriented

[23] Waddington, 1953, 1957, 177; Bateman, 1959.
[24] See Waddington, 1953, 1956, 1957.

with respect to the adaptive requirements of the organism (see Chapter 8). It is therefore desirable to find an interpretation of the observed results that is consistent with our present knowledge concerning the mutation process.

Stern has offered an interpretation involving the following postulations, all of which are consistent with known genetic facts.[25] (1) The ancestral fly populations were polymorphic for genetic factors affecting the character of the wings or thorax. (2) These genetic factors were unexpressed in the standard environment but were brought to expression in the abnormal environment. (3) When the flies were reared in environments in which the genetic factors were expressed, selection could alter their frequencies. (4) The selection of genes with a strong mutant phenotypic expression in an abnormal environment had the collateral effect of building up genotypes which produced the mutant phenotype also in a normal environment. Bateman independently made similar suggestions and proposed that the types of genetic factors being favored by selection may be penetrance modifiers, which determine incomplete penetrance at the beginning of the experiment but bring about a more complete penetrance later on.[26] The following paragraphs are an elaboration on these ideas.

Let us now pick up the train of thought developed in the preceding section. We saw that penetrance or expressivity modifiers which enhance the phenotypic expression of some major gene will be subject to control by selection. We may now consider, as an additional complication, the influence of the environment on the modifying effects of modifier genes.

The action of penetrance and expressivity modifiers of our hypothetical major allele a_1 is affected by environmental conditions, as is the action of all genes. The modifier genes may exert their modifying effects on development in one environment (E_2), but not in another (E_1). A given set of modifier genes (M^+), having an enhancing effect on the expression of the allele a_1, will then produce different phenotypes in the different environments E_1 and E_2. Conversely, in environment

[25] Stern, 1958, 1959; also Mayr, 1959, 8.
[26] Bateman, 1956, 1959. Waddington seems to be adopting a similar interpretation, especially in his more recent discussions. For a more complete account of Waddington's views than needs to be given here the reader is referred to Waddington, 1957.

E_1 where the modifier complex M^+ is inactive, the phenotype engendered by M^+ may be indistinguishable from that engendered by a different set of modifiers M^-. It follows that selection, acting directly on phenotypic differences between individuals, will be ineffective in discriminating between the M^+ and M^- genotypes in environment E_1, but not so in environment E_2, where it can raise or lower the frequency of the alternative modifier complexes M^+ and M^-.

Selection in environment E_2 for the enhancers of penetrance or expressivity (M^+) of some trait is of course related primarily to the expression of that trait in that environment. But selection for modifying factors acting in a certain direction in one environment (E_2) may have consequences as regards the character expression in other environments (E_1, etc.). We have assumed above that the modifier complex M^+ enhances the expressivity or penetrance of the major allele a_1 in environment E_2 but not in E_1. It is clear that selection carried out in environment E_2 for the phenotypic trait brought to expression by the modifiers M^+ will at first increase the frequency of these modifiers and of the genotypes carrying them in the population. Now observe that the same selective process can next go on to assemble in the population a new and different set of modifier genes (M^{++}) which has an even stronger enhancing effect on the development of the phenotypic character in environment E_2. And the new and stronger modifier complex M^{++} may, as a by-product of its greater strength of action, produce the phenotypic trait not only in environment E_2 but also in environment E_1.

Thus a certain major allele (a_1) together with a certain modifier complex (M) in a particular environment (E) will produce a given phenotype (P). The relations between these factors may be recapitulated in the form of a series of equations:

$$a_1 + M^+ + E_2 \rightarrow P_m \quad \text{(i)} \qquad a_1 + M^- + E_2 \rightarrow P_n \quad \text{(iv)}$$
$$a_1 + M^+ + E_1 \rightarrow P_n \quad \text{(ii)} \qquad a_1 + M^{++} + E_2 \rightarrow P_m \quad \text{(v)}$$
$$a_1 + M^- + E_1 \rightarrow P_n \quad \text{(iii)} \qquad a_1 + M^{++} + E_1 \rightarrow P_m \quad \text{(vi)}$$

In summary, the same normal phenotype (P_n) may be produced by different modifiers in different environments [eqs. (ii) and (iv)]. And the mutant phenotype P_m may be produced by a modifier complex of a certain strength in one environment but not in another [eqs. (i) and (ii)]. However, a stronger modifier complex may be capable of producing the mutant phenotype P_m in a wider range of environments [eqs. (v) and (vi)].

Here E_1 stands for the standard environment and E_2 for an abnormal environment containing ether or heat in Waddington's experiments. P_n stands for a normal type of thorax or wing and P_m for a bithorax or crossveinless condition of these body parts. The potentially mutant and potentially non-mutant flies are both normal in phenotype in the standard environment [eqs. (ii) and (iii)]. In the abnormal environment, however, these potentialities come to expression, and a mixture of normal and aberrant flies appears in the population [eqs. (i) vs. (iv)]. In the abnormal environment selection discriminates between genotypes on the basis of the allelic differences in their modifier genes which affect the expression of the character.

Selection for a strong expression of the trait in the new environment leads first to the establishment of the modifier complex M^+ which produces the character in this environment, [eq. (i)], then to the assembling of a new modifier complex M^{++} which determines an even stronger expression in the same environment, [eq. (v)]. The modifier complex M^{++} has, however, the capacity to cause the mutant trait to be formed also in the normal environment, [eq. (vi)].

The model outlined above, hypothetical though it is, shows us how selection for a given set of modifier genes in one environment may lead to genotypes having unpredictable phenotypic expressions in other environments. The evolutionary aspects of such shifts in expressivity, resulting from selection for different modifier alleles in different environments, have been relatively little investigated as yet.[27] These evolutionary aspects are worth considering. The new potentialities of phenotypic expression, which can be created inadvertently by selection in one environment, may sometimes have important consequences in evolution, as where a population becomes prepared or "preadapted" in an ancestral environment for entering a new environment.

BUFFERS AGAINST SELECTION

Natural selection involves an interplay between organism and environment. We have assumed in the discussion up to this point that

[27] But see *inter alia* Stern, 1958, 1959; and Bateman, 1956, 1959.

the environment is active and discriminates between genetically different forms of the organism. But in practice the different genotypes are going to react in more than passive ways to environmental selection. Since natural selection in its negative aspect is a wasteful process, entailing the death of many individuals, the reactions of these individuals to the environmental factors causing the selective pressures will be such as to reduce the losses.

The various adjustments of individual organisms to the environment which were described in Chapter 6—phenotypic modifications, physiological homeostasis, habitat selection—all help to reduce the number of genetic deaths caused by selection in its negative aspect.[28] To the extent that these phenotypic adjustments cushion the organism against natural selection, they slow down the genotypic changes that are synonymous with evolution. This buffering against the effects of selection is, however, advantageous to the organisms concerned, for their reproductive potential is increased by the capacity to make adaptive phenotypic responses to environmental stresses. All existing genotypes have the ability to make such phenotypic responses.

In the final analysis a certain proportion of genetic deaths cannot be avoided in an organism subject to competition and selection. The loss of a number of individuals may be a burden to the population. This burden, though unavoidable, can be mitigated in various ways. The loss of a given number of eggs, seeds, embryos, or juveniles is less burdensome to a population than the loss of the same number of adult individuals. Competition between individuals is frequently more severe during early stages of development, and the individuals are frequently more susceptible to selective elimination at such stages, than in the adult phase. The burden of genetic deaths in many plants and animals is thus often shifted from the adults to some embryonic or juvenile stage of growth (see also Chapter 10).[29]

As noted earlier in this chapter, the extra high cost of selection for characters controlled by separate unlinked genes may be kept within limits tolerable to the population if only a moderate number of alleles are proceeding toward fixation at any one time, and if the whole selective process is spread out over many generations.

[28] Also Waddington, 1960.
[29] Haldane, 1957.

STABILIZING SELECTION VERSUS
PROGRESSIVE SELECTION

Any population of organisms exists in a certain environment and must be fitted or adapted to live successfully in its particular habitat. If the environment remains stable and if the population has already arrived at a high state of adaptedness, the main effect of selection will be to eliminate such peripheral variants or off-types as arise by mutation, gene immigration, or recombination. A certain range of genotypes of proven adaptive fitness is thereby preserved from generation to generation. This form of selection, known as stabilizing selection, does not bring about evolutionary changes, but rather maintains an existing state of adaptedness.[30]

Bumpus measured wing spread, weight, and other characters for a sample of English sparrows, *Passer domesticus*, overcome by a snowstorm in North America. He found that the sparrows that survived conformed closely to the mean for the species, whereas the sparrows that died from the storm showed a greater range of variation in the characters measured.[31] In a well-adapted species like the English sparrow, deviations from the norm are suppressed by stabilizing selection.

If the environment changes, however, some of the peripheral variations may prove to have a higher adaptive value than some of the old normal genotypes in the new environment. Selection now takes a different form, wherein some of the standard genotypes become eliminated and some of the new mutant or recombination types are preserved. The population maintains its fitness in a changing environment by undergoing the appropriate changes in its genetic composition. If the environmental changes are oriented in a particular direction over a long period of time, as when the climate becomes progressively colder and moister with the advance of an ice age, the genetic composition of the population may undergo a directed change in successive generations.

The form of selection which operates in a progressively changing environment is known as progressive selection.[32] The result of progressive

[30] Schmalhausen, 1949, 78 ff.; Simpson, 1953a, 148 ff. (as centripetal selection). Lerner, 1958, 6 ff.
[31] Bumpus in 1899; see Huxley, 1943, 448.
[32] Schmalhausen, 1949, 73 ff.; Simpson, 1953a, 148 ff.; Lerner, 1958, 6 ff. The terms dynamic selection, linear selection, and directional selection are also used.

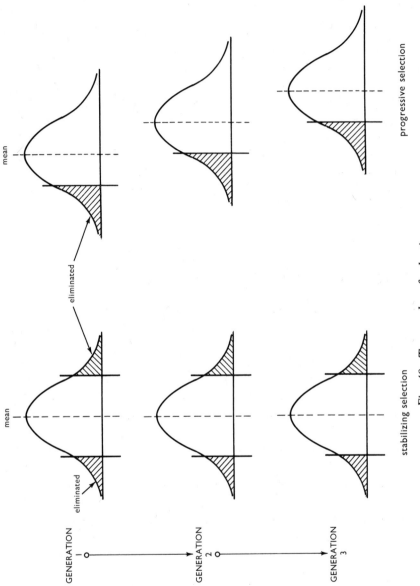

Fig. 40. Two modes of selection

selection is a systematic change in allele frequencies by which the population keeps pace with the changing environmental conditions.

The two modes of selection are portrayed graphically in Fig. 40. The genetic variation in a population is assumed to have a normal distribution with numerous individuals near the mean and a few individuals at the extremes for each variable and measurable character. This variation can accordingly be represented as a normal curve. Under stabilizing selection the peripheral variants on both extremes of the normal curve are eliminated generation after generation. The reproduction preferentially of the individuals possessing characteristics near the mean for the population results in the preservation of a constant modal condition through time. Under progressive selection, on the other hand, the elimination of the genetic variation is one-sided. The mean characteristics of the population therefore shift progressively during a succession of generations (Fig. 40).

It is evident that progressive selection is a process by which a state of adaptedness is reached, while stabilizing selection is a process by which the state of adaptedness, once reached, is maintained.

SELECTION EXPERIMENTS

The theory of selection has been verified in numerous experiments.[33] The subject in one such experiment was the small mosquito fish, *Gambusia patruelis*, which can be pale gray or black.[34] A pale-colored fish blends into a light background but a black fish stands out in the same setting; conversely, a black fish is harder to see against a black background than a pale-gray fish (Fig. 41). Sumner reasoned that coloration harmonizing with the background would help to protect the fish against their normal predators, fish-eating birds. To test this assumption he built two tanks 15 feet long by 8 feet wide and painted one of them black on the inside and the other pale gray. Each tank was filled with water and with equal numbers of black mosquito fish and pale mosquito fish. The fish were now exposed to attacks by the Galapagos penguin. After the birds has been allowed to feed for a short time they were removed and the surviving fish in each color class were counted.

[33] Reviews in Dobzhansky, 1951a, ch. 4, and 1955a, chs. 5–6.
[34] Sumner, 1935.

It was found that the penguin in the pale tank had eaten 176 pale-gray fish and 278 black fish, or 39 percent of the original number of pale fish as compared with 61 percent of the black fish. The penguin in the black tank had taken 78 black fish to 217 gray fish, the mortality of the two color classes being 26 percent and 74 percent, respectively. A similar differential in mortality occurred in another series of experiments where night herons instead of penguins were used as predators. The experiment shows that a coloration blending in with the background does not

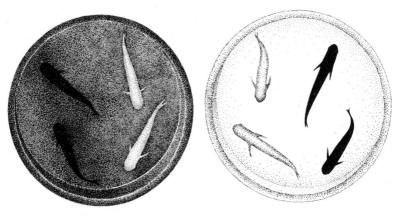

Fig. 41. The black and pale-gray phases of the mosquito fish, Gambusia patruelis, *as seen against black and light backgrounds*

Redrawn from Sumner, 1935.

protect a fish completely, but does give it an advantage over a fish without concealing coloration. For the experiment as a whole the birds ate nearly twice as many contrastingly colored fish as concealingly colored fish.

Several selection experiments with corn, *Zea mays*, have been carried on at the Illinois Agricultural Experiment Station for over half a century. The starting point of one experiment was a single variable population of corn grown in 1903. In the original population the average height of the ears above the ground ranged from 43 to 56 inches in different individuals. The individuals with the ears borne closest to the ground were chosen as parents for the next generation. In the second generation the individual plants with the lowest ears were again chosen to reproduce the population. In this way a selection line for ears low on the stem was set up and continued generation after generation. The result by 1927 was a population of corn plants in which the ears were an average of 8 inches above the ground. The new phenotypic trait was hereditarily fixed in the population as shown by the fact that progeny of the 1927 population bred true to type. Since corn is an annual crop, producing one generation each year, the change in the genotypic composition of the population corresponding to the above described phenotypic change from medium-high ears to very low ears took place in 24 generations of artificial selection.[35]

Selection was carried out concurrently in another line derived from the same foundation population of 1903 for ears borne high on the stem. In this case, the plants with the highest ears were used generation after generation to reproduce the strain. In 1927 at the end of 24 generations of selection the derived plants produced their ears an average of 120 inches (10 feet!) above the ground.[36]

Another selection line was set up for high content of protein in the grains. A sample of 163 ears was harvested from the foundation population of 1896 and the protein content was determined for a few rows of kernels in each ear. The remaining kernels on the 24 ears with the highest percentage of protein were then planted in a plot, where the resulting plants were allowed to interbreed only with one another. The process was repeated in each following year when the protein production of the different individuals was again sampled and the 24 (or in later years, the 12) ears highest in protein were chosen for reproduction. The hereditarily determined protein content of the grains gradually increased during 60 generations of artificial selection from an average of 10.9 percent to an average of 19.4 percent (Fig. 42).[37]

Selection was carried out concurrently for low protein content of the grains in another line derived from the same original population. The progress of selection in the low line is shown in Fig. 42; the average protein content decreased in this line from 10.9 percent to 4.9 percent in 60 generations.[38] The low line and the high line represent two

[35] Bonnett, 1954.
[36] Bonnett, 1954.
[37] Winter, 1929; Woodworth, Leng, and Jugenheimer, 1952; Leng, 1960.
[38] Winter, 1929; Woodworth, Leng, and Jugenheimer, 1952; Leng, 1960.

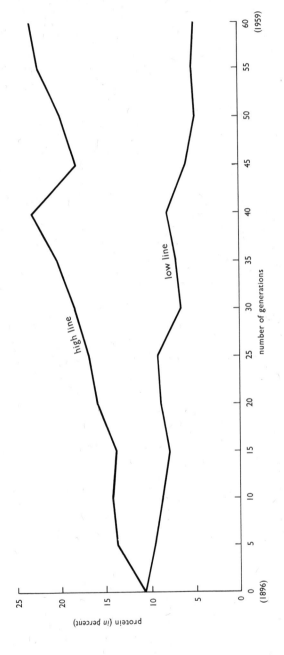

Fig. 42. Progress of selection for the protein content in the kernels of corn, Zea mays, during 60 generations
From data of Woodworth, Leng, and Jugenheimer, 1952.

populations derived from the same foundation stock which have diverged markedly in a measurable phenotypic trait and in the genotypes determining this trait. This divergence has been brought about by the orienting force of artificial selection, which has taken one course in one line and another course in the other line.

Are the differences between the two extracted lines due entirely to selection? There is no need to assume that only one evolutionary factor is involved. During 60 years time, gene mutations or recombinations could have arisen affecting the protein composition of the grains. The occurrence of new genetic variations for low protein content might explain why selection for this character made more progress during the last 33 years of the experiment than during the first 27 years.

A selection experiment carried out by Dobzhansky and Spassky was designed so that the evolutionary changes in the populations would be brought about by natural causes rather than by the will of the experimenters.[39] The foundation population was a strain of *Drosophila pseudoobscura* having less than the normal degree of viability; the actual viability of the strain was 29 percent of a certain standard known as "normal viability." A number of males and females were placed in a bottle containing food material and allowed to lay unlimited numbers of eggs, which soon resulted in the container becoming overpopulated. A random sample of the offspring was then transferred to a new culture bottle and allowed to breed again. A line was continued in this fashion for 50 generations (which took four years, as compared with 50 years for the same number of generations in corn). The average viability of the flies at the end of the experiment was compared with that at the beginning. It was assumed that the crowded conditions in the bottles and the strong competition for food would lead to an increase in the frequency of any mutant types possessing greater vigor than their siblings. This expectation was realized. The viability of the strain rose from 29 percent to 90 percent in the course of 50 generations.

The experiment was repeated using six other strains which, like the first one, were initially less viable than normal flies. The results based on a series of tests are more reliable statistically than the results of an experiment which is run only once. It was found that other strains

[39] Dobzhansky and Spassky, 1947.

underwent improvements in viability too: one from 60 percent to nearly 100 percent, another from 30 percent to over 80 percent, and so on. The viability rose in five of the seven independent lines.

Seven other lines were carried through 50 generations in the same way except that the male parents of each new generation were treated with x-rays before they were introduced into the fresh culture bottle. The purpose of the x-ray treatment was to increase the number of new mutations. A gain in viability was exhibited by six of the seven irradiated lines. Some of the viability gains were quite large, as from an initial 29 percent to a final 103 percent in one line, and from 65 percent to 115 percent in another line.

The rise in frequency of mutant genotypes of superior vigor in 11 of the 14 lines went hand in hand with an elimination of the weaker types under the conditions of intense competition for food in crowded culture bottles. Another series of lines in which the cultures were kept from becoming overcrowded and the competition between individuals was less severe did not show similar improvements in viability at the end of 50 generations. A differential mortality did not occur in the latter so-called control lines nearly as much as in the 14 crowded populations. This is another way of saying that selection was at work in the crowded lines but not in the controls. The form of selection involved, moreover, was natural selection, for although the flies lived in glass bottles on a laboratory shelf, the observed changes in their hereditary characteristics were due to interactions within the populations themselves and not to any direct intervention by the men running the experiment.

SELECTION IN NATURAL POPULATIONS

The peppered moth, *Biston betularia*, which occurs widely in England, flies at night but spends the day at rest on the trunk or branches of trees, where it is vulnerable to the attacks of insectivorous birds. For ages the populations of *Biston betularia* consisted of speckled gray individuals which blend well with the lichen-covered bark they use as a perching place, and the predation by birds was reduced by this concealing coloration. There is in addition a melanic form of the moth with black wings and body which arises as a result of a rare dominant mutation in a single gene **C**. The melanic form is conspicuous against the background of gray bark (Fig. 23) and is quickly and preferentially caught by birds when on such backgrounds, as has been shown experimentally.[40]

In the middle of the nineteenth century the conversion of certain areas in England from an agricultural to an industrial economy brought about extensive changes in the countryside. Soot from factories covered the bark of trees where the peppered moth had always perched. On the new black substratum the gray moths are conspicuous and the melanic forms are concealed (Fig. 23). Experiments have shown that insectivorous birds capture gray individuals much more frequently than black ones in soot-polluted areas.[41] The relative adaptive values of the two color forms of the moth are thus reversed in the industrial environment.

The melanic form of *Biston betularia* was first noticed as a rare mutant in populations of this moth around Manchester, England, in 1850. In the next few decades it became more frequent and has by now almost completely replaced the gray form in this and other industrial areas in England. It has not, however, become established in rural districts in western England, where the speckled gray type continues to thrive as in times past.

It has been calculated that the replacement of the gray phenotype (cc) by the black phenotype (Cc or CC) in the Manchester area corresponds to a rise in the frequency of the melanic allele (C) from about 1 percent in 1848 to 99 percent in 1898. This represents a very rapid evolutionary change compressed into a 50-year period of rapid environmental change and enacted before the eyes of human observers. The cause of this change in gene frequency is beyond any reasonable doubt natural selection, acting to preserve the melanic allele because of the advantage it confers to its carriers in a new environment. Some other kinds of moths underwent similar transformations from gray or brownish forms to a black form during the same period of industrialization in England.[42]

The hamster, *Cricetus cricetus*, a gopherlike rodent in Russia, also produces a mutant black form in addition to the normal gray form.

[40] Kettlewell, 1956.
[41] Kettlewell, 1956.
[42] Reviews by Ford, 1955, ch. 5; Sheppard, 1959, 68 ff.; Owen, 1961.

The geographical distribution and relative frequency of the black color phase are known with certainty because of the importance of this animal in the fur trade. In the eighteenth century the black hamster appeared in Russia, and in the nineteenth century it spread into the Ukraine. There are several areas of mixed forest-steppe vegetation along the Dnieper and other rivers where black hamsters constitute up to 27 percent of the population. To the north in the boreal forest and to the south in the open steppe, few or no black hamsters occur. From government records on almost two million pelts turned in by trappers, Gershenson was able to trace the increase in frequency of the black allele. The selection coefficients calculated from these data turned out to be rather high, indicating a strong selection for the black forms in certain regions.[43] Just what the adaptive advantage of the black hamsters over the gray forms is in these regions has not been explained. We know that selection has greatly increased the frequency of the melanic allele in some hamster populations, but we do not yet know why.

When the insecticide DDT was first introduced in recent years it proved fatal to the common housefly, *Musca domestica*. DDT was widely used in sprays and liquid washes to control this common household and barnyard pest. Within a few years it was noticed in different parts of the world that houseflies were not succumbing as much as they had at first to this poison. Laboratory tests in which regulated amounts of DDT were applied directly to the bodies of flies showed that some individuals and strains could tolerate stronger doses than others. The reduced effectiveness of DDT as an insecticide is concluded to be due to the increase in frequency of the DDT-resistant genotypes in fly populations which are systematically exposed to this new environmental factor.[44]

Some disease bacteria have gone through a parallel history in relation to the antibiotics developed during World War II and brought increasingly into medical use since the end of the war. Small doses of penicillin, streptomycin, and other antibiotics were at first effective in controlling bacterial infections in human patients. Soon, however, it was observed that many infections were not brought under control by the application of these same antibiotics. Laboratory tests have

[43] Gershenson, 1945; Dobzhansky, 1951a, 140–42.
[44] King and Gahan, 1949; see also Johnston, Bogart, and Lindquist, 1954.

confirmed the existence of genetically different strains of the same kind of bacterium, some of which are resistant and others susceptible to antibiotics.[45]

The susceptible strain of *Micrococcus pyogenes*, for example, ceases to grow when cultured in minute concentrations (.2 micrograms, or 2 ten-millionths of a gram, per cubic centimeter) of streptomycin, whereas the resistant strain tolerates dosages many times greater (125,000 micrograms per cubic centimeter). The properties of the resistant strains are due to mutations which arise spontaneously at a given rate in bacterial populations. When antibiotics are applied to the population the few resistant genotypes quickly increase in frequency and become established in the bacterial colonies to the exclusion of the susceptible types. This evolutionary change has now been repeated in laboratory experiments. Striking increases in the resistance of *Micrococcus pyogenes* to streptomycin occurred when this bacterium was cultured on a succession of 23 petri dishes containing nutrient medium with increasingly large concentrations of streptomycin.[46]

Bacteria resistant to antibiotics and houseflies resistant to DDT tend to disappear and be replaced by the susceptible types when the poison is no longer applied and the environment reverts to the original condition. The explanation of this reversal lies in the fact that the mutations producing resistance also bring about some other phenotypic effects. Penicillin-resistant strains of bacteria are found to be less virulent, to have a reduced ability to synthesize some of the vitamins and other products they require for growth, and to grow more slowly than the susceptible strains.[47] There is evidence that DDT-resistant houseflies are slower growing than the DDT-susceptible types. In the non-toxic environments the resistant bacteria or flies are at a selective disadvantage compared with the susceptible forms. Susceptible genotypes arising in a predominantly resistant population either through reverse mutations from drug resistance to susceptibility or through immigration of susceptible individuals from neighboring colonies will be increased in frequency by selection in a poison-free environment just as the resistant types spread in a poisoned environment.

[45] Dobzhansky, 1955a, 92 ff.
[46] McVeigh and Hobdy, 1952.
[47] McVeigh and Hobdy, 1952.

ARTIFICIAL SELECTION

For thousands of years primitive man practiced an unplanned selection in various useful plants and animals which he had domesticated. If he wanted to grow a certain plant for food or raise a certain animal to help him in his hunting forays it was only natural that he should choose for breeding in each generation those individuals which best served his purpose and should discard the unsuitable individuals. This systematic but unplanned process brought about vast transformations in the hereditary characters of many kinds of animals and plants from the wild ancestral stocks to the domesticated forms in existence at the beginning of the historical period. This process of selection has been continued and intensified in modern times with the development of scientific breeding practices.

The modern short-legged, barrel-bodied hog is very different from the smaller, more slender, razor-backed wild pig of China from which it was derived. The modern specialized breeds of dogs, such as the toy dogs, dachshunds, terriers, setters, and others, are all derived from wolflike ancestors in various parts of the world.

The transformations which have occurred in corn during the thousands of years it has been cultivated as a food plant have been studied by Mangelsdorf and other investigators.[48] The evidence consists partly of fossils and partly of primitive characteristics still preserved in some living varieties of corn.

Although corn belongs to the grass family, it is strikingly different from all other grasses including the cultivated cereals in a number of ways. The flowers of corn are unisexual, containing either the pollen-bearing stamens or the grain-producing ovaries. The male and female flowers, moreover, are grouped in different inflorescences on different parts of the plant. The male flowers are borne at the top of the stem in the so-called tassel. The female flowers are grouped in a spike of their own, the ear, which arises on a short lateral branch in the middle of the stalk. The corn ear, as is well known, consists of a stout central axis, the cob, bearing many rows of naked grains, and enveloped by a sheath of husks. Nothing like the corn ear is found anywhere else in the grass family.

[48] Mangelsdorf, 1958a; Mangelsdorf and Reeves, 1959.

The ancestor of cultivated corn is believed to have been a wild grass existing several thousand years ago in the American tropics. It was a low stocky plant with several stalks arising from the base. The tops of the plants bore tiny spikes measuring only two or three centimeters long, and not enveloped by a husk. The spikes contained both male flowers at the tips and female flowers with grains in the basal part. The latter arose on short side branches from the slender central axis and were surrounded by bracts or chaff as in other grasses and in modern pod corn. The grains were capable of breaking off from the spike and being dispersed individually, as are the seeds of grasses generally. The kernels of primitive corn were tiny and hard as in most wild grasses and in modern popcorn.

Remains of a cultivated grass with the foregoing primitive characteristics have been found in a cave in New Mexico which was inhabited for thousands of years by an agricultural people. The ancient corn remains found in Bat Cave, New Mexico, are estimated by the radioactive carbon method of dating to be 5,600 years old. It is believed that the wild ancestors of this plant were first brought into cultivation for food in Central America or the Andean region or perhaps independently in both areas. If the new crop plant had spread as far as the southwestern United States by the time of the Bat Cave deposits about 5,600 years ago, the initial step in domestication must have taken place earlier.

Figures 43 and 44 illustrate some of the stages through which the corn plant and the corn ear have probably passed in their evolution. A trend is shown in Fig. 43 from small plants with several main stalks arising from the base and numerous small ears borne high on the stems to tall plants with a single stalk and a single large ear borne in the middle of the stem.

The stages in the evolution of the corn ear shown in Fig. 44 are partly hypothetical. The spike of Tripsacum (Fig. 44a), a wild grass related to corn, has female flowers below and male flowers above. Synthetic pod-popcorn (Fig. 44b) is a type of corn derived from hybridization between popcorn and pod corn. The latter two varieties (Fig. 44d, e) possess different primitive characteristics which were combined in the ancestral forms of corn and which can be combined artificially by hybridization today. Synthetic pod-popcorn, like

Tripsacum and the ancestral corn, has female flowers in the lower part and male flowers in the upper part of the same ear, and the kernels break off readily from the spike. Fossil pod-popcorn from Bat Cave, New Mexico, about 5,600 years old, has kernels which are tiny and hard as

Fig. 43. Stages in the evolution of the shoot of the corn plant, Zea mays

    (a) Pod popcorn   (c) Flint corn (New England)
    (b) Popcorn   (d) Dent corn (corn belt)

Rearranged from Mangelsdorf, 1958*a.*

in popcorn, and are borne on long side branches and enclosed by chaffy bracts as in pod corn (Fig. 44c). In dent corn, as grown today in the corn belt, the kernels are numerous, large, naked, and tightly attached to the large cob (Fig. 44f). Furthermore, the ear consists entirely of female flowers and is enclosed by a husk, a leafy envelope not found in primitive corn.[49]

[49] Mangelsdorf, 1958*a.*

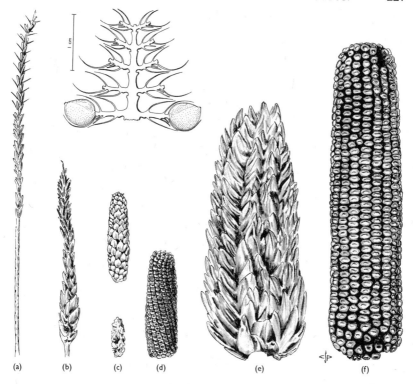

Fig. 44. Stages in the evolution of the corn ear

(a) Tripsacum
(b) Synthetic pod popcorn
(c) Fossil pod popcorn from Bat Cave; fossil cob below, reconstruction of whole ear
   in center, and diagram of structure of ear above.
(d) Popcorn
(e) Pod corn
(f) Dent corn

All to same scale.

Rearranged and redrawn from Mangelsdorf, 1958a.

The primitive corn grown by the Bat Cave peoples and the grasslike plants grown by the still earlier agriculturalists in the American tropics were amenable to cultivation and yielded some edible seeds. However, numerous improvements were possible. Gene mutations are known in modern corn varieties of a primitive type which cause the plants to become single-stalked and robust, the ears to develop in the middle part of the stalk, the ears to develop to a large size and to consist of grain-bearing flowers only, a husk to grow up around the ears, the chaff surrounding the individual grains to disappear, and the grains to be large and relatively soft. All of these changes would be considered desirable by humans using the plant as a source of food, for more and larger grains would be produced on each ear and would be retained there until harvested. Mutant phenotypes with the desirable features would have been preserved and propagated preferentially by the early American corn-growers. This unconscious selection resulted in extensive changes in the characteristics of the plants and in the genotypes controlling these characteristics during a relatively short period about 6,000 years long.

SELECTION AS A CREATIVE FORCE

We have examined cases where selection in combination with mutations in single genes has brought about relatively simple evolutionary changes. Such simple changes are exemplified by the rise of black peppered moths from speckled gray ancestors, or of poison-resistant flies or bacteria from non-resistant strains. In the evolution of corn and other domesticated plants and animals from their wild ancestors, we are confronted with changes of a much greater magnitude, changes involving a complex of characters determined by a combination of genes. A black peppered moth is still a peppered moth, but modern corn differs profoundly in many ways from its grasslike ancestors of 6,000 years ago.

In Chapter 5 we saw that the major adaptations of organisms consist of harmoniously working combinations of characters. We learned in Chapter 8 that sexual reproduction is a mechanism for assembling the combinations of genes that determine these character complexes. Natural selection, as we now see, is a mechanism for preserving and establishing the gene combination which produces a favorable character complex.

The complex structures of organisms, like an ear of corn or a giraffe's neck, are highly improbable phenomena. Their development by random processes alone is inconceivable. And no evidence has been found for any internal guidance of the evolution process toward the development of more perfect structures and adaptations. However, natural selection is a creative force, as Weismann pointed out long ago.[50] It is, as Fisher and J. Huxley have stated, a mechanism for bringing about highly improbable events.[51]

[50] Weismann, 1902, 11.
[51] Fisher, 1930; Huxley, 1943, 474–75, 1945.

TEN

Levels of Selection

In my Father's house are many mansions.   JOHN 14:2

THE PRIMARY FORM of selection in nature, Darwinian selection, takes place between individuals in a population. But competition and selection can and do occur between reproducible biological units lower or higher than individuals on the scale of complexity. The purpose of the present chapter is to survey the operation of natural selection at various levels of organization.

SINGLE-GENE SELECTION RECONSIDERED

The single-gene model of selection was set forth in the preceding chapter as a simple but abstract introduction to the workings of natural selection. Sister individuals in some groups of higher plants, particularly among self-pollinating and homozygous forms, may occasionally differ with respect to a single gene. Where this gene is the only variable factor in the alternative genotypes, Darwinian selection is in effect selection for single-gene differences. This situation must be relatively uncommon in the majority of plants and animals, however, where sister individuals usually differ in many major genes and many modifying factors.

Before we dismiss the single-gene model of selection as a useful but unrealistic abstraction, however, we must recall that the highest structural unit of life postulated to exist during the primitive stage of evolution was the naked or seminaked genic particle (see Chapters 3 and 4). For the first eon or so in the history of life, during the stage of primitive saprophytic particles, evolution must have been guided predominantly by selection at the genic level.

Recent theories of the origin of life, as expounded by Oparin, Wald, Calvin, and others, hold that natural selection must have come into play at the molecular level of organization. In the primitive waters containing energy-rich carbon compounds, molecules and macromolecules would unite spontaneously to form colloidal aggregates, and the resulting particles would compete with one another for raw materials and energy sources. Some molecules and particles by virtue of an especially favorable internal arrangement might be able to acquire the new molecules and chemical energy required for self-maintenance and reproduction more rapidly than others. The molecular and colloidal species which maintained and reproduced themselves most successfully emerged as the dominant types in the period of chemical evolution, and became the ancestors of the genic particles in the ensuing period of early organic evolution (see Chapter 3).[1]

In Chapter 3 it was suggested that the self-reproducing molecular and colloidal particles on the borderline between non-living and living would reach a saturation point and enter into a strong competition sooner in small water bodies than in the vast seas. Selection at the macromolecular and genic levels would therefore have a headstart in small but permanent ponds and lakes, and would lead to the evolution of efficient genic particles sooner there than in the oceans. On theoretical grounds we would expect that the effective site of the origin of life was not the sea, as is frequently assumed, but medium-small water bodies which were large enough to have permanence but small enough to become overcrowded in a brief time with competing self-perpetuating molecules.

SELECTION AT THE CHROMOSOME LEVEL

The heterozygote a_1a_2 is expected, on the basis of Mendelian theory and much empirical evidence, to produce the two classes of gametes a_1 and a_2 in a 1:1 ratio. This 1:1 ratio depends on the mechanics of the meiotic process by which the gametes are formed. Normally at meiosis the homologous chromosome segments containing a_1 and a_2, respectively, separate and go to daughter poles, each of which develops into a gamete. This is not invariably the case however.

In some strains of *Drosophila pseudoobscura* the males with the heterozygous constitution XY for the sex chromosomes produce all daughters (XX) and no sons (XY). The altered sex ratio is not due to

---

[1] Oparin, 1938; Wald, 1954; Calvin, 1956.

preferential mortality of the XY types in the zygote or early embryo stages, nor is it due to a superior functioning of the X sperms over the Y sperms in fertilization. The possibilities of preferential zygotic mortality and of gamete selection can be excluded in this case. Cytological studies reveal that spermatogenesis in some lines is modified in such a way as to alter the gametic ratio from the expected 1X:1Y to an observed 100X:0Y. The single X chromosome present in the mother cell at the beginning of meiosis divides twice and becomes incorporated into the four daughter nuclei and hence into each of the gametic products of meiosis, while the Y chromosome present in the mother cell is usually left out of the daughter nuclei and eventually degenerates in the cytoplasm. As a result a full complement of viable sperms is produced by the sperm mother cells, but these are almost entirely X sperms. The aberration of meiosis which brings about this effect is determined by the allele sr of the "sex-ratio" gene **Sr** on the X chromosome.[2]

Analogous cases of unequal gamete formation resulting from aberrations of meiosis have been described elsewhere in Drosophila and in corn.[3]

It will be noted that an altered gametic ratio can change the proportions of different types of gametes in the gamete pool and can lead to an increase in frequency of certain alleles in the population. A non-Mendelian segregation of chromosomes at meiosis, in other words, can directly alter allele frequencies in a population. This factor in evolutionary change has been called meiotic drive.[4]

As a result of meiotic drive an allele can increase in frequency, even though it has harmful effects, if it is regularly included in more than half of the successful gametes produced by a heterozygote.[5] In *Drosophila melanogaster* the "segregation-distorter" gene **SD** on chromosome II causes unequal segregations of this chromosome at meiosis in the males. Heterozygous males (sd/+[sd]) produce many more sd sperms than +[sd] sperms, and transmit the sd chromosome to the majority of their progeny. A recessive lethal allele is linked with the

[2] Sturtevant and Dobzhansky, 1936.
[3] See Sandler and Novitski, 1957; and Swanson, 1957, 323–28.
[4] Sandler and Novitski, 1957.
[5] Sandler and Novitski, 1957; Hiraizumi, Sandler, and Crow, 1960.

sd allele in some natural populations of *Drosophila melanogaster* in Wisconsin. The sd allele and its linked lethal are found in fairly high frequencies in the natural populations, and rise in frequency spontaneously in experimental populations. In one population cage the allele frequency of sd underwent changes from .03 to .10 during 212 days; in another cage this allele changed from an initial frequency of .02 to an ultimate level of .76 in 303 days. These increases in the frequency of sd in both the experimental and natural populations are presumably checked when counter-selection against the lethal factor becomes as strong as the meiotic drive.[6]

It would seem that the reported cases of meiotic drive can be viewed as the differential reproduction at meiosis and the differential inclusion in the gametes of alternative homologous forms of a chromosome. Meiotic drive, in other words, is interchromosomal selection.[7] The evolutionary effects of meiotic drive form an interesting subject, which is just beginning to be explored.

GAMETE SELECTION

Between the formation of the gametes and the formation of the zygotes there intervenes a period during which the gametes carry out the functions leading to fertilization. Different classes of gametes may not be equally capable of performing these functions. Under competition between the gametes, some may produce proportionately more zygotes than others. The occurrence of gamete selection can be demonstrated by showing that the initial frequencies of different classes of gametes in the gamete pool are statistically different from the frequencies in which they are respectively represented in the fertilized eggs.

The pollen grains of seed plants effect fertilization by forming a pollen tube which grows through the female ducts—the style in flowering plants, the space between cone scales in cycads and conifers—to the ovule and then within the ovule to the egg. Pollen grains formed by one heterozygous plant or by genetically different plants frequently differ in genes or chromosome segments affecting their ability to grow rapidly and successfully through the female ducts to the egg. Since the

[6] Hiraizumi, Sandler, and Crow, 1960.
[7] Also Stalker, 1961, 177.

pollen is produced in excess of the amount required for fertilization, a competition between genetically dissimilar pollen grains and pollen tubes normally occurs. Under these conditions some classes of male gametes may be more effective than others in producing seeds.

The endosperm in the kernels of corn (*Zea mays*) may be sugary or starchy depending on the allelic form of a gene **Su**. The dominant allele Su determines starchy, and the recessive allele su sugary endosperm. Crosses of true-breeding starchy parents (Su/Su) with true-breeding sugary parents (su/su) thus usually produce F_2 progeny segregating into starchy and sugary types in a 3:1 ratio; and the backcross progeny of the F_1 usually segregate in a 1:1 ratio.[8]

The gene **Su** is linked with a gene **Ga** controlling the rate of pollen-tube growth, the allele Ga bringing about rapid growth and the allele ga poor growth of the pollen. A cross was made between rice popcorn (with the constitution Su Ga/Su Ga) and sugary corn (su ga/su ga), and the resulting F_1 hybrid (Su Ga/su ga) was self-pollinated to produce an F_2, and was crossed with the parental types in various combinations to yield backcross progeny. The backcross Su/su ♀ × su/su ♂ yielded 1,374 starchy and 1,397 sugary kernels. The backcross Su/su ♀ × Su/Su ♂ gave 207 heterozygotes and 213 homozygotes for S in the B_1 generation. In these cases, the expected 1:1 backcross ratio is realized. In the F_2 derived by selfing the F_1 hybrid, however, the observed numbers of starchy and sugary kernels deviated markedly from the expected 3:1 ratio as shown below:

	Starchy kernels	Sugary kernels
Observed	3,085	596
Expected	2,761	920

There is an excess of the starchy type and a deficiency of sugary kernels in the F_2 generation. A deviation from a 1:1 Mendelian ratio was also found in the backcross progeny from the cross Su/Su ♀ × Su/su ♂.[9]

Thus Mendelian ratios are obtained when the F_1 is pollinated with one type of pollen alone. The Su-carrying pollen and su-carrying pollen are both functional. But when a mixture of the two classes of pollen is placed on a style, distorted ratios are obtained in the next generation.

[8] Mangelsdorf and Jones, 1926.
[9] Mangelsdorf and Jones, 1926.

Although both classes of pollen are able to accomplish fertilization, they are not equally effective at this task, and the differences between them are revealed under conditions of pollen competition. The selective differences between Su and su pollen are not due to the **Su** gene itself, which merely serves as a convenient morphological marker of the action of the linked gene **Ga**.[10]

Altered ratios attributable to differences in the effectiveness of different classes of pollen grains have also been observed in cotton (Gossypium), lima beans (Phaseolus), and other plants.[11]

A selective elimination of certain classes of pollen is also clearly revealed by studies of monosomic types of wheat (*Triticum vulgare*) and tobacco (*Nicotiana tabacum, N. rustica*). Monosomics are individuals lacking a single chromosome. If normal wheat plants have 21 pairs or 42 chromosomes, monosomic types have 20 II + 1 I or 41 chromosomes; if the tobacco species mentioned above have normally 24 II, the monosomic forms have 23 II + 1 I. The monosomic chromosome may be any chromosome in the complement; thus a plant containing the A chromosome in single dose but the B, C, D, ... in the diploid condition can be distinguished as monosomic A; while a plant with the chromosomal constitution AA B_ CC DD ... would be monosomic B, and so on. Extensive series of different monosomic types have been obtained and studied in wheat and tobacco. While each monosomic type has its peculiar morphological and genetical features, there also exist, as regards the hereditary transmission of the monosomic condition, some features which are common to all monosomic types and which are relevant to our present discussion.

The gametes produced by a monosomic plant are of two types with respect to the monosomic chromosome: normal haploid (n) gametes and deficient (n − 1) gametes. The two sorts of gametes could theoretically arise in equal frequencies. In actuality a majority of n − 1 gametes is invariably produced. Thus wheat plants monosomic for chromosome F produced functional eggs in the proportion 71 no F: 29 F; and monosomic type G of wheat produced deficient and haploid eggs in the ratio 73 no G: 27 G.[12] Similar percentages of n − 1

---

[10] Mangelsdorf and Jones, 1926.
[11] Roman, 1948; Finkner, 1954; Bemis, 1959.
[12] Nishiyama, 1928.

gametes are found for other monosomic types in wheat[13] and Nicotiana,[14] and occur in both female and male gametes.[15] The excess of n − 1 gametes over 50 percent is explained by the fact that the monosomic chromosome frequently lags during the meiotic division and consequently does not become incorporated in half of the daughter nuclei.

Owing to the excess of n − 1 eggs produced by a monosomic parent, the cross monosomic ♀ × diploid ♂ usually yields progeny more than 50 percent of which are monosomic. In *Nicotiana rustica*, for example, two monosomic types used as females and pollinated by normal males produced 63 percent and 71 percent monosomic progeny (Table 10).

Table 10. Relative numbers of monosomic and diploid progeny produced from reciprocally different crosses between monosomic and diploid parents[a]

	Progeny		
Cross	Monosomic	Diploid	Total
Monosomic D♀ × diploid ♂	65	38	103
Diploid ♀ × monosomic D♂	0	94	94
Monosomic G♀ × diploid ♂	77	31	108
Diploid ♀ × monosomic G♂	1	104	105

[a] Lammerts, 1932.

The reciprocal cross, diploid ♀ × monosomic ♂, by contrast, yields a greatly reduced proportion of monosomic progeny. Table 10 shows that two crosses of diploid ♀ × monosomic ♂ in *Nicotiana rustica* gave 0 percent and 1 percent of monosomic progeny, respectively.[16] These large differences between the transmission of the monosomic condition through the ovules and through the pollen are characteristic features of the breeding behavior of monosomic types generally in both tobacco and wheat.[17]

To summarize, a monosomic plant produces over 50 percent, and usually about 75 − 80 percent, of n − 1 gametes on both the male and

[13] Sears, 1944.
[14] Lammerts, 1932; Clausen and Cameron, 1944.
[15] Clausen and Cameron, 1944.
[16] Lammerts, 1932.
[17] Nishiyama, 1928; Lammerts, 1932; Clausen and Cameron, 1944; Sears, 1944.

female side. A monosomic female pollinated by a diploid male gives rise to monosomic and diploid progeny approximately in proportion to the frequency of n — 1 and n ovules. But a monosomic male parent, which forms more than 50 percent n — 1 pollen grains, yields a low percentage of 2n — 1 progeny in crosses with normal females. The pollen tubes produced by a monosomic plant enter into a strong competition, which is not shared in an equal degree by the ovules. This pollen-tube competition results in a selective elimination of the male gametes which are deficient for one chromosome.[18]

It might be predicted that where the pollen competition is relieved somewhat, in a sparsely pollinated flower, a greater variety of pollen types could function in fertilization than under conditions of dense pollination. Some preliminary evidence suggests that this may be the case. Flowers of cotton (*Gossypium hirsutum*) were pollinated with known numbers of pollen grains ranging from 5 to 1,000. The progeny of flowers pollinated with less than 300 grains showed marked variation from plant to plant. This individual variation became progressively greater as the number of pollen grains became fewer.[19]

The pollen of an F_1 hybrid of tomato (*Lycopersicum esculentum*) was applied in two different concentrations on the stigmas to make F_2 and B_1 seeds. Some flowers were pollinated sparsely with fewer pollen grains than ovules, thus eliminating pollen competition, while other flowers were pollinated heavily, creating a competitive condition among the pollen grains. Various quantitative characters, such as leaf length, plant height, flower number, and fruit weight, were measured in the F_2 and B_1 progeny. The results differed according to the method of pollination. Although the mean for each character was the same in progenies derived from dense or sparse pollinations, the variation around the mean as expressed in the coefficient of variation was lower in all characters under conditions of dense pollination. The extreme large or small classes were more poorly represented in the progeny produced by pollination with superabundant pollen grains. Where the pollen competition is severe, therefore, the pollen grains carrying gene alleles determining extreme phenotypes tend to be eliminated.[20]

[18] Nishiyama, 1928; Lammerts, 1932; Clausen and Cameron, 1944; Sears, 1944.
[19] Ter-Avanesjan, 1949.
[20] Matthews; see D. Lewis, 1954.

GENOTYPIC DIFFERENCES IN GAMETIC OUTPUT

In the preceding section we considered the relative efficacy of alternative classes of gametes as a factor altering allele frequencies in a population. It was assumed that one heterozygous individual a_1a_2 or two homozygous individuals a_1a_1 and a_2a_2 would produce a_1 and a_2 gametes in equal numbers, but owing to an inherent functional superiority of, say, the a_1 gametes, these would bring about more fertilizations and contribute their allele to more zygotes than would the a_2 gametes.

We now have to consider another possibility. It is conceivable that the individuals in a population may differ genotypically in ways affecting the relative output of the a_1 and a_2 gametes. Thus the homozygote a_1a_1 might produce a greater number of gametes than the genotypes a_1a_2 or a_2a_2. In such cases a selection between genotypes, that is, Darwinian selection, will take place for the important characteristic of gamete productivity.

A factor determining the frequency of any allele in a population of plants is the relative amount of pollen produced per flower and the relative number of flowers produced per plant by the genotypes carrying that allele.[21] If a_1a_1 individuals have a greater total output of pollen than a_1a_2 or a_2a_2 individuals, the a_1 allele will inevitably rise in frequency relative to the a_2 allele. In this way poor pollen producers will eventually become extinct, and the population will come to consist of genotypes which have a more or less uniformly high rate of pollen production. It is probable that the capacity for producing very large amounts of pollen, which is found in so many species of plants, is a result in part of competition between genotypes with respect to participation in pollination.

SELECTION WITHIN SUBGROUPS OF A POPULATION

The breeding populations of most animals and of many plants are composed of morphologically and physiologically different male and female individuals. The adaptive requirements and the adaptations which meet these requirements are, in general, the same in the two sexes. For example, male and female deer of the same species must be adapted for living in the same kind of terrain and the same climate, for eating the same types of food, for escaping from the same predators,

[21] Stephens, 1956.

and so on. Male and female deer do in fact share the common adaptive characteristics of the species. They differ of course in their organs of reproduction, and, in addition, they differ in various other characteristics such as body size and the presence of antlers.

It is well known that the males in most species of deer are larger and stronger than the females, and possess antlers which are lacking in the females. The large body size and the antlers of male deer are not among the necessary adaptations of the species as a whole, or else they would be expected to occur in the female sex too, nor are they necessary for the primary male functions of sperm production and impregnation. The origin of these secondary sexual characters in the males cannot, therefore, be attributed to selection either for general fitness to the environment or for coadaptations between the sexes related directly to reproduction.

Yet such sex-limited characters exist, not only in deer, but widely throughout the animal kingdom. Among insects, crabs, fish, reptiles, birds, and mammals there are numerous species in which various characters of size, strength, aggressiveness, and weapons are specially developed in the male sex. The evolution of such characters in the males of many animal species requires an explanation. Darwin raised and solved this problem in *The Descent of Man* in 1871.

Darwin pointed out that the special characteristics of the males in many species of animals, while superfluous for fertilization per se, are connected indirectly with reproduction insofar as they help their possessors to secure a mate.

Now, as Darwin observes[22]:

It is certain that amongst all animals there is a struggle between the males for the possession of the females. This fact is so notorious that it would be superfluous to give instances. Hence the females have the opportunity of selecting one out of several males, on the supposition that their mental capacity suffices for the exertion of a choice. In many cases special circumstances tend to make the struggle between the males particularly severe. Thus the males of our migratory birds generally arrive at their places of breeding before the females, so that many males are ready to contend for each female. . . .

We are naturally led to enquire why the male, in so many and such distinct classes, has become more eager than the female, so that he searches for her, and plays the

[22] Darwin, 1871, ch. 8.

more active part in courtship. It would be no advantage and some loss of power if each sex searched for the other; but why should the male almost always be the seeker?

The eagerness of male animals, which leads them to compete for females, follows from the respective specializations of the two sexes. The male sex is specialized for finding and fertilizing the female, the female for providing nourishment and protection to the fertilized egg and developing embryo. In the sessile and hermaphroditic lower aquatic animals, as in plants generally, the female gametes and the organs which contain them remain attached to the organism and thus close to the sources of food materials, while the male gametes travel in one way or another—by swimming, floating, drifting in the air, or attaching themselves to flower-visiting animals—to the female organs. In some lower animals the females alone are sessile and the males are mobile and travel as individuals to the females. Those males possessing a strong eagerness to mate would find more females and leave more progeny than males lacking in sex drive, and this eagerness of the males would become a characteristic of the species. Moreover it would be inherited by the higher and fully mobile animals which arose from the species of lower animals with strong male sex drive.[23]

The male eagerness, which developed in the first place as a concomitant of the functional differentiation of the sexes, had the further consequence among higher dioecious motile animals of fostering a competition between males for females. And this competition led to the evolutionary development of other secondary sexual characters in the males. For, under conditions of competition between genetically different male individuals, those males possessing weapons which gave them an advantage over other males in finding and holding mates would transmit their hereditary characteristics preferentially to future generations. This relative advantage of different males of the same species with respect to reproduction was termed sexual selection. It is the process responsible for the diffusion throughout the species in many animal groups of special male characteristics of strength or ornamentation which are related neither to the general struggle for existence nor to fertilization specifically, but to the securing of female mates.[24]

[23] Darwin, 1871, ch. 8.
[24] Darwin, 1871, ch. 8.

To quote Darwin again[25]:

There are many . . . structures and instincts which must have been developed through sexual selection—such as the weapons of offence and the means of defence—of the males for fighting with and driving away their rivals—their courage and pugnacity—their various ornaments—their contrivances for producing vocal or instrumental music—and their glands for emitting odours, most of these latter structures serving only to allure or excite the female. It is clear that these characters are the result of sexual and not of ordinary selection, since unarmed, unornamented, or unattractive males would succeed equally well in the battle for life and in leaving a numerous progeny, but for the presence of better endowed males. We may infer that this would be the case, because the females, which are unarmed and unornamented, are able to survive and procreate their kind. . . . When we behold two males fighting for the possession of the female, or several male birds displaying their gorgeous plumage, and performing strange antics before an assembled body of females, we cannot doubt that, though led by instinct, they know what they are about, and consciously exert their mental and bodily powers.

Just as man can improve the breeds of his game-cocks by the selection of those birds which are victorious in the cock-pit, so it appears that the strongest and most vigorous males, or those provided with the best weapons, have prevailed under nature, and have led to the improvement of the natural breed or species. A slight degree of variability leading to some advantage, however slight, in reiterated deadly contests would suffice for the work of sexual selection; and it is certain that secondary sexual characters are eminently variable.

The operation of sexual selection can be seen in a clear form in polygamous animals, where the strongest and most attractive males collect and guard a harem of females, while their weaker brothers possess few or no mates. Some birds and many mammals are polygamous. Among birds polygamy is found in chickens, pheasants, and peacocks; among mammals in deer, cattle, sheep, most antelope, elephants, seals, sea lions, walruses, baboons, and gorillas. Secondary sexual characters are particularly well developed in the males of these animals. Thus in the polygamous chickens, pheasants, and peacocks the cocks are notably larger, more pugnacious, and better decorated than the hens, while in the monogamous partridge, grouse, and ptarmigan the differences between the sexes are relatively slight. The mountainous size and strength of bull gorillas, walruses, and sea lions are well known, as are the antlers or horns of the males in many ungulates and the aggressive behavior of male baboons. The sexes are, by contrast, nearly equal

[25] Darwin, 1871, ch. 8.

in size and strength in monogamous wolves and in certain monogamous species of monkeys, in the members of the cat family which form matriarchal families, and in colonial but non-polygamous rodents.[26]

Secondary male characters, though especially well developed in polygamous animals, are found in animals with other breeding systems as well. How then does sexual selection operate in a non-polygamous species? Darwin considered the case of a species consisting of approximately equal numbers of males and females, all of which form pairs and produce progeny. The selective advantage of the superior males is not so great in this case, since even the worst-endowed males eventually find mates and leave offspring; nevertheless the superior males may have a slight selective advantage. The strongest and best-armed males could begin to breed earliest in the season. Moreover they would pair with the most vigorous females which would also be ready to breed early in the season. "Such vigorous pairs, " as Darwin remarks, "would surely rear a larger number of offspring than the retarded females, which would be compelled to unite with the conquered and less powerful males."[27]

Furthermore, the courtship of animals does not consist simply of males conquering their rivals and thereby gaining possession of the females apart from the choice of the latter. On the contrary[28]:

The females are most excited by, or prefer pairing with, the more ornamented males, or those which are the best songsters, or play the best antics; but it is obviously probable that they would at the same time prefer the more vigorous and lively males, and this has in some cases been confirmed by actual observation. Thus the more vigorous females, which are the first to breed, will have the choice of many males; and though they may not always select the strongest or best armed, they will select those which are vigorous and well armed, and in other respects the most attractive. Both sexes, therefore, of such early pairs would as above explained, have an advantage over others in rearing offspring; and this apparently has sufficed during a long course of generations to add not only to the strength and fighting powers of the males, but likewise to their various ornaments or other attractions.

The effects of sexual selection, as Haldane and Huxley have noted, may be advantageous to certain individuals within a species, but at the same time disadvantageous to the species as a whole. The long train of

[26] Darwin, 1871, ch. 8.
[27] Darwin, 1871, ch. 8.
[28] Darwin, 1871, ch. 8.

a peacock cannot improve its chances of escape from predators. The heavy burden of antlers that must be borne by the males of the larger deer species is not an unmixed blessing.[29]

When we consider groups as different as insects, crabs, birds, and mammals, or even different genera and families within these groups, and when we consider characteristics as different as body size, horns, colorful plumage, and musical ability, we are bound to be impressed with the fact that the secondary sexual characters of male animals form an extremely varied and heterogeneous assemblage. Many, but not all, of these secondary sexual characteristics confer an advantage on some males over others in reproduction. Those male characteristics which do relate to success in mating can have come into existence by intrasexual selection as Darwin suggested. Darwin's theory of sexual selection has the great merit of explaining, by a single mechanism, the origin of a very large and diverse group of phenomena in the animal kingdom.

The value of this theory is not diminished by the recognition that it cannot explain the origin of *all* secondary male characters. As Wallace later pointed out in a brilliant analysis of the problem, Darwin's theory of sexual selection provides a satisfactory explanation of the development of characters in males related to battle for females (these are the examples emphasized in the present discussion), but not for the development of ornamentation and song in the male sex.[30]

Some characters of male animals serve other functions besides the winning of a victory in an intermale competition. Many of the distinctive color patterns, songs, and forms of courtship behavior in birds, for example, serve primarily as species-specific recognition marks which promote intraspecific matings and prevent interspecific hybridization. Secondary male characters such as these have no doubt been developed by modes of selection other than sexual selection.[31]

Another subgroup frequently found within animal populations is the litter. Haldane reported that in mice about one-quarter of the embryos die during pregnancy owing to insufficient space and nourishment in

[29] Haldane, 1932, 120; Huxley, 1943, 484.
[30] Wallace, 1889, ch. 10.
[31] Wallace, 1889, ch. 10. For a statement regarding the present status of the theory of sexual selection, see Huxley, 1938.

the uterus for the whole litter. This intrauterine competition probably leads to a selection for rapid embryonic growth. Rapid developmental rates have important effects on the adult characteristics of the animals; the process of learning, for example, requires a prolonged juvenile period. In an animal which produces many young at each birth, therefore, the diffusion throughout the species of genes for rapid growth, as a result of intrauterine selection, may set restrictions on the potentialities of the species for developing greater intelligence.[32]

THE HETEROZYGOUS GENOTYPE

Natural selection is a force causing one allele or genotype, say a_1, to increase in frequency, while the alternative alleles or genotypes a_2 and a_3 decrease in relative abundance in the population. It might be assumed that the ultimate result of these upward and downward trends in allele or genotype frequencies is the fixation of a_1 and the concomitant extinction of a_2 and a_3 in the population. The complete replacement of one allele or genotype by another is indeed a common result of selection in nature. But it is not the only necessary result of the action of this evolutionary force.

Individuals heterozygous for a single gene or for several or many genes distributed throughout the genotype frequently exhibit heterosis, being superior to the corresponding homozygotes in viability, fecundity, homeostasis, or some adaptive trait. Where the heterozygotes (a_1a_2) possess an adaptive advantage over the corresponding homozygotes, (a_1a_1 or a_2a_2), and are consequently preserved as an important constituent of the breeding population, neither a_1 nor a_2 will go to extinction, since the favored individuals in successive generations will be the carriers of both alternative forms of the gene or genotype.

The normal individuals comprising the breeding populations in many plants and animals, as for example corn, Oenothera, Drosophila, chickens, and man, are known to be heterozygotes. Although these individuals are undoubtedly homozygous for many genes and chromosome segments, they are heterozygous in many other parts of their respective genotypes. The result after thousands of generations of selection has not been the fixation of single forms of the genes or genotypes in a homozygous condition—has not been racial purity—but

[32] Haldane, 1932, 124; Huxley 1943, 525.

instead has been perpetual heterozygosity of individuals and consequently a permanent polymorphism in the populations. This situation has been thoroughly investigated in Drosophila.

Populations of various species of Drosophila the world over consist of flies which carry numerous deleterious recessive alleles in their chromosomes. These recessive factors can be detected by collecting normal flies in nature and inbreeding them in the laboratory so as to make particular chromosomes of the complement homozygous.

Chromosome II of *Drosophila melanogaster* has been put into a homozygous condition by inbreeding flies taken from natural populations in North America and Israel. It is found that between 25 percent and 60 percent of the natural occurring second chromosomes in the populations carry genetic factors which are lethal or semilethal when present in homozygotes. About 41 percent of the second chromosomes and 32 percent of the third chromosomes in Brazilian populations of *Drosophila willistoni* are lethal or semilethal in the homozygous condition; another 57 percent and 49 percent of the second and third chromosomes, respectively, produce various defects and constitutional weaknesses when present homozygously. Those individual chromosomes of *Drosophila willistoni* which are free of recessive lethals or semilethals, thus usually carry other deleterious recessive factors. Recessive subvital alleles of this sort are found in a high frequency, ranging from 41 to 98 percent, of the autosomal chromosomes in natural populations of *Drosophila pseudoobscura* and *D. persimilis* in western North America.[33]

The deleterious recessive alleles occur on different chromosomes of the complement and at different sites on a single chromosome. In *Drosophila pseudoobscura* and *D. persimilis*, for example, very few of the lethals tested, only 2 out of 68, were allelic.[34] Many separate genes thus contribute to the concealed polymorphism of the flies.

Despite their large and diversified load of deleterious factors, the flies are normal and healthy, and the populations in which they occur are flourishing. This is because the various lethal, semilethal, and subvital alleles are recessive, and the flies composing the natural

---

[33] Reviews and references to the voluminous literature are given by Dobzhansky, 1951*a*, 65 ff., and 1955*b*.
[34] Dobzhansky and Spassky, 1953.

populations are heterozygous for the chromosomal regions in which these deleterious recessive factors occur.

Some of the deleterious recessive alleles in wild populations of Drosophila exist in closely linked blocks located in particular regions of particular chromosomes. These blocks of genes, or supergenes, occur in chromosomal segments which have undergone inversions. The linear order of the genes may be ABC DEF GHI in one chromosome, and ABC FED GHI in another which differs from the former by an inversion in one segment (see also Chapter 8). A fly can carry either type of chromosome in homozygous condition, thus (−C DEF G−)/ (−C DEF G−) or (−C FED G−)/(−C FED G−), or it can carry both chromosome types in the heterozygous state, (−C DEF G−)/ (−C FED G−).

Drosophila pseudoobscura in western North America and Mexico possesses 16 different types of inversions on chromosome III. Most populations of this species are polymorphic for two or more of these inversion types. Thus the populations in the Sierra Nevada and San Jacinto Mountains of California contain third chromosomes of the types designated as ST, AR, CH, TL, SC, and PP. The inversion types ST, AR, and CH together comprise the majority of the third chromosomes in these California populations, while the chromosomal types TL, SC, and PP are infrequent here, though abundant in other parts of the distribution area of the species.[35] (The symbols are abbreviations of the vernacular names by which these alternative forms of chromosome III are known, thus: ST = Standard, AR = Arrowhead, CH = Chiricahua, TL = Tree Line, SC = Santa Cruz, PP = Pikes Peak.)

In the larval stage of *Drosophila pseudoobscura* the different inversion types, and the various homozygotes and heterozygotes for these inversions, can be identified cytologically by the characteristic banding patterns and pairing configurations of the giant salivary gland chromosomes (see Fig. 45). It is therefore possible at this stage to classify the flies according to their genetic constitution for the inverted region of chromosome III by direct observation under the microscope. This procedure enables the geneticists to measure accurately the proportions of the different chromosome types (ST, AR, CH, etc.) and of the different diploid combinations of these (ST/ST, AR/AR, ST/AR, etc.) in

---

[35] Dobzhansky and Sturtevant, 1938; Dobzhansky and Epling, 1944.

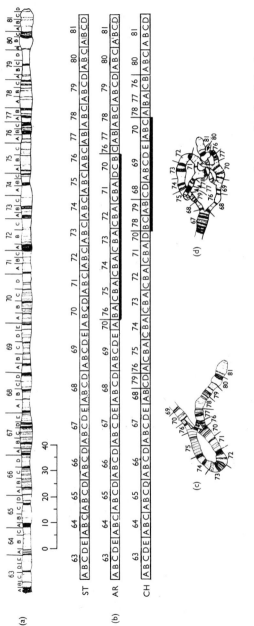

Fig. 45. Chromosome III of Drosophila pseudoobscura *as seen in the salivary glands of the larvae*

(a) The standard form (ST) of chromosome III. The arbitrary subdivisions of this chromsome into numbered sections and lettered subsections are shown. The 40-micron scale indicates the magnification.

(b) The length and position of the inversions Arrowhead (AR) and Chiricahua (CH) in relation to the sections of the standard chromosome.

(c) Type of chromosome pairing in the inversion heterozygote ST/AR.

(d) Chromosome configuration in the heterozygote ST/CH.

(a), (c), (d) From Dobzhansky and Sturtevant, *Genetics*, 1938. (b) From Dobzhansky and Epling, *Carnegie Inst. Wash. Publ.* No. 554, 1944.

samples taken from either natural or experimental populations. And by sampling a population at different intervals of time, the geneticist can detect and measure any changes in frequency of the alternative chromosome types and genotypes.

A sample of adult flies, which have already passed through the larval stage and no longer possess giant chromosomes, cannot of course be identified as to genotype by direct cytological observation. The procedure followed in practice, therefore, is to take adult flies from a population and let each fly produce F_1 larvae in the laboratory. The F_1 larvae are then determined for chromosomal constitution. From the examination of the chromosomes in a family of larvae derived from any individual adult fly, reliable inferences can be made concerning the type or types of third chromosomes carried by the parental individual.

A great deal of information concerning the composition of populations of *Drosophila pseudoobscura* with respect to chromosome III has been obtained by Dobzhansky and his co-workers. Extensive data on the frequencies of the alternative types of chromosome III in both natural and artificial populations have been published in numerous papers and summarized in several reviews.[36] Data on the frequencies of the various diploid combinations of these third chromosomes in wild populations are given in papers by Dobzhansky, Epling, and others.[37]

The array of genotypes to be found in natural populations of adult flies can be exemplified by a population at Keen Camp in the San Jacinto Mountains. Samples were collected from this population by Epling, Mitchell, and Mattoni at successive dates during the summer of 1952.[38] The contents of these samples are given in Table 11. It will be noted that 5 inversion types occur commonly enough in the Keen Camp population to be picked up in a sample of 112 flies. Furthermore, all but 1 of the 15 possible diploid combinations of the 5 inversions are represented in the samples. The different diploid genotypes vary greatly in their relative abundance however. In early summer the homozygote

---

[36] Dobzhansky, 1949; 1951*a*, ch. 5; da Cunha, 1955.
[37] Dobzhansky and Epling, 1944; Dobzhansky and Levene, 1948; Epling, Mitchell, and Mattoni, 1953.
[38] Epling, Mitchell, and Mattoni, 1953.

ST/ST and the two heterozygotes ST/CH and ST/AR are most common in the sample; in midsummer CH/CH and CH/AR also become frequent; and these 5 genotypes preponderate over all others for the season as a whole.[39]

Table 11. The numbers of adult flies with different genotypes for the inverted segments of chromosome III in samples taken from a wild population of Drosophila pseudoobscura *at Keen Camp in the San Jacinto Mountains, California*[a]

Genotype	May, 1952	July, 1952	May–August, 1952
Homozygotes			
ST/ST	15	10	32
CH/CH	2	16	27
AR/AR	1	3	10
TL/TL	1	0	1
PP/PP	0	0	0
Heterozygotes			
ST/CH	26	19	58
ST/AR	26	16	58
ST/TL	6	1	9
ST/PP	8	3	13
CH/AR	4	18	33
CH/TL	7	2	10
CH/PP	3	0	4
AR/TL	3	1	6
AR/PP	9	1	12
TL/PP	1	0	1
Total homozygotes	19 (17%)	29 (32%)	70 (26%)
Total heterozygotes	93 (83%)	61 (68%)	204 (74%)
Size of sample	112	90	274

[a]Epling, Mitchell, and Mattoni, 1953.

Knowing the frequencies of the different classes of gametes in the gamete pool, it is possible to calculate the expected frequencies of the genotypes resulting from their random union by the Hardy-Weinberg formula, as shown in Chapter 7. Let us analyze in this way the sample of flies collected by Epling, Mitchell, and Mattoni at Keen Camp in

[39] Epling, Mitchell, and Mattoni, 1953.

May. The frequency of the various chromosome types in the gamete pool can be calculated from the data given in Table 11 to be

ST	.429	(p)
CH	.196	(q)
AR	.196	(r)
TL	.085	(s)
PP	.094	(t)
Total	1.000	

The expected frequency of the diploid combinations of these chromosomes is then given by the formula $(p + q + r + s + t)^2$.

Solving this equation for p^2, which represents the genotype ST/ST, leads to the expectation of a frequency of .184 for this genotype. The observed frequency of ST/ST individuals in the population, however, is only 15/112 or .134, and is thus considerably less common than predicted. The other homozygous types, CH/CH and AR/AR, are also less frequent in the population than expected. Solving the same equation for $2 pq$ and $2 pr$, representing the heterozygotes ST/CH and ST/AR, yields expected frequencies of .168 for each of these types. The actual frequencies of .232 ST/CH and .232 ST/AR are much greater than the expected .168.

All 5 inversion homozygotes in the Keen Camp population in May would be expected from the Hardy-Weinberg law to occur in the combined frequency of .276. The observed frequency of all homozygous individuals combined was .170. All heterozygous types combined should be present in a frequency of .724, but actually attain a frequency of .830. In this population at this season of the year, therefore, the heterozygotes are more abundant and the homozygotes are less frequent than would be expected on the basis of the Hardy-Weinberg formula for a population at equilibrium.

The deviations from the theoretical equilibrium condition in this population could be attributed to several causes. The sample of flies studied might not reflect the true proportions in the population. But statistical tests show that the observed deviations in this case are extremely unlikely to arise from errors of sampling.[40] Departures from

[40] Epling, Mitchell, and Mattoni, 1953.

the predicted equilibrium frequency of genotypes could also result from non-random mating in the population. Independent evidence indicates that this is not a significant factor in *Drosophila pseudoobscura*.[41] Finally, it must be remembered that the Hardy-Weinberg equilibrium is based on the assumption that the various genotypes are all equally well adapted to their environment. There is no reason to assume that this premise is valid in this case. If the heterozygotes have a higher fitness than the homozygotes, the observed deviations from the predicted equilibrium condition could be accounted for.

The possible role of selection in altering the genotype frequencies in a wild population can be tested by comparing the frequencies of chromosomal heterozygotes and homozygotes in an egg sample with those in a sample of adult flies. Dobzhansky and Levene have shown that the eggs deposited by adult flies in natural populations in the San Jacinto Mountains and several other localities contain heterozygotes and homozygotes for inversions on chromosome III in the approximate proportions demanded by the Hardy-Weinberg formula. Among adult flies in the same natural populations, however, the heterozygotes are more frequent, and the homozygotes less so, than in the egg stage. Therefore, between the egg and adult stages, a differential mortality must occur in favor of the heterozygous genotypes.[42]

This conclusion can be confirmed experimentally in artificial populations maintained in population cages in the laboratory. If the eggs deposited in the population cage include inversion heterozygotes and homozygotes in the proportions expected from the Hardy-Weinberg rule, but the population of adult flies developing from the same batch of eggs contains an excess of heterozygotes and a deficiency of homozygotes, it can be concluded again that a differential mortality has taken place during the lifetime of the flies so as to favor selectively the heterozygous individuals.[43]

Dobzhansky performed such an experimental test with a San Jacinto Mountain strain of *Drosophila pseudoobscura*. The parental population of flies in the cage contained ST and CH chromosomes in known proportions. From the known gametic frequencies it was possible to

[41] Dobzhansky and Levene, 1948.
[42] Dobzhansky and Levene, 1948.
[43] Dobzhansky, 1947*a*.

calculate the Hardy-Weinberg ratios of homozygotes and heterozygotes expected among the fertilized eggs and among the adult flies in the next generation.

Samples of eggs taken out of the cage and raised to the larval stage under optimal conditions were determined for chromosomal constitution. It was found that the relative numbers of ST/ST, ST/CH, and CH/CH types were in fairly close agreement with the expected Hardy-Weinberg ratios. But samples of adult flies from the same cage, left to grow to maturity under conditions of competition, deviated markedly from the equilibrium proportion of genotypes. In a sample of 255 adults, the expected and the actual numbers of each genotypic class were:

	ST/ST	ST/CH	CH/CH
Expected	81	122	52
Observed	57	169	29

The figures reveal a significant excess of heterozygotes and deficiency of homozygotes among the adult flies. The alteration of the ratios between the egg samples and the adult samples points to a selective process favoring the ST/CH heterozygotes over the homozygous types.[44]

In a parallel experiment conducted in the same way, but using a population polymorphic for AR and CH chromosomes, it could be shown that here too an excess of heterozygotes CH/AR and a deficiency of homozygotes CH/CH and AR/AR arose between the egg and adult stages.[45]

The properties of the heterozygote ST/CH which endow it with a superior fitness compared with either homozygote have been identified more precisely in later studies. In laboratory populations containing ST and CH chromosomes from the San Jacinto Mountains, and maintained at 25°C, it was found that the ST/CH individuals developed more rapidly, were more viable in the larval stage, were longer-lived, and laid more eggs than the ST/ST and CH/CH flies. The heterozygotes are thus superior in several different respects, namely, in rate of development, viability, longevity, and fecundity.[46]

[44] Dobzhansky, 1947a.
[45] Dobzhansky, 1947a.
[46] Moos, 1955.

Where the inversion heterozygotes are superior in fitness to the inversion homozygotes, selection in favor of the former will operate to preserve a balanced chromosomal polymorphism in the population, and no single inversion type will become extinct. Studies of both natural and artificially caged populations show that the common inversion types are indeed permanent components of the population.

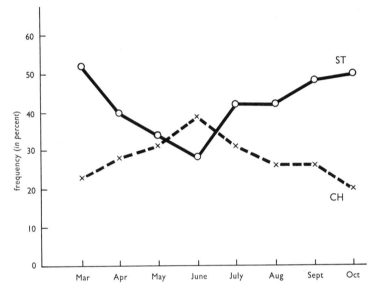

Fig. 46. Changes in frequency of chromosomes carrying two inversion types in natural populations of Drosophila pseudoobscura in the San Jacinto Mountains during the advance of the season from March to October

Averages for comparable dates during the years 1939 to 1946. From data of Dobzhansky, 1943, 1947b.

In natural populations of Drosophila pseudoobscura in the Sierra Nevada and San Jacinto Mountains, the ST, CH, and AR forms of chromosome III undergo regular cyclical changes in frequency during the course of each year. In the San Jacinto Mountains, the frequency of ST chromosomes in the population decreases from March until June, then increases again with the advance of summer and fall. As shown in Fig. 46, the average frequency of ST is 52 percent in March, but 28 percent in June, and 50 percent again in October; during the winter from October to March no appreciable changes take place. The reverse changes occur in the frequency of CH. This inversion type rises from a frequency of 23 percent in March to a maximum of 39 percent in June, then decreases in relative abundance during the hot season from July to September (Fig. 46). The AR inversion remains relatively constant from month to month in the San Jacinto Mountains.[47]

In populations in the Yosemite region of the Sierra Nevada the seasonal changes involve the ST and AR chromosomes. Here ST rises in frequency with the advance of summer, reaching its maximum in August and September, while AR, which is most abundant in May and June, diminishes during the summer to its low point in September. These trends are apparently reversed during winter hibernation to restore the proportions which exist at the beginning of summer.[48]

Systematic changes in frequency ending, not in the replacement of one inversion type by another but in a balanced polymorphic condition, are also found to occur in caged populations maintained in the laboratory under certain conditions. Two separate experiments were performed with flies from the San Jacinto Mountains polymorphic for AR and CH. The initial population in cage 44 had 29 percent AR and 71 percent CH chromosomes. In successive generations the frequency of AR rose rapidly, as shown in Fig. 47, until it attained a frequency of about 75 percent, at which point it leveled off and underwent no further changes. The complementary changes took place in the frequency of CH chromosomes in this population cage from the original high frequency of 71 percent to the ultimate level of 25 percent. Type AR did not replace CH, but instead an equilibrium point was attained where both chromosome types continued to exist in the population.[49]

Population cage 38 was started with a very different proportion of AR and CH chromosomes, namely, with 84 percent AR and 16 percent CH, as contrasted with the ratio of 29 AR:71 CH in the original generation of cage 44. Nevertheless, after a few generations of natural selection in cage 38, the same equilibrium point of 75 percent AR and 25 percent CH was reached (See Fig. 47).[50]

[47] Dobzhansky, 1943, 1947b.
[48] Dobzhansky, 1948a.
[49] Dobzhansky, 1948b.
[50] Dobzhansky, 1948b.

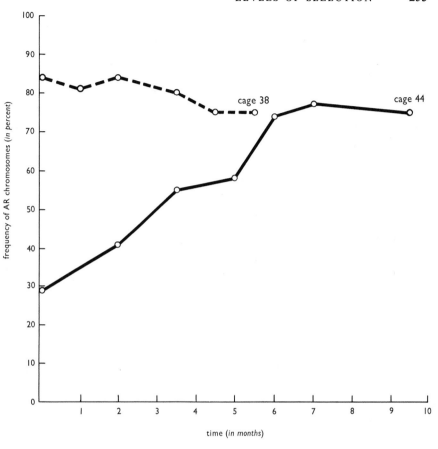

Fig. 47. Changes in frequency of AR *chromosomes in two caged populations of* Drosophila pseudoobscura

The populations in both cages were polymorphic for AR and CH. The proportions of AR and CH were very different in each cage at the start of the experiment but became similar after a few generations of selection. From data of Dobzhansky, 1948*b*.

In the light of what we have already learned regarding the relative selective values of flies heterozygous or homozygous for inversions, the permanent chromosomal polymorphism in the wild and caged populations can readily be accounted for as a result of the adaptive superiority of the inversion heterozygotes.

The inversion heterozygotes do not *always* exhibit an adaptive superiority over the corresponding homozygotes. This is true of some natural populations at some seasons of the year. Let us return to Table 11 summarizing the proportion of inversion heterozygotes in the Keen Camp population of *Drosophila pseudoobscura* during different months of 1952. In the May sample analyzed by Epling, Mitchell, and Mattoni, the proportion of heterozygotes actually observed (83 percent) was significantly greater than that expected from the Hardy-Weinberg formula (72 percent). In the July sample, on the other hand, the actual frequency of heterozygotes (.678) agrees closely with the expected frequency (.681). This close agreement suggests that the heterozygotes do not have a selective advantage in this population at this season.[51]

Similar results are obtained in some of the cage experiments. It is found, for example, that the selective advantage of the inversion heterozygotes is manifested only under certain prescribed conditions. It is necessary, in the first place, that the cages be overcrowded so that a competition exists for food and space.[52] At warm temperatures (21°–25°C) ST chromosomes rise and CH chromosomes decrease in frequency until an equilibrium point determined by the relative adaptive values of the three genotypes is reached. In cages kept at cool temperatures (16°C) such changes do not take place. The heterozygotes are adaptively superior at warm temperatures but not at cool temperatures.[53]

The relative adaptive values of the ST homozygotes, CH homozygotes, and ST/CH heterozygotes also vary according to the kind of yeast which the adult flies or larvae use for food. In a laboratory population of flies fed with one type of yeast the ST/CH heterozygotes may be superior to the homozygotes, while in a similar laboratory population supplied with a different type of microorganism for food the heterozygotes may be equal or inferior to the homozygotes in adaptive value.[54] Thus, at 21°C, the ST/CH heterozygote is adaptively superior when feeding on the yeast Kloeckera, but is not superior on a food substratum of the yeast Zygosaccharomyces.[55]

[51] Epling, Mitchell, and Mattoni, 1953.
[52] Levine, 1952.
[53] Wright and Dobzhansky, 1946; Dobzhansky and Spassky, 1954.
[54] da Cunha, 1951.
[55] Dobzhansky and Spassky, 1954.

These observations remind us that the heterosis of inversion heterozygotes, like any other expression of gene action, is capable of being
modified by environmental conditions. As Dobzhansky and Spassky
state[56]: "It is an oversimplification to believe that once a genotype is
heterotic it should always be heterotic."

The selective advantage of inversion heterozygotes under certain
environmental conditions and the resulting balanced chromosomal
polymorphism are by no means unique features of the single species
Drosophila pseudoobscura. These phenomena are known in about 30
wild species of Drosophila and in some other Diptera.[57] Inversions
occur commonly in the heterozygous state in various plant species such
as *Paeonia californica* and *Paris quadrifolia.* Permanent heterozygosity
for translocations is as common or more so among higher plants, being
characteristic of the *Oenothera biennis* group, *Datura stramonium,* and
Rhoeo discolor.[58] Permanent heterozygosity for genes not associated
with any cytologically visible structural rearrangements of the chromosomes, as found in corn and chickens, may well be even more widespread
in the living world.

Where the heterozygotes enjoy a selective advantage over the homozygotes, and the population consequently remains polymorphic, the
action of progressive selection is probably to bring about a series of
shifts from one heterozygous pair of alleles to another, thus

$$a_1a_2 \rightarrow a_2a_3 \rightarrow a_2a_4.$$

The sequence of events in the population is then not the replacement of
one allele by another, but the replacement of one diploid combination
by another.[59]

DISRUPTIVE SELECTION

A population inhabiting a heterogeneous environment is apt to be
subjected to somewhat different selective pressures by the different facets
of its habitat. Each of the several selective pressures will tend to preserve a different variant form of the organism which is best adapted to

[56] Dobzhansky and Spassky, 1954.
[57] Review by da Cunha, 1955.
[58] Dobzhansky, 1951*a*, ch. 5, for brief review with references.
[59] Lerner, 1954, 113–14.

one particular facet of the environment. Insofar as the various phases of the heterogeneous environment continue to exert different strong selective pressures on a single interbreeding population occurring throughout this environment, a diversity of polymorphic forms may continue to coexist in the population. In theory, a heterogeneous habitat may set up a series of diverse selective pressures, known as disruptive selection, and the latter may in turn create and preserve a series of genotypes in a state of polymorphic equilibrium.[60]

Permanent balanced polymorphism certainly exists in many natural populations. Furthermore, in *Drosophila willistoni* and other species, the polymorphism reaches its greatest development in populations inhabitating the most diverse environments in the tropics.[61] But whether this polymorphism can be attributed to disruptive selection, alone or in combination with other factors such as selection for heterosis, remains to be clarified.

Again, under experimental conditions strong disruptive selection can maintain two or more separate classes of genotypes in a population in spite of random interbreeding.[62] But as yet we do not know very much about the effectiveness of disruptive selection under natural conditions. Clarke and Sheppard have recently presented some evidence pointing to the occurence of disruptive selection as between mimetic and non-mimetic females in natural populations of the butterfly, *Papilio dardanus*, in Abyssinia.[63]

THE POPULATION AS A UNIT OF SELECTION

The population system of many species consists of a series of more or less disjunct colonies. These colonies, semi-isolated from one another by uninhabited or sparsely inhabited areas, comprise the regular breeding units of the species. They are also the decentralized storehouses of variability and the separate fields of evolution of the species.

The tendency of selection is to maintain and increase the fitness of the individuals in each population separately. If for any reason this

---

[60] Mather, 1953, 1955.

[61] da Cunha and Dobzhansky, 1954; da Cunha, Dobzhansky, Pavlovsky, and Spassky, 1959.

[62] Thoday and Boam, 1959; Streams and Pimentel, 1961. See the discussion of the interaction between selection and migration in Chapter 9.

[63] Clarke and Sheppard, 1962.

tendency of selection is thwarted in some colonies by absence of the necessary genetic variations, or strong counter-mutation, or prevalence of random changes in allele frequencies, etc., those colonies may dwindle and become extinct. While this result is fatal for the particular colonies concerned, it is not necessarily disastrous for the species as a whole, so long as other colonies exist in which the selective process can operate successfully to establish and maintain fitness. In this case the populations containing genes which engender high fitness replace the less fit populations.

The replacement of one population by another may be a direct process, as where the poorly adapted colony dies out completely and its territory is subsequently occupied by representatives of a well-adapted colony. Or the replacement may come about gradually by the immigration of superior individuals or gametes from the well-adapted colony into the less fit colony, followed by their subsequent rise in frequency in their new home. The end result is the same in either case: a spreading and increase of the genes of the better adapted population and a concomitant decline in abundance of the genes of the poorly adapted population. Where separate breeding populations differ in their average level of genetically determined fitness, selection can favor the spread of one population relative to another.

The subdivision of a species into semi-isolated colonies is, as Wright has emphasized, a particularly favorable situation for rapid evolution in a changing environment. By comparing the subdivided population with two contrasting types of population structure, we can see the evolutionary advantages of the former.[64]

In a species consisting of a few small populations, most alleles will be fixed at a frequency of 100 percent, often without regard to their selective value, with the result that the amount of genetic variation is insufficient and the types available are inadequate to satisfy the requirements of adaptation. This situation will be discussed further in the following chapter. In a species consisting of very large continuous populations with free interbreeding, selection is relatively ineffective in raising the frequency of rare recessive alleles, including most new mutations. Furthermore, the favorable alleles that do become established in one part of the large continuous population are usually swamped out,

[64] Wright, 1931, 1943 *passim*.

and the favorable gene combinations that become assembled are usually broken apart by interbreeding with neighboring segments of the same population.[65]

But a species consisting of many semi-isolated colonies of intermediate or small size possesses an adequate but decentralized store of genetic variations in a form where selection can get a hold on them. The primary mode of selection—Darwinian selection between individuals of one population—will take place in each separate colony. If the Darwinian selection fails to produce the required adaptedness in some colonies, there is still a good chance that it will do so in other colonies. The latter can then replace the former and can become the characteristic representatives of the species. The primary effects of Darwinian selection are reinforced by the second-order effects of interpopulation selection.[66]

The results of interpopulation selection can be seen in a concrete form in the distinctive adaptations of the worker caste in social ants and bees. In the honeybee (*Apis mellifica*), for example, the important functions of food-getting, defense and maintenance of the colony, and rearing of the broods are carried out by the workers, which possess bodily and psychic adaptations for performing these functions. Being sexual neuters, the workers do not reproduce as individuals, and consequently have no opportunity to pass the genes determining their adaptive characteristics on to the next generation. That task is performed by the sexually fertile but economically useless queens and drones. If the queens and drones do not carry genes making for adept and efficient worker bees, the hive will not thrive and may be eliminated by competition from other hives containing better adapted workers. The chief adaptations by which the colony lives reside in the worker bees. The unit of selection which has brought about those adaptations is the colony as a whole.

As discussed in the preceding section, the populations of many plants and animals are composed mainly or entirely of heterozygotes, which have a higher fitness than the corresponding homozygotes.[67] Now

---

[65] Wright, 1931.

[66] Wright, 1931.

[67] For an excellent discussion of the integrative properties in populations composed of heterozygotes, see Lerner, 1954.

heterozygotes do not breed true to type. An individual heterozygous for any given gene or chromosome segment is expected to produce some progeny which are homozygous for that gene or segment as well as some heterozygous progeny. In a population of permanent heterozygotes, therefore, some inferior homozygous zygotes are regularly formed as the price the population must pay in order to maintain a high fitness based on heterosis.

This price can, however, be reduced by maintaining a series of multiple alleles in the population.[68] For any gene or supergene present in two allelic forms, the maximum proportion of heterozygotes that can be produced by random mating between the two classes of gametes is 50 percent, and all the rest of the zygotes are homozygotes. With three alleles and hence three classes of gametes, this maximum proportion of heterozygotes can be raised to 67 percent, with a corresponding reduction in the number of homozygotes to 33 percent. Random mating in a population polymorphic for four or five alleles can lead to the production of as few as 25 percent or 20 percent of homozygotes, respectively. It is therefore advantageous for a population basing its fitness on a heterozygous condition to contain many multiple alleles, as a means of lowering the incidence of ill-fitted homozygotes produced by random mating. Selection would favor the populations possessing large numbers of multiple alleles over those possessing few alleles of each gene or supergene.

The frequency of heterozygous genotypes can also be raised above .50 by non-random mating, particularly by various systems of negative assortative mating, or preferential mating between unlike genotypes. Thus if like homozygotes do not mate, but mating is otherwise at random, the heterozygotes at equilibrium will have a frequency of .555. And, if homozygotes mate with unlike homozygotes, and heterozygotes with heterozygotes, the frequency of heterozygotes will eventually become .67.[69]

The cost to the population of a system of permanent heterozygosity can also be met in part by an increase in the number of independently assorting, heterozygous genes.[70] An individual heterozygous for a

[68] Dobzhansky, 1955b, 10.
[69] Naylor, 1962.
[70] White, 1958.

single gene **A** yields progeny half of which are homozygous for **A**. But if the individual is heterozygous not only for **A** but also for an unlinked gene **B**, only one-quarter of its progeny will be complete homozygotes. Half the offspring which are homozygous for **A** will be heterozygous for **B**, and vice versa. With larger numbers of independent heterozygous genes, the chances are increased that any given zygote will be heterozygous for at least one or a few genes. If, therefore, the different heterozygous genes **A**, **B**, **C**, etc., are more or less interchangeable, so that heterozygosity in one gene (**B**) compensates to a certain extent for homozygosity in another gene (**A**), the addition of new unlinked heterozygous genes or supergenes to a polymorphic system may help to increase the average fitness of the population.

But if the foregoing assumption does not hold true, and if the heterosis depends on heterozygosity simultaneously in all genes, then permanent polymorphism in a multiplicity of independent genes will lower the proportion of complete heterozygotes and hence of heterotic individuals produced by the population. Lerner points out that if heterozygosity were rigidly enforced in seven independent genes, only $(.5)^7$ or less than 1 percent of the population would have offspring.[71] The reproductive rates in domestic animals are not high enough to tolerate selection of such intensity. Yet polymorphism does seem to be maintained in populations of poultry and other animals for several independent genes or supergenes. How, Lerner asks, can we reconcile the empirical evidence with the expected reproductive disadvantage of multiple polymorphism?[72]

The theoretical difficulty in accounting for successful reproduction by a multiple polymorphic system arises from the assumption that heterosis requires complete heterozygosity in all genes. This theoretical difficulty disappears, indeed is replaced by a theoretical advantage, on the assumption that heterosis can be produced by heterozygosity in just one or a few of the many polymorphic genes. The existence of successfully reproducing populations which are permanently polymorphic in several or many unlinked genes suggests, therefore, that the latter assumption is most likely to be valid. It is probable, in other words, though not certain, that heterosis can be attained through different

[71] Lerner, 1958, 71–72.
[72] Lerner, 1958, 71–72.

pathways, and as a result of the action of different heterozygous genes which can compensate for one another to a certain extent.[73]

Heterosis does not arise from just any combination of two dissimilar alleles. The interaction between a pair of alleles may produce either favorable or unfavorable phenotypic results. Natural selection within and between populations will act to preserve those alleles which in combination give rise to heterotic individuals, while eliminating the alleles that lack good combining ability.

The superiority of the heterozygotes between the ST, CH, and other inversions in *Drosophila pseudoobscura* is subject to an interesting geographical limitation. It will be recalled that these chromosome types have a widespread geographical distribution in western North America. The heterosis of ST/CH is not, however, an intrinsic property of any pair of ST and CH chromosomes derived from any geographical area in which these chromosome types occur. In fact, an individual containing a ST chromosome from one locality and a CH chromosome from some other widely distant locality usually does not exhibit heterosis and may even be inferior to the homozygotes. Whereas the heterozygote ST/CH derived from ST and CH chromosomes indigenous to the San Jacinto Mountains is heterotic under certain conditions, as noted in the preceding section, the same heterozygous type produced by combining ST chromosomes from the San Jacinto Mountains with Mexican CH chromosomes is not heterotic.[74]

The differences in performance of different ST/CH heterozygotes can be explained by assuming that the genic contents of the ST and CH inversions vary throughout the geographical range of the species. The ST and CH chromosome segments which coexist in any local population, and more specifically the gene alleles in those segments, have been selected for their ability to produce high fitness in the heterozygous state. But the California type of ST chromosome and the Mexican type of CH chromosome do not combine regularly in any natural population. Therefore these particular forms of the two inversions have not been exposed to selection for the mutual adjustments between their gene alleles that bring about heterosis.[75]

[73] See also Lerner, 1954, 66–68, for a discussion of this problem.
[74] Dobzhansky and Levene, 1951.
[75] Dobzhansky and Levene, 1951.

The two types of chromosomes can, however, *become* coadapted by selection in experimental populations. Some population cages started with mixtures of California ST and Mexican CH chromosomes, which at first yielded heterozygotes exhibiting no heterosis, were reexamined after 15 generations of natural selection and found then to contain heterotic heterozygotes. Between the early and the 15th generations, variant forms of the ST and CH chromosomes with a favorable combining ability had replaced alternative forms without favorable joint effects.[76]

The ST and CH chromosomes from different geographical areas become coadapted sometimes, but not invariably, in caged populations containing new mixtures of them. Among six replicate experimental populations composed of ST and CH chromosomes, which were not mutually adapted so as to engender heterosis at the beginning of the experiment, heterosis based on coadaptation of these chromosomes developed eventually in two cages, but did not develop in the other four.[77] The effectiveness of selection in bringing about the coadaptation is, of course, contingent upon the existence of suitable genetic variations within each population. The chromosomes in some caged populations may include the raw materials from which selection can build up coadapted combinations, while other populations lack the variations necessary for the development of heterozygote superiority.

The average fitness of different Drosophila populations no doubt varies. It has in fact been found that laboratory populations of *Drosophila pseudoobscura* which are polymorphic for AR and CH produce more individuals and a greater total weight of flies from a given amount of food, and are more stable in these respects under varying environmental conditions, than monomorphic populations.[78] If the average fitness is generally higher in the populations in which coadaptation develops and lower in those where it does not, the former populations would be able to reproduce more successfully than the latter. Under conditions of competition the more fit populations would tend to replace the less fit ones, so that in time the features of the former would come to be distributed throughout the whole species. Coadapted

[76] Dobzhansky and Levene, 1951.
[77] Dobzhansky and Pavlovsky, 1953.
[78] Beardmore, Dobzhansky, and Pavlovsky, 1960.

alleles, along with multiple alleles and unlinked heterozygous genes, are in fact found throughout the area of *Drosophila pseudoobscura*. The wide distribution attained by these features throughout the species can be explained as a result of selection in favor of polymorphic populations having the highest average fitness.

ON THE INTERRELATIONSHIP BETWEEN DARWINIAN SELECTION AND INTERPOPULATION SELECTION

The basic form of selection is, as we have seen, the differential reproduction of genetically different individuals within a population. This is Darwinian selection. A differential survival and reproduction also takes place in nature between different breeding populations of one species and between different species. For the purpose of the present discussion we will refer to the differential reproduction of populations of any magnitude, from local breeding groups to non-interbreeding species, collectively as interpopulation selection.

The relationships between Darwinian selection and interpopulation selection are complex and are not fully understood. Undoubtedly the two modes of selection work together and in balance. Apparently interpopulation selection acts as a censor on the effects of Darwinian selection. The severity of the censoring action of interpopulation selection varies, however, with the degree of interpopulational competition in the environment. The purpose of the following section is to develop this idea briefly.

The number of genetic deaths which may be involved in making the complete substitution of one allele for another may be many times the number of individuals present in any one generation. If this substitution comes too fast the population may be unable to stand the cost of the Darwinian selection.[79] A value of s = 1.0, for example, corresponds to complete lethality of the carriers of one allele, and a value of s = .5 means that the carriers of one allele are semilethal. Changes in the relative frequencies of the alternative alleles will be very rapid, theoretically, where the value of s is high. In practice, however, the population might not be able to support the loss by selection of a large number of individuals in each generation. On its way to becoming better adapted too rapidly the population might become extinct and hence

[79] Haldane, 1957.

never arrive at its goal. In other words, a population which is too poorly adapted to its present conditions may have already lost its chance to *become* adapted.

It might be argued that an organism capable of multiplying its numbers in reproduction should be able to support the loss of half or more of its individuals each generation through intense selection and still survive as a population. Under actual conditions in nature, however, the population does not usually exist in an empty environment, but is surrounded by other competing populations ready and able to move into its territory and replace it as soon as it becomes decimated in numbers. One kind of weed in a field or one kind of Drosophila in a forest can hold its own against competing kinds of weeds or Drosophilas only so long as the former continues to occupy its niche with about as many individuals as the niche will hold. If the niche is vacated to any considerable extent as a result of the selective elimination of many poorly adapted individuals produced by one population, it will be occupied by the individuals of a competitor population.

The average fitness of a population can be improved by Darwinian selection only within certain limits. As long as the population is better adapted to its particular niche than other competing populations it can maintain and improve its level of adaptedness on the basis of inter-individual selection. If, however, it becomes relatively less fit for its conditions than neighboring populations, the chances are good that the ill-adapted population will be replaced by a better adapted one. If this happens, we can say that the defunct population *might* have developed a better state of adaptation by Darwinian selection in time, but that under the circumstances it was not granted the time in which to do so.

The existing array of natural populations is maintained at a high level of fitness not only by Darwinian selection but also by the second-order effects of interpopulation selection. As an example, Darwinian selection would tend to increase the proportion of self-centered and aggressive individuals in a population of social animals. The end result of this process would be a group in which habits of cooperation are poorly developed. If this population comes into conflict with another in which team work is well developed, the former is apt to come out second best. The spread of individualism by the basic form of selection is thus counter-balanced by interpopulation selection, which favors the spread of cooperativeness, and in a real situation the outcome will be a balance between the opposing tendencies.

One of the factors affecting the balance between Darwinian selection and interpopulation selection in any particular case is the strength of the competition between different populations for a limited amount of territory, food, and raw materials. This is a factor which varies widely in different parts of the world and at different times in earth history. Let us consider the forms of selection which will operate in populations living in two contrasting situations: first, in an environment densely inhabited by many other populations, and, second, in an uninhabited environment.

In a saturated environment interpopulation selection will play a preponderant role. Darwinian selection occurs, but its effects are modified by selection between populations under the conditions of strong intercolonial and interspecific competition. A poorly fitted population, and hence one which must sustain the loss of a high proportion of individuals on the way to becoming well adapted, is likely to be replaced *in toto* by a better adapted population already in existence. Whatever effects Darwinian selection might have had in the ill-adapted population are canceled when this population dies out as a whole. The net result of strong competition and selection between populations is a community of well-adapted populations.

The various well-adapted populations continue to be subject to Darwinian selection, involving the death of a relatively small proportion of poorly adapted individuals. This process cannot be avoided. The cost of individual selection can, however, be mitigated. The loss of a given number of eggs, seeds, or juvenile forms is less disastrous to a population than the loss of the same number of adult breeding individuals. The burden of individual selection can, so to speak, be shifted from the adults back to some juvenile stage.[80] The stage of development at which a mutant gene expresses itself is subject to modification by other genes in the complement. A population can therefore become composed of genotypes which cause new mutations or old alleles to express their advantageous or disadvantageous effects at relatively early developmental stages, when the selective elimination of them

[80] Haldane, 1957.

reduces the reproductive potential of the population to a minimal extent. This result will be favored by interpopulation selection.

The potentialities for evolutionary change by Darwinian selection are quite different where the pressure of competition between populations is relieved. Occasionally in history it happens that a population is confronted with an unsaturated environment and has few or no competitors, predators, or parasites. A new island might rise above the sea and be exposed for colonization by terrestrial plants and animals. Or glaciers and ice sheets may retreat at the end of an ice age, opening up extensive areas of bare ground. Out of the numerous kinds of organism in the world fauna and flora that are potentially capable of inhabiting the new environmental niches thus opened up, only a minority will exist in bordering communities within easy dispersal range of the new territory.

For a while these first pioneers will have the territory pretty much to themselves and will be relatively free of competition Under exceptional conditions such as these, the pioneering populations can tolerate a large number of deaths concentrated in a few generations. In other words, they can afford to pay the high price of Darwinian selection that goes with rapid evolutionary change.[81] For a relatively brief period, until the new environment becomes fully populated again, the first arrivals can undergo a rapid evolution to a new state of adaptation and at the same time increase in numbers.

Annual herbaceous plants of the genus Clarkia grow in a series of local colonies in favorable spots in the foothill region of California. Lewis has observed that local colonies of Clarkia frequently become extinct. In a series of years with increased winter rains, the Clarkias may be crowded out by grasses; while in a series of drought years a given population may dwindle in numbers and eventually disappear, although the habitat remains open and potentially available. Now one of the prevailing trends in the evolution of Clarkia has been adaptation for ever more arid conditions; there has been no corresponding trend toward increased tolerance of moist conditions. In order to explain the observed short-range and the inferred long-range changes in Clarkia, Lewis suggests that wet phases of a climatic cycle result in the total

[81] Haldane, 1957.

extermination of Clarkias by more mesic types of vegetation, whereas dry phases of the climatic cycle sometimes bring about "catastrophic selection" in favor of new genotypes adapted for greater aridity.[82]

In terms of the framework presented in this section, Clarkias can be said to face severe interpopulation competition from better adapted grasses in a moist environment, but not in an arid one. Consequently, under arid conditions, strong Darwinian selection can take place in Clarkia populations and can lead occasionally to the rapid evolution of new adaptive types. But under moist conditions the effects of Darwinian selection within Clarkia populations are obliterated by the effects of interpopulation selection, when the Clarkias are supplanted by grass.

CONCLUSIONS

The term natural selection subsumes a wide variety of complex phenomena. It refers to the differential and non-random reproduction of any reproducible biological unit whatsoever. One way to analyze the complex phenomena of selection into their constituents is to classify them according to the type of biological unit, the members of which reproduce at different rates.

An intergene selection and even an intermolecular selection probably occurred in primitive evolution. We find gamete selection and even chromosome selection exemplified in higher plants and animals. Selection takes place between subgroups within populations of higher animals, thus between males (sexual selection) or litter mates (intra-uterine selection). Of primary importance in higher organisms is selection between genotypically different individuals of a population, or Darwinian selection. Selection discriminates, finally, between different local populations, different species, and even different biotic communities, as will be shown in Chapter 14.

The phenomena of selection can also be analyzed into components on the basis of the processes in the life of a biological unit which affect the rate of reproduction. Darwinian or individual selection, for example, may be an outcome of differential viability, differential mortality, differential longevity, differential fertility, differential

---

[82] Lewis, 1962. Lewis applies the concept of "catastrophic selection" to a process of decimation of Clarkia populations and their revival from a few survivors, which would be described in this book as the combined action of selection and drift.

fecundity, differential success in mating, and so on. These components can be broken down still further; thus differences between individuals in viability may be expressed in either an embryonic, a juvenile, or an adult stage, and may be due in the final analysis to differences in resistance to disease, in tolerance of extreme climatic conditions, in ability to heal wounds or rejuvenate lost parts, etc.

Again, selection can be classified according to whether it operates in a stable or a changing environment, or in a relatively homogeneous or highly heterogeneous environment. In a stable environment stabilizing selection maintains the existing state of an organism or population. This is probably the most common form of selection in nature. In a changing environment progressive selection is a factor for changes in the population. In a heterogeneous environment, finally, disruptive selection may act to maintain in the population an array of different polymorphic types adapted to different facies of the habitat.

Notwithstanding the diversity of phenomena involved, the goal of selection is always the same—adaptation. Now adaptation is by no means a simple matter. The environment to which an organic unit must adapt is a complex of many different factors, physical, social, and biotic. Each separate factor may carry out its own selective processes separately. The adaptations created by selection for one aspect of the total environment are not necessarily useful, and may even be detrimental, in relation to other facets of the environment. Furthermore selection is opportunistic in that it brings about adaptations to existing environmental conditions. Such adaptations may or may not be valuable to their possessors in future environments. The collective processes of natural selection, while they promote the formation of adaptations of diverse kinds, do not guarantee evolutionary success in the long run under what Darwin termed the complex conditions of existence. Indeed, for every gene allele, genotype, or species that is preserved by natural selection on account of its adaptive properties, many sister alleles, genotypes, or species are exterminated by the same process.

The workings of natural selection are extremely intricate, and despite much study since the time of Darwin and Wallace, are still not fully understood. In order to attain a better understanding as to how selection really works in nature, as opposed to how it *could* operate in theory, we need much more quantitative information than we have at present. What is the actual number of genetic deaths in a population under various normal or abnormal conditions? How many such genetic deaths can the population tolerate and still survive under various different conditions? In a word, we need to be able to quantify some of the components involved in the reproductive process, the interplay of which determines the success or failure of a population in the struggle for existence.

ELEVEN

Genetic Drift

> The fact is, that a selective advantage of the order of 1 per cent., though amply powerful enough to bring about its evolutionary consequences with the utmost regularity and precision when numbers of individuals of the order of 1,000,000 are affected, is almost inoperative in comparison to random or chance survival, when only a few individuals are in question.
> R. A. FISHER[1]

THE HARDY-WEINBERG LAW states that a selectively neutral allele will tend to remain at a constant frequency from generation to generation. The constant frequency referred to in this formula is one of statistical averages. In actuality the allele will fluctuate above or below the statistical average in different generations due to purely chance deviations in the reproduction of different genotypes. We can predict that a penny flipped into the air repeatedly will come up heads 50 percent of the time. In any actual series of trials, however, we are apt to get 6 heads out of 10 flips, or 45 heads in 100 flips, or some other deviation from the expected average of 50 percent heads. The chance fluctuations in allele frequencies within a biological population are known as genetic drift.

THE EFFECTS OF SAMPLING IN TWO HYPOTHETICAL POPULATIONS DIFFERING IN SIZE

Drift will not make much difference in a large population. In a small population, by comparison, a chance excess or deficit of one genotype may cause a given allele to shift radically in frequency or even in extreme cases to reach fixation (100 percent frequency) or extinction (0 percent).

Let us consider two populations, one large and the other small, composed of interbreeding individuals of the genetic constitutions

[1] Fisher, 1930, 77.

a_1a_1, a_1a_2, and a_2a_2. We will suppose that the proportions of these genotypes are such that the initial frequency of the allele a_2 is 50 percent in each. The large population consists originally of 250 a_1a_1, 500 a_1a_2, and 250 a_2a_2 individuals (N = 1,000, frequency of a_2 = .50); the small population consists of 6 a_1a_1, 12 a_1a_2, and 6 a_2a_2 individuals (N = 24, frequency of a_2 = .50). The a_1 and a_2 genes are further assumed to have equal adaptive values. Now let some accident befall 6 a_1a_1 and 12 a_1a_2 individuals in each hypothetical population before they reproduce. In the small population this accident would lead to the extinction of the a_1 allele, and consequently to the fixation of the a_2 allele; N will become 6, and the frequency of a_2 will become 1.00. In the large population the same accident will have negligible effects, N becoming 982 and the frequency of a_2 .503.

The change in the frequency of the a_2 allele from .500 to .503 in one generation in the large population does not represent any systematic trend. In the next generation a high proportion of carriers of the a_2 allele could accidentally die before reproducing, so that the allele frequency is reversed to .500 or .496 or some like value. In the small population, by contrast, the accidental death of the carriers of one allele may lead to a new allele frequency of .000. This represents an irreversible change (apart from recurrent mutations or the immigration of mutant genes from other populations).

In the foregoing hypothetical example we have considered the extreme case where one allele drifts into complete fixation and another into extinction in a small population. As we shall see later, this fate is the usual *ultimate* result of drift in small isolated breeding groups. Before the allele frequencies reach the values of 100 percent or 0 percent, however, they may drift to new levels differing markedly from those found in the original population. If in the previous example of a small population composed of 6 a_1a_1, 12 a_1a_2, and 6 a_2a_2 individuals, 6 a_1a_1 individuals accidentally die before reproducing (instead of 6 a_1a_1 + 12 a_1a_2 as postulated earlier), the allele frequency of a_2 will have drifted from .500 to .667 and that of a_1 from .500 to .333 in one operation. This represents a significant change in the genetic composition of the population in itself. In the next generation the allele a_2 could, of course, shift up to .750 or 1.00 or back down to .500 or to some other value by chance.

The conclusion that an allele may shift in frequency in a population as a result of a purely random component in reproduction means that some changes in the genetic composition of populations and hence some evolutionary changes may be brought about by chance alone. The random changes in allele frequencies will nearly always be minor in large populations. In small breeding groups, by contrast, many of the fluctuations in gene reproduction can be expected to alter significantly the allele frequencies. Drift, as a factor in evolution, is thus expected to be operative primarily in small populations.

THE GAME OF CHANCE

The influence of chance on the genetic composition of a small population during a succession of generations may be illustrated by the following example. Let us assume that a population is originally composed of 100 individuals and is originally polymorphic for 4 alleles of a gene **A**, namely, a_1, a_2, a_3, and a_4. These alleles can be combined in 10 different kinds of genotypes as follows:

$$a_1a_1 \quad a_1a_3 \quad a_2a_2 \quad a_2a_4 \quad a_3a_4$$
$$a_1a_2 \quad a_1a_4 \quad a_2a_3 \quad a_3a_3 \quad a_4a_4$$

There are 10 mature individuals of each genotype, and so N is 100. Each allele has a frequency of .250 in the original population under these conditions. We will assume that the different alleles have equal selective values.

Suppose that the original population is drastically reduced in size from 100 to 3 individuals by some catastrophe which has an equal chance of befalling all individuals alike. Such accidents are common in nature. Ninety-seven individuals of some population of plants or animals, for instance, might lie in the path of an avalanche and be wiped out, while 3 other individuals which happen to be on the sidelines at the time survive.

The 3 survivors are likely to represent a biased sample of the alleles present in the original population. In order to find out what happens to the allele frequencies as a result of chance when the original population passes through a bottleneck of small size, we can make use of a pack of playing cards. We take 100 cards and write each genotype on 10 of them, then we shuffle the pack, and draw out 3 cards at random.

I have done this repeatedly. On the first trial I turned up cards standing for 3 individuals with the genotypes a_1a_3, a_2a_3, and a_3a_4. No alleles have been lost in the new reduced population so far. We will return to this case in a moment.

In many other trials I lost 1 or 2 alleles in passing from a full deck to a sample of 3 cards. On the fifth drawing all 3 surviving individuals turned out to be a_4a_4 types. In one stroke the a_1, a_2, and a_3 alleles were lost and the a_4 fixed in the new daughter population. The population was converted from a polymorphic to a monomorphic or invariable state for the gene **A** in a single generation as a result of purely random processes.

We have not considered the first drawing in sufficient detail. Although all 4 alleles are represented in the daughter population, their frequencies are changed. The frequency of the allele a_3, for example, was .250 in the parental population—it is now .500; and each of the other alleles has decreased in frequency from .250 to .167.

Now suppose that this derivative colony is repeatedly decimated and thus remains small for several generations. Any one of the alleles is apt to undergo violent fluctuations in frequency in any subsequent generation. Sooner or later 3 of the 4 original alleles will be lost by chance and the remaining allele will be fixed.

The three surviving individuals—a_1a_3, a_2a_3, and a_3a_4—produce gametes in the ratio: 1 a_1 : 1 a_2 : 3 a_3 : 1 a_4. If these gametes unite at random they will produce the 10 possible genotypes in the following proportions:

1 a_1a_1	6 a_1a_3	1 a_2a_2	2 a_2a_4	6 a_3a_4
2 a_1a_2	2 a_1a_4	6 a_2a_3	9 a_3a_3	1 a_4a_4

But owing to a second accident only 3 of these survive to reproduce.

We can use our deck of cards to follow out the fate of the different alleles in the subsequent history of the colony. To do this we must reconstitute the deck to conform to the new proportion of genotypes. Then we simply draw 3 cards at random again.

In drawing these cards I turned up 1 individual of the genotype a_2a_4 and 2 individuals with the genotype a_3a_3. The allele a_1 has now dropped out, a_2 and a_4 are low in frequency, and a_3 has risen to a frequency of .667.

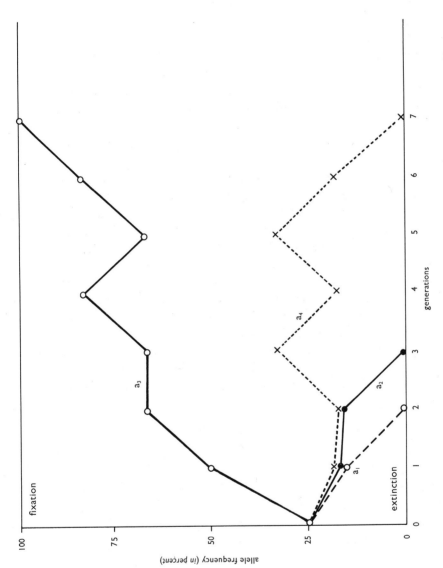

Fig. 48. Random changes in the frequencies of 4 "alleles" during 7 "generations"
The changes occurred as a result of drawing successive small samples at random from a deck of cards.

It will be of interest to see what might happen to this population if it passes through a third bottleneck of small size. We draw again from a newly reconstituted deck with the following results: 2 a_3a_3 and 1 a_4a_4. Now the a_2 allele is extinct in this population.

In successive generations the remaining alleles, a_3 and a_4, fluctuated as shown in Fig. 48, until the colony finally became monomorphic in

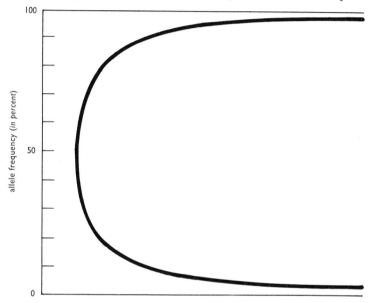

Fig. 49. *The expected distribution of the frequencies of selectively neutral alleles in a small isolated population*

The curve shows that a majority of the selectively neutral alleles either become fixed or lost in a small isolated population. Note that the curves in Fig. 48, which are based on a series of actual samplings, approximate in their general outline the theoretical curve shown here.

Based on Wright, 1931.

Generation 7. This was the particular sequence in one series of drawings. Anyone making random choices out of a deck of cards would obtain results different in detail, but the principle would remain the same. In a small breeding population a majority of the alleles fluctuate widely in frequency from generation to generation, and eventually drift into fixation or loss (Fig. 49).

It is evident that if the alleles determine particular phenotypic traits, the visible characters of a population might change as a result of a random component in gene reproduction whenever the population is reduced to a small number of breeding (hence inbred) individuals. If the gene **A** controls flower color in a plant, and if the various alleles determine a series of shades from blue to white, like deep blue, light blue, pale blue, and white, the large and polymorphic parental population will comprise a variable mixture of individuals having the different shades. As a result of drift a fragment of this population may become homogeneous for one flower color. If drift occurs repeatedly in different segments of the original population, a series of derivative colonies might arise which are characterized by different flower colors in pure form. One daughter colony might be all white, another all deep blue, and still another all light blue.

THE FIXATION OF GENE COMBINATIONS BY DRIFT

We have considered the fixation of particular alleles of a single gene **A** by drift in small populations. The alleles of other independent genes **B, C**, etc., may be fixed in a similar way. If the ancestral population is polymorphic for the alleles a_1 and a_2 of the gene **A** and b_1 and b_2 of an independent gene **B**, producing the various homozygous and heterozygous combinations of these alleles, a small derivative colony might become monomorphic for particular alleles of each gene. One homozygous allele combination, for instance $a_1a_1b_2b_2$, could be established by drift in one small derivative colony; another homozygous allele combination, $a_2a_2b_1b_1$, might be fixed in another daughter colony; still another colony could become composed uniformly of $a_2a_2b_2b_2$ genotypes; and so on. Inbreeding and drift in small populations can speed up the fixation, not only of single genes, but also—and perhaps more importantly—of gene combinations (see also Chapter 16).

THE INTERACTION OF DRIFT AND SELECTION

Whether drift will occur or not in a real population depends on the interrelationships between various factors: the number of breeding individuals (N), the selective value of the allele (s), mutation pressure (u), and gene flow (m). The relations between these factors have been worked out mathematically by Sewall Wright.[2]

[2] Wright, 1931.

Let us consider first the relations between population size, selection, and drift. It is sometimes stated that selection and drift are opposing forces; it is more accurate to regard them as interacting forces. Wright has pointed out[3]:

> The genes in a population may be put into 3 classes with respect to the roles of selection and random sampling. One class of segregating genes in any population may be expected to be almost wholly dominated by selection in one way or another, another class almost wholly by accidents of sampling [i.e., drift], while an intermediate class . . . will show important joint effects.

We have stated further in the introductory paragraphs that drift prevails in small populations and have implied that it is inoperative in large populations. This again is an oversimplification. Drift could occur in a population where N is in the thousands if s, m, and u are very low. Bentley Glass has pointed out that a population of 10,000 breeding individuals would be enormous for an allele having a selective value of s = .2 but would be "small" for an allele whose selective value is s = .0001.[4] Thus one and the same population may be too large for drift to occur in one gene, but below the critical size for drift with respect to some other gene. When we say that a population is "small," we mean that it is relatively small or, more specifically, that N is low in relation to s.

The relationships are summarized quantitatively in the following equations worked out by Wright[5]: (a) selection predominates when $N \geq 1/4s$; (b) drift predominates when $N \leq 1/2s$; (c) both selection and drift occur in between. If we arrange the different proportions between N and s on a linear scale, selection predominates over one range, drift over another, and the two ranges overlap, as shown in Fig. 50.

We can visualize these relations better by assigning concrete values to N. In order to do this we shall assume arbitrary but realistic values for s, and then solve for N in the equations $N = 1/4s$ and $N = 1/2s$.

Assume first that an allele has a selective value of .01. The frequency of this allele will be controlled by drift if N is 50 or less, by selection if $N = 25$ or more, and by both factors operating jointly in the overlapping range between $N = 25$ and $N = 50$ (see Fig. 51). In other

[3] Wright, 1948, 281, quoted from the journal *Evolution*.
[4] Glass, 1954.
[5] Wright, 1931.

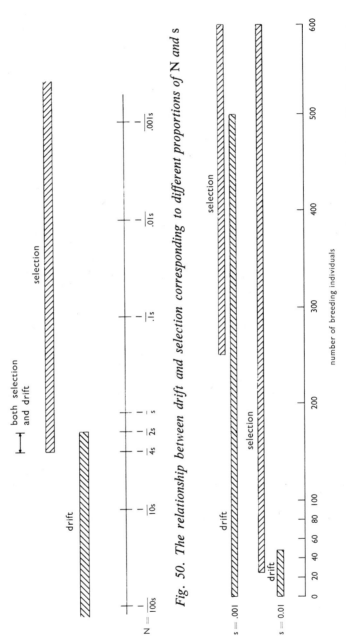

Fig. 50. The relationship between drift and selection corresponding to different proportions of N and s

Fig. 51. The range in population size in which drift is expected to prevail or to act jointly with selection for two values of s

words, an allele with a 1 percent selective advantage can drift into fixation or extinction in a population composed of 50 or less breeding individuals (N = 50).

If the allele is more nearly neutral in selective value than in the foregoing example, it can be lost or gained by chance in a population of larger size. It is simple to calculate that if s = .001, drift can occur in a population of as many as 500 breeding individuals, and predominates over selection below N = 250 (see Fig. 51).

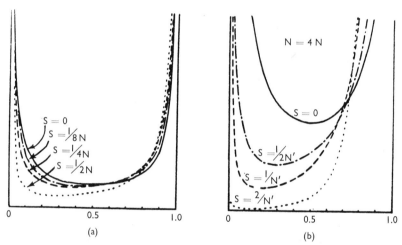

Fig. 52. The expected distribution of the frequencies of alleles with neutral or slight selective values in populations of different sizes

(a) In a small population
(b) In a population 4 times larger with the same absolute intensity of selection
Wright, 1931.

Wright has expressed the relations between N, s, and drift in a more general way in the curves shown in Fig. 52.[6] The symmetrical U-shaped curve corresponding to s = 0 in a small population (Fig. 52a) is the same as that shown in Fig. 49. As noted earlier, this curve indicates that selectively neutral alleles will usually drift into frequencies of either 0 percent or 100 percent in a small isolated population. Under conditions of slight selective pressure (s = 1/4N or 1/2N, etc.) in the same small population, the curve is skewed slightly. The frequency of

[6] Wright, 1931.

the allele is controlled partly by drift and partly by selection in this case. The chance component governing the allele frequency is reflected in the approximation of the skewed curve to the symmetrical curve expected when $s = 0$. The non-chance component is reflected in the deviation of the skewed curve from the symmetrical U- shape.

Most alleles tend to be either fixed or lost in a small population whether $s = 0$ or $1/2N$. But if $s = 0$, the alleles have an equal chance of drifting to fixation or loss; whereas if $S = 1/2N$, the favored allele will be fixed more often than it will be lost (Fig. 52a).

The central curve in Fig. 52b expresses the idea that in a larger population, a selectively neutral allele is not likely to drift to frequencies of either 0 percent or 100 percent. If the allele has even a slight selective advantage in a population of larger size, its frequency will be controlled mainly by selection. The predominating effect of selection over drift when N is relatively large is indicated by the remaining three curves in Fig. 52b.

THE EFFECTS OF MIGRATION AND MUTATION ON DRIFT

The random fixation of variations in a population depends not only on the relationships between N and s, but also on the influence of two additional factors, mutation pressure (u) and gene flow (m). In general, the frequency of alleles is affected partly by drift when u and m are less than $1/2N$, and mainly by drift when u and m are less than $1/4N$ (Fig. 53).[7]

In other words, allele frequencies are controlled by gene flow when $N \geq 1/4m$; by drift when $N \leq 1/2m$; and by both drift and gene flow when the value of N lies between these critical points. Similarly, allele frequencies are affected by mutation pressure if $N \geq 1/4u$; by drift if $N \leq 1/2u$; and by both factors acting jointly in the intermediate range.

Unless a population is very small, therefore, recurring mutations or the immigration of genetically different individuals from another population can prevent the fixation or loss of an allele by drift. But if the population is very small in relation to the mutation rate and the amount of gene flow, significant changes in allele frequencies can occur due to accidents of sampling during reproduction.

[7] Wright, 1931.

EXPERIMENTAL STUDIES

The U-shaped curves shown in Fig. 52a tell us that most alleles will eventually be fixed or lost in any small isolated population. These curves can also be taken to mean that any given allele will tend to reach fixation or extinction in a majority of the small isolated populations of the species. These are two ways of expressing the same relationship. In certain types of experimental studies it is more convenient to compare many populations with respect to one gene than to trace the changing frequencies of many genes in one population.

In one of the few experimental studies on drift, Kerr and Wright observed the fate of a single gene in 96 very small laboratory populations of *Drosophila melanogaster*.[8] They made the foundation populations polymorphic for the allele "forked," which causes the bristles on the body of the flies to be short and bent, and for the normal allele of the same gene. They placed 4 females and 4 males together in a laboratory bottle and allowed them to breed. The parental individuals

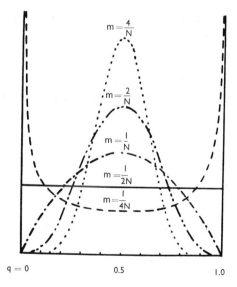

$$m = \frac{4}{N}$$

$$m = \frac{2}{N}$$

$$m = \frac{1}{N}$$

$$m = \frac{1}{2N}$$

$$m = \frac{1}{4N}$$

$q = 0$ 0.5 1.0

Fig. 53. Distribution of allele frequencies in a population with varying amounts of immigration

Wright, 1931.

[8] Kerr and Wright, 1954.

were selected so that the initial frequency of the allele f for forked was .50. The F_1 generation consisted as expected of numerous flies, but just 4 females and 4 males were chosen at random from among the numerous progeny as parents for the next or F_2 generation. The effective breeding size of the population was kept down to $N = 8$ by this process of random choice of parental individuals during the course of the experiment, which was continued for 16 generations.

This process was repeated in 96 different populations. That is, 96 laboratory bottles each received at the beginning of the experiment 4 male and 4 female flies. In each line the effective breeding size of the population was kept down permanently to 8 individuals. The allele frequency of f was .50 in each population at the beginning; no artificial selection was applied to any of the populations during the course of the 16 generations; and the allele frequency of f was determined for each line at the end of the experiment.

Only 26 of the 96 lines remained polymorphic to the last. In 29 lines the allele f reached fixation, and in 41 lines it became extinct by the 16th generation. These results show that drift does take place in small laboratory populations of Drosophila.

In an important experiment demonstrating the combined effect of selection and drift, Dobzhansky and Pavlovsky set up 20 replicate laboratory populations of *Drosophila pseudoobscura* and maintained them in a uniform environment for 18 months.[9] The foundation populations were all polymorphic for PP chromosomes from Texas and AR chromosomes from California in initial frequencies of 50 percent PP and 50 percent AR. Ten of these populations were started with large numbers of individuals (4,000 flies), and the other 10 with small numbers of founder individuals (20 flies). Although the flies rapidly multiplied to the maximum number possible within each cage, the lines descended from 20 founder individuals will be referred to for convenience as small populations, while the lines derived from 4,000 flies will be called large populations.

The PP and AR chromosomes are known to affect the fitness of the flies. A high fitness in the populations is often achieved through heterosis of the heterozygous combination PP/AR, in which case the

[9] Dobzhansky and Pavlovsky, 1957.

two classes of chromosomes are both preserved in a balanced polymorphic condition. Since in the present experiment the PP and AR chromosomes are derived from different geographical areas, and therefore have not become coadapted in nature, the development of a state of balanced polymorphism in the new artificial populations is by no means assured, and where a heterotic balance does arise the relative frequencies of the PP and AR chromosomes at equilibrium may differ from case to case depending on their genic constitution. The situation is the same as that where ST and CH chromosomes of different geographical origin are put together in a new population, as described in Chapter 10. Consequently, although the changes in frequency of PP and AR within the populations are controlled by natural selection, the final result of the selection in terms of equilibrium frequencies of the two chromosome types is unpredictable.[10]

In the present experiment the frequency of PP at the end of 18 months of natural selection varied from population to population as expected. The equilibrium frequency of this chromosome type ranged from 16 percent to 47 percent in the different populations. In the group of large populations this frequency value of PP exhibited a moderate cage-to-cage variance, the range of variation between populations being 20 percent to 35 percent. By comparison, the range of variation between small populations for the equilibrium frequency of PP was 16 percent to 47 percent. The variance between populations was 4.4 times greater for the 10 small populations than for the 10 large ones. Thus although selection controls the frequency of PP within each population, the operation of this factor is more indeterminate in the small populations. The differences between populations which are correlated with population size point to the operation of the additional factor of genetic drift. In short, the outcome of the experiment, as seen particularly in the 10 small populations, is due to the combined action of selection *and* drift.[11]

THE STUDY OF DRIFT IN NATURAL POPULATIONS

Drift would be expected to occur primarily in natural populations in which the effective breeding size either remains small or periodically becomes small. These conditions are met in many natural populations.

[10] Dobzhansky and Pavlovsky, 1957.
[11] Dobzhansky and Pavlovsky, 1957.

Some populations remain small in size from generation to generation. The cypress trees in California, for example, exist in small groves composed of scores or a few hundred individuals; many of these groves are restricted spatially to areas a few acres in extent and are separated from one another by many miles of unfavorable terrain where no cypresses exist.

In other cases a natural population may be large but has descended from a few migrant individuals. This is often the case in organisms which colonize new areas. A volcanic eruption may destroy all traces of preexisting life on a mountain; a retreating ice sheet may leave a stretch of bare ground behind it; a fire may sweep through a forest; or a tree may simply die of old age, exposing a spot of bare ground. The new ground becomes colonized by migrant plants and animals from surrounding areas. It is not likely that the colonizing individuals can bring with them a complete sample of the genetic diversity in the old ancestral populations. If the founders of the new population are few in number they could, by chance, represent only one or a few of the genetic variants in the ancestral population from which they came.

The direct result of drift is a change in the allele frequencies and hence a change in some phenotypic traits between the ancestral and the descendant populations. Such changes could be detected by observing a natural population over a period of time. It is easier in practice, however, to observe a series of related populations at one time. If drift has occurred, its effects can be seen in a divergence between different contemporaneous colonies derived from a common ancestral population. The evolutionary changes are observed in space rather than in time.

If we now examine natural populations of plants and animals which are permanently or periodically reduced in numbers, do we find the distinctive pattern of variation expected from the operation of drift? Do we find in short that each colony is relatively uniform genetically and that neighboring colonies are frequently different? Many plant and animal populations in nature do in fact conform to this pattern.

The California cypresses, mentioned earlier as existing in a series of small isolated groves, exhibit little individual variation within a single colony but perceptible differences between colonies. In some colonies, for example, the trees have a slender form and in others a pyramidal form. The bark is rough in some colonies, smooth in others. The color of the foliage varies from gray to bright green or bluish, and the globular seed-bearing cones from small to large in different groves. Each group of groves or in some cases each particular grove possesses its particular assemblage of phenotypic features.[12]

Gilia achilleaefolia, a herbaceous annual plant belonging to the Phlox family, occurs in a series of more or less isolated colonies in the California Coast Ranges. Some colonies are small, consisting of scores or hundreds of individuals; others are large with thousands or tens of thousands of plants. Since, however, a colony is usually founded by one or a few migrant individuals, which multiply in later generations, an opportunity exists for the action of drift during the period of colonization. This opportunity for drift is independent of the ultimate size attained by the population in its later history. The periodical reductions in population size in the Gilias have the same net effect on local variation as do permanently reduced numbers of individuals in the cypresses. Accordingly we find that any given colony of *Gilia achilleaefolia* is relatively uniform with respect to genetically determined characteristics of the flowers and other plant parts, but neighboring colonies often differ strikingly in these same characteristics (Fig. 54).[13]

A similar variation pattern is found in the genus Clarkia of the family Onagraceae, another group of herbaceous annual plants inhabiting the California foothill region. In the Clarkias as in the Gilias much more variation usually exists between colonies than within a colony. The variability of the species is broken up into a series of modes corresponding to the different discrete populations. These populations frequently originate from one or a few seeds dispersed into an uninhabited area from some neighboring population. The intermittent action of drift during the stage of population establishment would lead to the observed fixation of the variations in the different colonies.[14]

It is not necessary to assume that the morphological and genetic differences between adjacent colonies of cypress, Gilia, or Clarkia are due solely to the random fixation of variations by drift. In fact it is

[12] Wolf; see Grant, 1958.
[13] Grant, 1954a, 1958.
[14] Lewis, 1953.

unlikely that this is the case. We can see the effects of selection in the general adaptedness of different populations to their respective environments. These effects are channelized, however, within the limits of

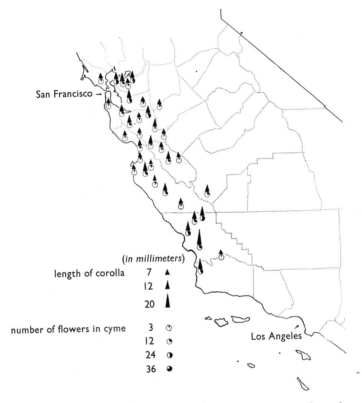

Fig. 54. Intercolonial variation in the length of the flower and number of flowers per head in Gilia achilleaefolia

The area of this species lies in central California between San Francisco Bay and Los Angeles.

Grant, 1954a.

separate small populations. We would expect the effects of sampling to be superimposed on the effects of selection. Some of the striking differences observed between neighboring colonies living in similar environments are probably brought about, partly at least, by the action of drift.

THE GEOGRAPHICAL DISTRIBUTION OF BANDLESS
SHELLS IN EUROPEAN LAND SNAILS

The land snail, *Cepaea nemoralis*, is widespread in western Europe. Habitats in which the snails can live and reproduce, such as beech woods, meadows, fields and hedgerows, are seldom continuous over large areas; the favorable habitats usually exist as a series of islands separated from one another by gaps up to several kilometers wide where few or no snails exist.

The snail populations are usually polymorphic for the pattern of bands on the shell. Most individuals have dark brown bands running around the shell, but some individuals lack bands (Fig. 55). As noted in Chapter 8, the difference between banded and bandless shells is due

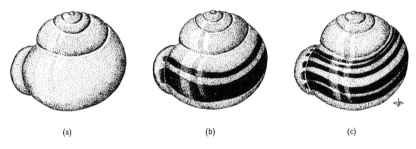

(a)   (b)   (c)

Fig. 55. The land snail, Cepaea nemoralis
(a) Bandless individual   (b), (c) Two types of banded snails

to the action of different alleles of a gene **B**, bandless shells being determined by the dominant allele b+. The observable phenotypic variability of the snail populations with respect to the character of banding depends on an underlying polymorphism in this gene.

The frequency of bandless shells varies in different populations. Lamotte determined the frequency of this characteristic for 826 populations in France.[15] He then divided the colonies into 3 size classes—large (3,000 to 10,000 individuals), small (500 to 1,000), and intermediate—and compared the colony-to-colony diversity among the large populations with that of the small populations (Fig. 56). He found that the allele frequency of b+ lies somewhere between 1 percent and 30 percent in a majority of the large colonies. Neighboring large

[15] Lamotte, 1951, 1959.

colonies, moreover, tend to be similar in allele frequency; thus if one large colony contains a certain percentage of bandless alleles, other large colonies in the same area are likely to exhibit similar frequencies. Along the Garonne River, for example, a series of large colonies several kilometers apart contained bandless snails in the following frequencies (in percent): 21, 12, 23, 18, 11, 11.

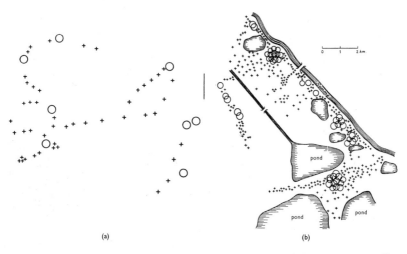

(a)   (b)

Fig. 56. Two types of population structure of the snail Cepaea nemoralis
in France

A circle represents 500 to 1,000 adults individuals in (a) and about 1,000 adult individuals in (b). A cross represents about 10 individuals.
(a) Semi-isolated small colonies connected by a few migrating individuals in Aquitaine
(b) Widespread large populations in the Somme Valley
Lamotte, 1951.

Among the small colonies, by contrast, the frequency of bandless alleles or bandless individuals fluctuates over a much wider range. Lamotte found a few small colonies in which the bandless allele had reached the fixation value of 100 percent, whereas none of the large populations were pure bandless. Furthermore, marked differences were sometimes evident in the genetic composition of neighboring small colonies. On the Ariège River in the Pyrenees Mountains, colonies with no bandless individuals were flanked by colonies with 20 and 21 percent bandless individuals, and colonies with 74 percent bandless snails were surrounded by colonies with 7, 4, and 32 percent bandless types. Figure 57 shows the haphazard distribution of the frequency of bandless types in a series of snail colonies along the Ariege River.

There are reasons for believing that the allele frequency of b+ in *Cepaea nemoralis* is controlled partly by natural selection.[16] If the genetic composition of the populations with respect to this gene were determined *solely* by natural selection, however, one would expect to find a similar range in frequencies in the different colonies irrespective of population size. This is not the case. If on the other hand the bandless allele sometimes reaches fixation as a result of random processes, one would expect to find monomorphic colonies primarily where N is small. Lamotte's observations agree with expectation according to the hypothesis that the relative frequencies of banded and bandless snails are sometimes controlled by genetic drift. In short, selection is probably involved, but this factor is insufficient in itself to explain all the observed facts about the local variability of *Cepaea nemoralis* in France.

GEOGRAPHICAL DISTRIBUTION OF THE BLOOD-GROUP ALLELES IN HUMAN POPULATIONS

Some examples of colony-to-colony divergence associated with small size of the breeding group are found in man. Human populations in the hunter stage of culture often consisted (or still consist in a few cases) of 200 to 500 breeding individuals. Some religious sects in modern civilizations also comprise small numbers of individuals who do not intermarry with outsiders. In such cases the small isolated group often differs abruptly and strikingly in various inherited traits from neighboring populations. The abruptness of the divergence in some small populations of *Homo sapiens* stands in marked contrast to the similarities or gradual changes which are typical of neighboring large populations of the same species.

A phenotypic characteristic which has been studied extensively in both large and small populations of man, and which can be related to specific genetic factors, is the biochemical properties of the red blood cells. It is well known that blood transfusions can be made successfully between certain individuals if they have the same type of blood or belong

[16] See Sheppard, 1959, 84–88, for a brief review of this aspect of the problem.

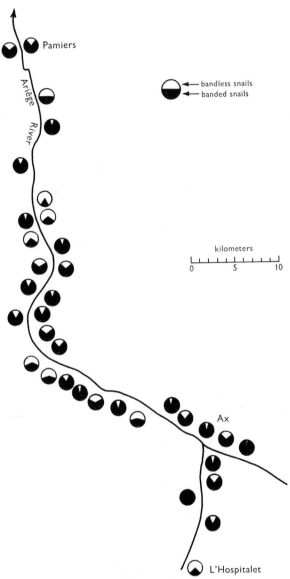

Fig. 57. The proportion of bandless and banded snails in a series of
colonies along the Ariège River in the Pyrenees Mountains

Note that adjacent colonies are frequently dissimilar in composition.

Redrawn from Lamotte, 1951.

to the same blood group, and that successful transfusions cannot be made between other individuals possessing different and incompatible blood types. This is because in the latter case the mixture of bloods leads to the clumping of the blood cells, which may prove fatal, whereas in the former case this unfavorable reaction does not occur.

The human species consists of individuals with different blood types, designated O, A, B, and AB. The serological types of blood are determined by the various homozygous and heterozygous combinations of a series of multiple alleles of the gene I. The principal alleles are i^O, i^A, and i^B. (Three slightly different allelic forms of i^A are known, but it is

Table 12. Proportion of 3 blood-group alleles in
2 human populations[a]

| | Allele frequencies (in percent) | | |
	i^A	i^B	i^O
Western Europeans	30	7	63
Mongolians	18	27	55

[a] Lundman, 1948.

not necessary to distinguish between them for the present discussion.) The genotype $i^O i^O$ determines blood of the type O. The blood group A is produced by the genotypes $i^A i^A$ and $i^A i^O$; blood group B by $i^B i^B$ and $i^B i^O$; and blood group AB by $i^A i^B$.[17]

Most human populations contain individuals belonging to 2 or more blood groups. The populations are thus polymorphic for the blood groups and for the alleles of the gene I. Any given human population, moreover, exhibits characteristic frequencies of the various blood groups and of the blood-group alleles. The different proportions of 3 blood-group alleles in 2 human populations are summarized in Table 12. It is seen that the allele i^B has a much higher frequency among the Mongolians than among Europeans, and that the i^A allele, conversely, is more frequent among Europeans than Mongolians.

These allele frequencies change in a gradual way along a geographical transect across Europe and Asia. The i^B allele, for instance, decreases regularly in frequency as one proceeds west from central Asia. The frequency of this allele ranges up to 32 percent in the Mongolian tribes,

[17] Stern, 1960, 176 ff.

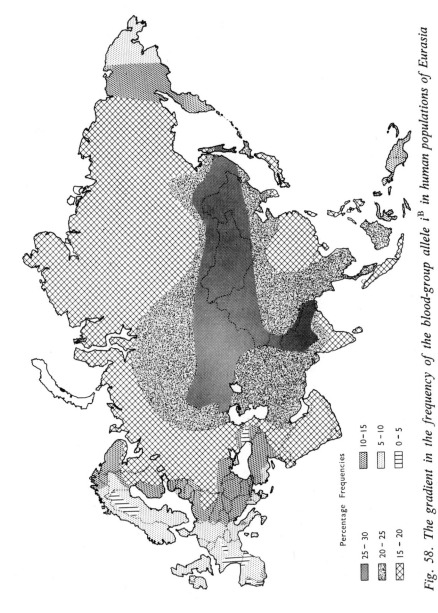

Fig. 58. The gradient in the frequency of the blood-group allele i^B in human populations of Eurasia
Drawn from data of Lundman, 1948; Mourant, 1954.

Percentage Frequencies

25 – 30
20 – 25
15 – 20
10 – 15
5 – 10
0 – 5

20 to 25 percent in the Ural Mountain region, 15 to 20 percent in Russia, 10 to 15 percent in southeastern Europe, 5 to 10 percent in western Europe, and decreases to less than 5 percent in Spain (Fig. 58).[18]

This gradient has been explained as a result of the repeated migrations of the Mongolians into western Asia and Europe. Interbreeding of the Mongolians with the native races to the west, among whom the i^B allele is believed to have been rare originally, would raise the frequency of this allele in the resident populations. The numbers of Mongolian invaders reaching any region varied inversely with the distance from their home territory. The gradient observed in the frequency of the i^B allele across Eurasia corresponds to a gradual decrease in the amount of interbreeding between the Mongolian tribesmen and other peoples at successively greater distances to the west.[19]

Small isolated human populations often present a variation pattern of a very different sort. A small group of Polar Eskimos numbering 271 or fewer individuals existed in complete isolation in northern Greenland for generations. When first contacted by another band of Eskimos from North Baffin Land, who spent several years trying to reach them, they believed that they were the only people in the world. The Polar Eskimos were found to differ markedly from the main populations of Eskimos in the frequencies of the blood-group alleles. The i^A allele varies from 29 to 40 percent in the larger Eskimo populations of Greenland; similar frequencies are found in Alaska and Baffin Land. In the Polar Eskimos, however, the frequency of this allele is only 9 percent (see Table 13).[20] Some smaller populations of Eskimos in Labrador and Baffin Land have lost the i^B allele entirely. The aboriginal tribes in south Australia were small in numbers and isolated from one another by differences in dialect. The i^A allele fluctuates greatly in frequency from one tribe to the next among these peoples.[21]

Among civilized peoples living in the same area, a small group of persons may be isolated from the surrounding population by social

[18] Lundman, 1948; Mourant, 1954.
[19] Candela; see Glass, 1954.
[20] Laughlin, 1950.
[21] Birdsell; see Glass, 1954.

barriers or religious beliefs. The Dunkers are a religious sect, the members of which have chosen to marry largely among themselves. The sect was founded in Germany in 1708 and immigrated to the United States in the early eighteenth century. A high degree of reproductive isolation has thus been maintained between the Dunkers and the surrounding German and American populations for generations. Some

Table 13. Frequency of the blood-group alleles in different native Eskimo populations[a]

Region	Allele frequencies (in percent)		
	i^A	i^B	i^O
Aleutian Islands	27	3	70
Alaska			
Bethel (Kuskokwin)	30	9	61
Nome	27	8	66
Point Barrow	30	7	63
Greenland			
South of Nanortalik, Julianhaab District, west Greenland	35	5	60
Nanortalik, Julianhaab District	27	3	70
Cape Farewell	33	3	64
Jacobshoven	29	5	66
Angmassalik, east Greenland	40	11	49
Thule, north Greenland	9	3	84
Baffin Land	25	0	75
Labrador	25–32	0	68–75

[a] Laughlin, 1950.

communities of Dunkers, moreover, are small; one in southern Pennsylvania contains about 90 individuals of reproductive age. Bentley Glass and his co-workers have determined the frequencies of blood groups and other traits in the Pennsylvania Dunkers and have compared these with the frequencies of the same traits in the non-Dunker populations of America and Germany.[22]

The frequencies of the blood-group alleles in the small Dunker community of Pennsylvania and in the racially similar large populations of western Germany and the eastern United States are summarized in Table 14. It is evident that the American and German populations have similar proportions of the blood-group alleles but that the Dunkers

[22] Glass, Sacks, Jahn, and Hess, 1952.

deviate strongly from their German ancestors and their modern American neighbors in the frequencies of these same genes. The i^A allele is significantly higher in frequency and the i^B allele has almost disappeared among the Pennsylvania Dunkers.

Other phenotypic traits studied in the Dunkers, such as type of ear lobe or hair, showed similar deviations from the ancestral or neighboring populations. In 5 different genes the allele frequencies found in the small breeding group diverged significantly from those typical of the surrounding populations.

Table 14. Proportions of 3 blood-group alleles in 1 small and 2 large human populations of the same racial stock[a]

| | Allele frequencies (in percent) | | |
	i^A	i^B	i^O
West Germans	29	7	64
Eastern Americans	26	4	70
Dunkers in Pennsylvania	38	2	60

[a] Glass, Sacks, Jahn, and Hess, 1952.

It has been discovered recently that persons with the genotypes $i^A i^A$ or $i^A i^O$ have a higher incidence of certain diseases, particularly stomach cancer and pernicious anemia, than members of the same human populations with the other possible genotypes for the 3 blood-group alleles. Ulcers of the stomach and upper intestine have a higher than average incidence among persons with the homozygous constitution $i^O i^O$. It is not likely, therefore, that the different alleles of the gene **I** are selectively neutral.[23] The frequencies of these alleles in human populations are probably controlled to some unknown extent by selection. The indication that selection plays a role in the control of the human blood-group alleles does not, however, entitle us to conclude that this is the only factor involved.

Some of the facts regarding the human blood groups, in particular the differences in the pattern of distribution of the **I** alleles in large versus small populations, are difficult to explain on the basis of selective agents alone. Where the breeding population is large in *Homo sapiens*, that is, where gene flow can occur more or less freely among numerous

[23] Mourant, 1959.

individuals, allele frequencies remain similar or change gradually over large geographical areas. Similar allele frequencies are exemplified by the i^A allele in different main populations of Eskimos, gradually changing frequencies by the i^B allele in the populations across Europe and Asia. On the other hand, where a small breeding group is segregated from a large parental population the alleles often shift abruptly in frequency. We see examples of this in the blood-group alleles of the Polar Eskimos, the south Australian tribes, and the Dunkers. The abrupt shifts in the genetic composition of some small human groups can be accounted for most satisfactorily by the hypothesis that drift is one of the effective evolutionary forces in such populations.

IMPORTANCE OF DRIFT IN EVOLUTION

Random variation in allele frequencies is predicted by the laws of probability and confirmed by experiments with laboratory animals. There is no reason to doubt that large random fluctuations in allele frequencies actually occur in natural populations under the appropriate conditions.

This does not tell us, however, how much importance we can ascribe to drift in the evolution of natural populations. The role of drift in nature is, in fact, a subject of debate among evolutionists at the present time. Some students argue that allele frequencies are controlled so much more effectively by selection than by drift as to render the latter factor insignificant in practice under natural conditions.[24] Other workers point out that some alleles producing deleterious effects on their carriers have become fixed by drift in small laboratory populations in the face of strong counter-selection.[25] It is probable, they say, that similar events sometimes occur in nature. Still other evolutionists emphasize the idea that selection and drift are interacting forces, the relative strengths of which vary from one situation to another. It is held by these students that some evolutionary changes in a population will be determined by selection, others by drift, and others by the joint action of both factors, depending on the balance between selection and drift in each particular case.[26]

[24] Fisher and Ford, 1947.
[25] Prout, 1954.
[26] Wright, 1948; Dobzhansky and Pavlovsky, 1957.

It is reasonable to suppose that some simple evolutionary changes may not represent any improvement in the fitness of the organism, but may be due instead to the random fixation of particular alleles in small populations. It is impossible to furnish a rigorous proof for this conclusion in any particular case, however, because it is impossible to be sure that the character established in the colony does not have some undetected value in adaptation.

In many cases both selection and drift may be involved. We find indeed among plants, animals, and man many cases of haphazardly shifting allele frequencies which are difficult to explain on the basis of mutation or selection alone, but which can be explained readily by invoking the additional factor of drift. The circumstantial evidence is strong enough in such cases to convince many evolutionists that drift is partly responsible for some observed evolutionary changes, and may be chiefly responsible for others.

Simpson has pointed out that an allele with a small selective advantage might require thousands of generations to replace the less favorable alleles as a result of selection working alone in a large population; but this same allele could easily be fixed in a few generations by drift in a small population.[27] Wright has expressed the same idea by means of the following comparison. If drift is occurring independently in 1,000 breeding populations each of which consists of 1,000 breeding individuals, the chance that at least one of these colonies will reach a new level of adaptedness for some gene is a million times greater than in a single interbreeding population of the same total size.[28] The species as a whole, in other words, can evolve a better adaptation sooner when the force of drift is added to that of selection.

Most of the adaptations of organisms depend not on single genes but on combinations of many genes. The new gene combinations are assembled by sexual reproduction. But how are they preserved? We saw in Chapter 9 that selection for gene combinations in a large breeding population is either very costly in terms of genetic deaths per generation or very slow in terms of number of generations required for fixation. Drift, being a random process, will tend to fix allele combinations without regard to their adaptive value. But the joint action of drift and

[27] Simpson, 1953a, 122.
[28] Wright, 1960, 463.

selection may facilitate adaptive evolution. As noted earlier in this chapter, new homozygous allele combinations can be established quickly by inbreeding and drift in small populations. If the new allele combination thus fixed is adaptively valuable, the population possessing it will be favored by selection. The evolution of adaptations can, theoretically, be brought about more rapidly by the joint action of drift and selection than by the operation of either one of these forces alone.

Part **4** *Evolution of Species*

TWELVE

Population Systems

Intercrossing plays a very important part in nature in keeping the individuals of the same species, or of the same variety, true and uniform in character. CHARLES DARWIN[1]

THE EVOLUTIONARY FORCES that produce and sort out genetic variations operate in a field of space and time. That field may be the local breeding population or colony. But the local population is a part of an aggregate of populations, a population system, which forms a breeding unit of large extent and long duration. The population system comprises a large-scale field in which evolutionary changes can take place.

The purpose of this chapter is to introduce and describe, with the aid of concrete examples, some of the principal types of population systems. The ways in which population systems are isolated genetically from one another will be described in Chapter 13, and their ecological relations will be discussed in Chapter 14. It is hoped that by dealing with the descriptive aspects of populations first, and then with their contemporaneous interactions, we will have laid a foundation for dealing with them later as stages in evolutionary divergence.

INTRODUCTION TO POPULATION SYSTEMS

Population systems of various orders of magnitude exist in nature. We can discern a series of ever more inclusive assemblages of local populations up to the limits of interbreeding. Naturalists and taxonomists have traditionally attempted to group the various population systems into two general categories, races and species. Races and species are supracolonial breeding groups of successively greater inclusiveness.

The local population or colony is, as we know, a community of

[1] Darwin, 1859, ch. 4.

interbreeding individuals occupying some niche in its habitat. Similar niches in neighboring localities are likely to be inhabited by separate but related colonies. Each colony represents a breeding unit in itself, for the individuals of which it is composed interbreed predominantly with one another because of propinquity. Nevertheless, some dispersal of individuals or gametes from one colony to another leads to occasional outcrossing on a wider scale. The neighboring breeding units in adjacent localities are likely to be similar in most of their genes, but different in some or many genes.

Interbreeding is more frequent among a series of local populations in the same general area than it is between inhabitants of widely distant areas. The regional cohort of populations is a race or, more particularly, a geographical race. Because interbreeding occurs much more commonly within a geographical region than between regions, and because of the more uniform mode of action of selection within (than between) geographical regions, each race tends to possess a distinctive and unique ensemble of genetic variations and phenotypic characteristics.

Races interbreed where they come into contact. This interbreeding leads to a more or less continuous intergradation between the races in their genetic constitutions and phenotypic characters. The sum total of the races that interbreed frequently or occasionally with one another, and that intergrade more or less continuously in their phenotypic characters, is the species. In other words, the hereditary variations are circumscribed within limits, and these limits are the boundary lines of species. Whereas the occurrence of interbreeding within a species is revealed by intergradations in various genetically determined morphological and physiological traits, the non-interbreeding between a species and its neighbors is indicated by a major discontinuity in the series of variations.

The presence or absence of interbreeding between populations is determined, among other factors, by the geographical and ecological relations between the populations. The terms allopatric and sympatric are useful to describe these relations. Populations that occur in different territories are said to be allopatric; those that coexist in the same area are called sympatric.[2] Allopatric populations may be

[2] Mayr, 1942, 148–49.

contiguous and adjacent, or disjunct and separated by geographical gaps.

In the case of population systems which occupy adjacent habitats—neighboring woods and meadows, or bordering upper and lower slopes in the same mountains—the geographical relations are partly allopatric and partly sympatric. The populations are separated spatially, and to that extent they are allopatric. But if the individuals or gametes belonging to the adjacent populations become dispersed over moderate distances, so that interbreeding between them is possible, they must be regarded as sympatric in the genetical sense.

Sympatric populations may live together with varying degrees of intimacy. In general, any two populations that occur within the normal range of dispersal of one another, so that opportunities for interbreeding arise normally, can be considered sympatric. There is an important difference, however, between populations that live in different habitats in the same area, and those that occupy different niches in the same habitat. The latter are subject to the various ecological interactions of competition or dependency that affect the members of a biotic community, whereas the former are not. Populations in neighboring habitats, like woodland and meadow, may be sympatric in the genetical sense alone. But two or more populations living in the same habitat, as within the meadow or within the woodland, are sympatric genetically *and* ecologically.[3] We shall refer to the latter situation as biotic sympatry[4] and to the former as neighboring or adjacent sympatry.

The array of population systems in the world is very diverse. The older naturalists in an attempt to classify the diverse observational data recognized two main types of population systems, namely, races (or varieties, subspecies, etc.) and species. As we shall see later in this chapter, the traditional set of general categories—races and species—had proved inadequate to deal with the facts of minor systematics and population biology already in Darwin's time. Some recent students have distinguished between geographical and ecological races (Rensch in 1934[5]), and between allopatric and sympatric species (Mayr in 1942[6]).

[3] Also Harper, Clatworthy, McNaughton, and Sagar, 1961.
[4] Or, in the terminology of Harper *et al.*, 1961, synecetic.
[5] Rensch, 1959, 47–50, and his earlier works cited therein.
[6] Mayr, 1942, 148–49.

These suggestions when combined give us four categories—geographical races, ecological races, allopatric species, and sympatric species—which provide a more satisfactory basis for the classification of population systems than does the older set of two categories.

Interbreeding and intergrading races may be allopatric or sympatric. Two or more allopatric population systems that interbreed where they come into contact are referred to as geographical races. Neighboringly sympatric population systems that interbreed and intergrade are known as ecological races. Ecological races may also be biotically sympatric in special cases, as we shall see later.

Non-interbreeding and distinct species may also be either allopatric or sympatric. Two or more population systems that inhabit the same territory without interbreeding comprise different sympatric species. Non-interbreeding population systems living in different territories, on the other hand, may be allopatric species.

Now the distinction between geographical races and sympatric species is usually clear enough in practice, but the distinctions between geographical races and allopatric species, and between ecological races and sympatric species, are frequently difficult to uphold in actual cases. Among the existing population systems, many fall clearly into one or another of the four named categories, indeed more population systems can be classified satisfactorily using the set of four main types than is possible with the older set of two types; yet even with these improvements in our method of dealing with populational phenomena many population systems remain which, owing to intermediate degrees of intergradation or intermediate ecogeographical relations, cannot be placed unequivocally into any single given category. Perhaps the recognition of still another category, the semispecies, for the population systems that are neither typical races nor distinct species will prove useful in many instances.

FROM POLYMORPHISM TO RACIAL VARIATION

Many local breeding populations, as noted in previous chapters, are polymorphic for various genes or supergenes and for the corresponding phenotypic traits. We have seen in Chapter 10, for example, that populations of *Drosophila pseudoobscura* normally contain several forms of chromosome III differing with respect to inversions. These inversion types, designated ST, CH, AR, etc., in their various homo-zygous or heterozygous combinations, affect the fitness of the flies under particular environmental conditions. The types of inversions and their relative frequencies are characteristic features of any population of *Drosophila pseudoobscura*.

It is of interest to compare the populations of *Drosophila pseudoobscura* in different geographical areas with respect to their composition for chromosome III. The different populations turn out to be alike insofar as they are all permanently polymorphic. But the average annual frequencies of the constituent inversion types change from population to population over moderate geographical distances. And over broader geographical areas changes become apparent in the types of inversions that are present at all in the populations.

Populations of *Drosophila pseudoobscura* near Mather in the Yosemite region of the Sierra Nevada, California, consist predominantly of ST, CH, and AR chromosomes in the average annual frequencies of 32, 38 and 19 percent, respectively. In addition there are 9 percent of TL chromosomes and small numbers of SC, OL, and PP chromosomes in the Mather populations. Populations of *D. pseudoobscura* living a few miles away from Mather at lower or higher elevations in the Sierras contain all the same inversion types. Two of them, however, namely, ST and AR, are present in characteristically different frequencies at each altitudinal station. Thus ST decreases in frequency from 41 percent at Lost Claim (below Mather), to 32 percent at Mather, to 26 percent at Aspen (above Mather). These trends extend over a larger linear and altitudinal transect from the foothills to the crest of the Sierra Nevada. As shown in Fig. 59, the frequency of AR chromosomes in the fly populations increases regularly with rises in elevation, while ST, which attains its maximal frequency in the foothill zone, decreases at higher elevations.[7]

Similar changes occur in *Drosophila pseudoobscura* over a broad front from southern California to southern Texas. The composition of the gamete pool at different localities along this transect of about 1,200 miles is shown graphically in Fig. 60. The polymorphic balance is seen to undergo a series of shifts from one geographical area to the next

[7] Dobzhansky, 1948a.

adjacent area on this transect. The ST chromosomal type declines in frequency from west to east and finally drops out entirely in south-central Texas. AR chromosomes, on the other hand, reach their maximal frequency in Arizona and New Mexico, and PP chromosomes

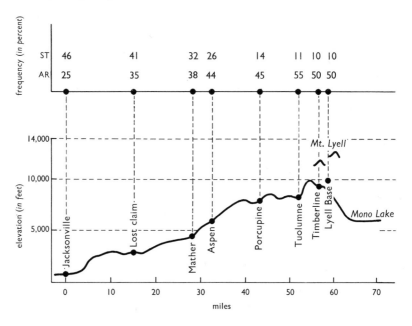

Fig. 59. Frequencies of two inversion types (ST *and* AR) *of chromosome III in populations of* Drosophila pseudoobscura *at different elevations in the Yosemite region of the Sierra Nevada*

The localities sampled are shown on the profile map. The percentage frequencies of ST and AR chromosomes in populations at each locality are given above the profile. Redrawn from Dobzhansky, 1948a.

which are absent or rare in the western part of the range become important constituents of the populations in the east.[8]

The geographical variation in the polymorphic balance for chromosome III of *Drosophila pseudoobscura* is shown on a still larger scale and in two dimensions in Fig. 61. Although all or nearly all of the populations are polymorphic in chromosome III, and some inversion types

[8] Dobzhansky and Epling, 1944.

are common to all or nearly all populations, the predominant inversion types vary from one geographical region to another. Thus ST is most frequent on the west coast, AR in the intermountain region, CH in

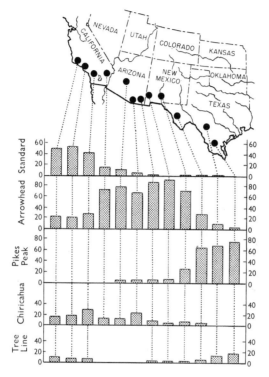

Fig. 60. Percentage frequency of different inversion types of chromosome III in a series of populations of Drosophila pseudoobscura *along a transect from southern California to southern Texas*

Dobzhansky and Epling, 1944.

northwestern Mexico, and PP in Texas. The rise in frequency of one inversion type in any geographical area is accompanied by decreases in alternative forms of chromosome III.[9]

Another example of a geographical gradient in allele frequencies, in this case in human populations, was described briefly in Chapter 11. We saw that different alleles of the gene **I** controlling the blood type

[9] Dobzhansky, 1951a, 138.

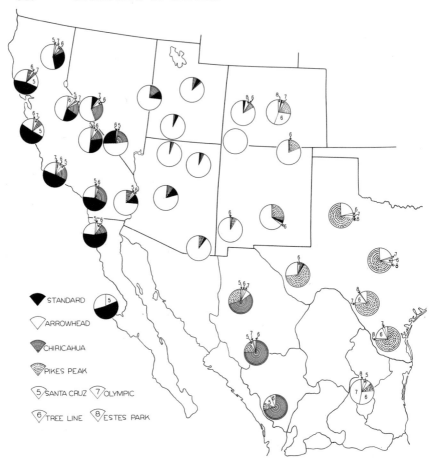

STANDARD

ARROWHEAD

CHIRICAHUA

PIKES PEAK

5 SANTA CRUZ   7 OLYMPIC

6 TREE LINE   8 ESTES PARK

Fig. 61. Relative frequencies of various inversion types in populations of Drosophila pseudoobscura *in the western United States and northern Mexico*

From Th. Dobzhansky, *Genetics and the Origin of Species* (New York, Columbia University Press, 1951).

vary in frequency across Eurasia. The allele i^B decreases gradually from a maximum frequency of 32 percent in human populations in Mongolia to less than 5 percent in parts of Spain, while the i^A allele undergoes increases in frequency along the same transect.

Let us next examine a group in which the genetic differences within and between populations are expressed in externally visible characters. Diplacus, a shrubby plant belonging to the Scrophulariaceae, is represented by a great variety of forms in southern California. As shown recently by Beeks, the various forms can be grouped for convenience into three main population systems: *Diplacus puniceus*, *D. longiflorus*, and *D. calycinus*. The characteristic appearance of the shoot and flowers of these three population systems is shown in Fig. 62. The distribution areas of the three entities are indicated in Fig. 63.[10]

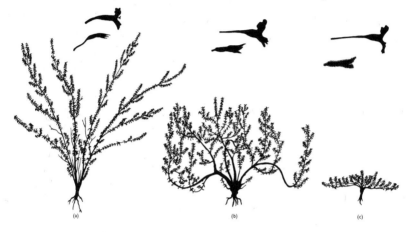

Fig. 62. The form of the plant body and flowers in three races or semi-species of Diplacus (Scrophulariaceae)

(a) *Diplacus puniceus.*   (b) *Diplacus longiflorus.*   (c) *Diplacus calycinus.*
Beeks, 1961, drawings by courtesy of Dr. Beeks.

Diplacus puniceus is a tall and erect shrub with red, short-tubed, long-pediceled flowers which grows in coastal hills and plains near the ocean. *Diplacus longiflorus* is a medium-tall, spreading shrub with salmon-orange flowers of intermediate length; it occurs on sun-beaten foothills in the chaparral zone of the interior southern California mountains. *Diplacus calycinus* grows in rocky places at higher elevations in these interior mountains and in the bordering areas of the Mojave Desert. It is a small, low, spreading plant with yellow, long-tubed, short-pediceled flowers. Intergrades occur between *D. puniceus*

[10] Beeks, 1961, 1962.

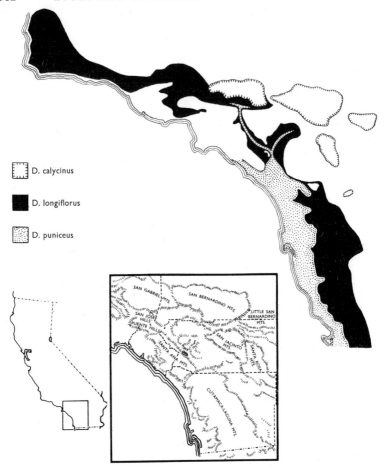

Fig. 63. Geographical distribution of Diplacus calycinus, longiflorus, *and* puniceus *in coastal southern California*

Beeks, 1961.

and *D. longiflorus* along their zone of contact in the western foothills, and again between *D. longiflorus* and *D. calycinus* where they meet at intermediate elevations in the interior mountains.[11]

The colonies of Diplacus are moderately to highly variable in several genetically determined characters of the shoot and flowers.

[11] Beeks, 1961, 1962.

Consider flower color for example. This character can be measured by comparing the corollas with a series of standardized colors; seven shades of red, orange, and yellow can be identified in this way. Colonies of *Diplacus puniceus* are mixtures of individuals with three, four, or even five shades of red, orange, and bronze. And the proportion of individuals belonging to any color group varies from colony to colony.

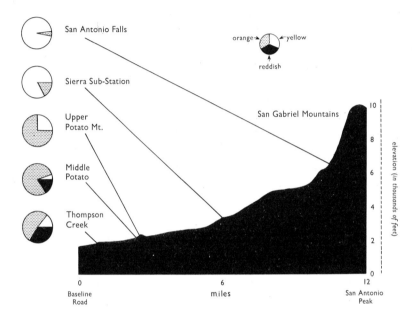

Fig. 64. Proportion of plants with red, orange, or yellow flowers in colonies of Diplacus on an altitudinal transect in the San Gabriel Mountains, California

Redrawn from Beeks, 1961.

Although *Diplacus longiflorus* is predominantly orange-flowered, individuals with cream-colored or reddish-orange flowers occur as minor constituents in many colonies of this population system. The proportion of the individuals belonging to each color group in a colony can be represented as a segment of a pie diagram, and a series of colonies can be represented by separate pie diagrams. Figure 64 shows how *D. longiflorus* intergrades with *D. calycinus* in the San Gabriel Mountains

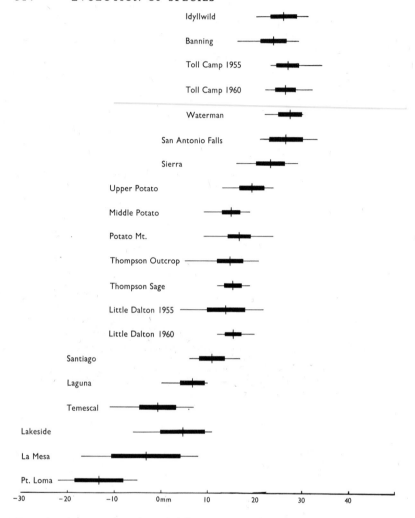

Fig. 65. Variation within and between populations of Diplacus for a quantitative character

The units on the scale express the difference between the length of the corolla tube and the length of the pedicel in millimeters. The colonies are arranged geographically from the coast (at the bottom) to the interior mountains (at top) of southern California. For each colony the length of the horizontal line gives the total range of variation, the verticle crossline the mean, and the horizontal bar the standard deviation of the mean for the character.

From Beeks, 1961.

by relative changes in the frequency of red-, orange-, and yellow-flowered individuals with rise in elevation.[12]

The colony to colony variation in a quantitative character, the difference between the length of the corolla tube and the length of the pedicel, is shown in Fig. 65, where the colonies are arranged on the graph so that the coastal ones are on the bottom and the interior montane ones are at the top. Although each colony is variable in this trait, as shown by the range and standard deviation for the individual measurements, there is a progressive shift in the mode of this variability along a geographical transect.[13]

The Diplacus populations vary simultaneously in several different morphological characters. Pooling the measurements of different characters on each individual plant, Beeks was able to express the aspect of the plant for all characters combined in terms of an aggregate index value. The index values range from 0 to 15. On this scale, a plant with the value of 0 would be an extreme form of *D. calycinus* (as illustrated in Fig. 62c), and a plant with a score value of 15 would be the extreme form of *D. puniceus* (Fig. 62a). Individuals with extreme index values arise in some populations. But the populations themselves always harbor a wider range of forms. The frequency of individuals with different index values in a population can be plotted as a bar graph (see Fig. 66). A series of bar graphs when related to the location of the colonies on the map profile shows a gradual shift in all characters combined from a puniceus combination of features on the coast to a calycinus complex of characters in the higher interior mountains.[14]

This intergradation occurs along most geographical transects but not on all. In certain areas in the San Gabriel Mountains, *D. longiflorus* and *D. calycinus* are neighboringly sympatric and morphologically distinct, and in parts of the Santa Ana Mountains *D. longiflorus and D. puniceus* coexist as non-intergrading sympatric neighbors.[15]

CONTIGUOUS GEOGRAPHICAL RACES

It is evident that "pure races" in the sense of homogeneous population systems which are qualitatively distinct from one another do not exist in Drosophila, Diplacus, or Homo; nor do such "pure races"

[12] Beeks, 1961, 1962.   [13] Beeks, 1961, 1962.   [14] Beeks, 1961, 1962.
[15] Beeks, 1961, 1962.

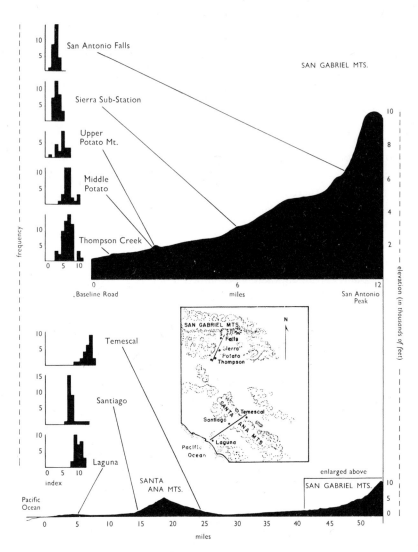

Fig. 66. Variation within and between populations of Diplacus along two transects in southern California for several characters combined

As indicated by the inset, one transect reaches from the coast to an interior valley at low elevations, and the other extends from the foothills to the upper slopes of the San Gabriel Mountains. The bar graphs show the frequency of individuals with different index values in each colony. The scale of index values ranges from 0, representing an extreme *calycinus* combination of characters, to 15, representing the most extreme character combination possible in *D. puniceus*.

From Beeks, 1961, with minor modifications.

316

exist in a variety of other sexually reproducing animals and plants that have been studied. Instead the racial variation in many organisms consists of geographical gradients in the relative frequencies of different alternative forms of genes or supergenes in polymorphic populations. With these facts in mind, Dobzhansky has offered the following definition[16]:

> Races may be defined as Mendelian populations [or breeding groups] of a species which differ in the frequencies of one or more genetic variants, gene alleles, or chromosomal structures.

But, as Dobzhansky points out, which, if any, of the racial groups should be named and recognized in taxonomic practice, and where, if at all, the dividing line should be drawn between races is a methodological problem quite separate from the description of geographical variation as a statistical phenomenon. To quote[17]:

> Statistically significant differences may occur between populations of localities only a dozen miles apart, and, in fact, between populations of stations only a fraction of a mile apart. These populations are, then, racially distinct. . . . Therefore, races are so numerous that they are practically uncountable, and they are not fixed but constantly changing. Although these facts are very important and must never be lost sight of in considering the nature of racial variability, the practical application of the race concept must be somehow restricted if this concept is to remain useful. . . .
>
> As we examine populations of many localities, it sometimes happens that the frequencies of gene and chromosome structures change gradually in a given geographic direction, so that the differences between the populations are proportional to the distances between the localities which they inhabit. Uniform geographic gradients of this kind may connect populations which are profoundly, qualitatively or quantitatively, different in genetic constitution. The end members of the chain of populations are racially distinct, but so are all the intermediate links. A systematist may or may not find it desirable to break the chain of populations into two or more sections and to designate them by racial or subspecific names. If he does so, the divisions are quite arbitrary. . . . On the other hand, the gene frequency gradients may be steep in some regions and relatively level in others. The differences between the populations are not proportional to the distances separating them; the species is broken up into more or less discrete arrays of populations. These arrays can be delimited, counted and named. This is not because the genetic differences between discrete population arrays are necessarily greater than between the end members of a continuous population chain, but solely because the discontinuities eliminate the

[16] Dobzhansky, 1951a, 138, quoted from *Genetics and The Origin of Species*.
[17] Dobzhansky and Epling, 1944, 138–40, quoted from *Carnegie Inst. Wash. Publ.* No. 554.

arbitrariness of drawing the dividing lines between groups of populations. It is convenient to refer to the discrete arrays of populations as races.

In nature, geographic gradients are seldom quite uniform or quite discontinuous. Intermediate situations occur more frequently than the extremes. Within a single species, two races may be easily separable while another two races show only a slight discontinuity. Furthermore, there may be discontinuities of different orders. A species may be split up into two parts showing a considerable break between them. but within one or both of these parts there may exist several minor discontinuities. One investigator may choose to distinguish two major races, and another may describe several minor ones. Which course is adopted is a matter of expedience, judgment, and the conventions which prevail among students at a given time.

Dobzhansky notes that the populations of *Drosophila pseudoobscura* could be grouped into four main geographical races: a Pacific coast race; an intermountain race in the region east of the Sierra Nevada and west of the Rocky Mountains; a race in the Rocky Mountains and Texas; and a Mexican race (see Fig. 61). Or, since important secondary racial differences exist within some of these main geographical groupings, one could subdivide *Drosophila pseudoobscura* into seven or more races.[18]

DISJUNCT GEOGRAPHICAL RACES

We have examined some cases of intergradation between populations living in different areas. The occurrence of a series of intergrading races in nature presupposes the existence of a more or less continuously inhabitable and intergrading territory. In widely separated areas or in different habitats we may then find well-differentiated races; but intermediate areas or ecological zones exist which are inhabited by races with intermediate characteristics.

If, on the other hand, the environment does not vary continuously through space, we would not expect to find a continuous series of intergrades in the population system living in this environment. Instead, we would expect the pattern of geographical variation to exhibit discontinuities coinciding with the gaps in geographical distribution. The environmental conditions conducive to the development of discontinuous geographical variation are frequently met with in nature; these conditions are found in island archipelagos, in mountain ranges, in series of spatially separated lakes, and elsewhere. The population

[18] Dobzhansky and Epling, 1944, 140.

systems living in such places frequently exhibit, as expected, a discontinuous pattern of geographical variation.

A good example is provided by the flycatcher, *Monarcha castaneoventris*, in the Solomon Islands. The races from a central group of islands, including Guadalcanal Island and others, have a black head

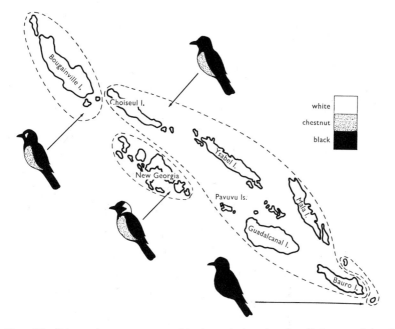

Fig. 67. Discontinuous geographical variation in the Solomon Island flycatcher, Monarcha castaneoventris

The different races have different patterns of black, chestnut, and white as shown.

From E. Mayr, *Systematics and the Origin of Species* (New York, Columbia University Press, 1942) with modifications.

and back and a chestnut belly (see Fig. 67). On other surrounding islands the birds differ from the aforementioned black and chestnut color pattern in the following ways. To the west in the New Georgia group a race is found with a large white patch on the back of the head; to the northwest on Bougainville Island there is a race with a white spot in front of the eye; and at the southeastern end of the archipelago on

Ugi and St. Anna Islands a solid black race occurs (Fig. 67). Intermediate forms are not found between these well-marked races. The absence of intergradation is associated with the disjunct distribution of their respective areas.[19]

Population systems broken into disjunct and discontinuously varying subgroups are common among terrestrial animals and plants living on islands, among aquatic animals living in isolated lakes and streams, and among the plant and animal inhabitants of isolated mountain tops or separate mountain ranges.

With regard to the variation pattern, there is an important difference as well as a similarity between small disjunct races and large contiguous races. Geographical variation can be described in general as a series of shifts in the relative frequencies of alleles and allele combinations within the populations along a geographical transect. In the case of contiguous races, the shifts in frequency of a particular allele or allele combination take the form of gradients. In the case of small disjunct races, on the other hand, the shifts in frequency are likely to be large and abrupt, and the frequencies themselves may approach or attain complete fixation or extinction. Thus a particular allele (a_2) or allele combination (a_2b_2) might change in frequency from 25 percent to 31 percent to 41 percent to 55 percent in a series of contiguous races, but in a series of disjunct races might change in a single step from 0 percent to 100 percent. Owing to gene flow between adjacent populations, in other words, contiguous races are generally polymorphic, as noted in the preceding sections, but in the absence of free gene flow and with the operation of drift, small disjunct races can be and frequently are monomorphic and uniform.

If the morphological differences between disjunct populations are relatively slight, the populations can be treated as geographical races. If the disjunct populations are very different they may be treated as allopatric species. Between these extremes there exist many intermediate degrees of differentiation between the allopatric and non-intergrading populations in which a non-arbitrary assignment to either alternative category, geographical race or allopatric species, is not possible. In such cases it seems desirable to bring the undefinable status of the population systems out into the open by calling them

[19] Mayr, 1942, 81–82.

allopatric semispecies.[20]   The assemblage of allopatric semispecies is known as a superspecies.[21]

ECOLOGICAL RACES

Neighboring habitats that are differentiated in any way, thus adjacent woods or meadows, sandy or clay fields, north- or south-facing slopes, or foothills and valleys, may be occupied by different habitat races. An example is furnished by *Gilia achilleaefolia*.

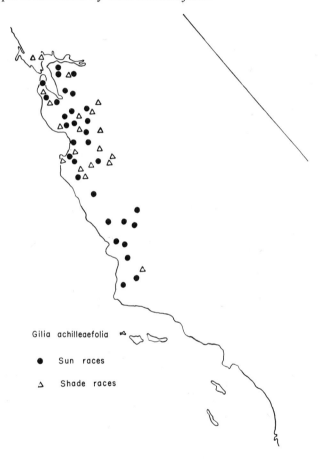

Fig. 68. Ecological races of Gilia achilleaefolia

Grant, 1954*a*.

[20] Following Mayr, 1940.   [21] Mayr, 1942, 169.

The area of distribution of *Gilia achilleaefolia* in the foothills of the California Coast Ranges is shown in Fig. 68, where each dot on the map represents a known colony. Although no two colonies are exactly alike in their morphological characters, as is indicated graphically by Fig. 53 (in Chapter 11), the diverse variations do fall into two major racial groupings. Many populations of *Gilia achilleaefolia* have large

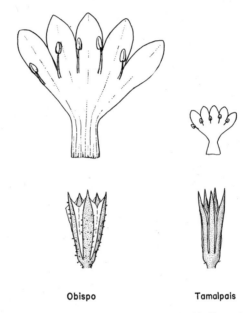

Obispo   Tamalpais

Fig. 69. Flowers of the sun race achilleaefolia *(left) and the shade race* multicaulis *(right)* of Gilia achilleaefolia

Calyx × 5; corolla (spread open) × 2.5.
Grant, 1954a.

flowers clustered in capitate heads (*G. achilleaefolia achilleaefolia*); other populations have small flowers borne in loose, few-flowered cymes (*G. achilleaefolia multicaulis*) (Fig. 69). Flower size and other racial traits in *Gilia achilleaefolia* are genetically determined. The extreme large- and small-flowered forms shown in Fig. 69 are connected by a complete series of intermediates.[22]

The two races occur in different habitats in the same foothill country,

[22] Grant, 1954a.

the large-flowered race (*achilleaefolia*) in open grassy slopes, and the small-flowered race (*multicaulis*) in shady oak woods or thin redwood groves. Both types of habitats are frequent throughout the California foothills, and the sun and shade races lie in juxtaposition as do their respective habitats. It is evident from the map in Fig. 68 that, unlike the situation in *Drosophila pseudoobscura*, the two races of *Gilia achilleaefolia* are both distributed in a scattered fashion throughout the whole area of the species.

As noted earlier, the sun race and shade race intergrade completely in nature. This intergradation takes place, not over any single geographical front, but within numerous separate localities where transitional habitats between shaded woods and grassy slopes are found. There is consequently no morphological discontinuity within *Gilia achilleaefolia* which could serve as a basis for the division of this population system into two sympatric species.[23]

Artificial hybridizations carried out in the experimental garden show that representatives of the different ecological races cross freely, in some combinations at least, to produce fertile and vigorous F_1 and F_2 offspring. The F_1 hybrids and most of the F_2 progeny are intermediate in morphology between the parental types. The intergradation found in nature indicates that some interbreeding between races takes place also under natural conditions. Wherever the distribution of a suitable range of habitats brings extreme parental types into contact and permits the establishment of their intermediate progeny, a series of intergrading forms can develop.[24]

The two races of *Gilia achilleaefolia* differ in still another respect, their breeding system, the sun race being cross-pollinated by bees and the shade race being largely self-pollinating.[25] These differences in mode of pollination reduce the amount of interbreeding between the two races in localities where they are adjacently sympatric. The small-flowered *multicaulis* sets seeds with its own pollen and is rarely visited by bees, which are more strongly attracted to the large showy nectariferous flowers of *achilleaefolia*. The breeding barrier between *achilleaefolia* and *multicaulis* is not absolute, as the presence of intergrading forms testifies, but it is no doubt a factor in the ability of the two races to live in adjacent habitats.

[23] Grant, 1954a.   [24] Grant, 1954a.   [25] Grant, 1954a.

In *Clarkia purpurea* (Onagraceae), another annual plant of the Pacific slope of North America, morphologically different but inter-fertile races, some of which are bee-pollinated and others self-pollinating, may even coexist in the same habitats. This coexistence of the races is apparently temporary, however, for many populations contain an array of intermediate and recombination products derived from occasional crosses between the biotically sympatric races.[26] Such pollination races, as we may perhaps call them, consisting of different outcrossing and self-pollinating forms that are not segregated geographically, seem to be fairly common in many groups of annual plants in the California flora.

Altitudinal races are a special case of ecological race found frequently in mountainous country. Populations of plants and animals generally differ genetically and phenotypically at different elevational zones on a mountain range. In *Drosophila pseudoobscura* in the Sierra Nevada, as we saw earlier, AR chromosomes increase and ST chromosomes decrease in frequency with rise in elevation (Fig. 59). Altitudinal races are well illustrated again by *Diplacus longiflorus* and *calycinus* in the San Gabriel Mountains (Figs. 64 and 66).

Among animals with very narrow food preferences, as for example many parasites and monophagous insects, different races may develop which are specialized for living on different hosts. These so-called biological races frequently reproduce in their feeding sites. If their respective hosts occur in the same locality, therefore, the biological races may be adjacently sympatric without interbreeding freely.[27]

The human louse (*Pediculus humanus*) consists of two interfertile biological races, the head louse (*capitis*) and the body louse (*vestimenti*), which differ in their preferred location on the host and in several morphological characters. The racial differences in morphology are not environmental modifications, but have a genetic basis, as is shown by the fact that head lice transplanted to the body retain their original characteristics during the first generation. However, individuals of race *capitis* when transferred to the body, while remaining *capitis* in form themselves, give rise to progeny which in later generations come to have the characters of race *vestimenti*.[28]

[26] Lewis and Lewis, 1955, 300, 308.   [27] Brief review in Mayr, 1942, 208–11.
[28] Alpatov *et al.*; see Levene and Dobzhansky, 1959, for a review of their findings.

This change from one race to another is accompanied by a high mortality rate, which suggests that natural selection may be involved. It has been proposed as a possible explanation that the louse populations are permanently polymorphic for genetic factors affecting their fitness in alternative parts of the human body. A small group of head lice invading the body, or of body lice finding their way to the head, could thus be transformed by selection in a few generations to the race best adapted to the new habitat. The races can either breed true to type or change rapidly from one to the other depending upon the circumstances.[29]

Seasonal races have been described in the herring (*Clupea harengus*) in the eastern Atlantic Ocean and the North Sea. A number of geographical races of the herring occur in these waters, some of which spawn in the same localities. The bank herring and the spring herring, for example, breed in the same spawning grounds in the southern North Sea, but have different breeding seasons. The spring herring breeds from March to May and the bank herring from August to October. Consequently, although fish belonging to the two races mingle in the same schools in their area of overlap, they retain their separate racial characteristics.[30]

A species of tarweed, *Madia elegans* (Compositae), which grows in the California foothills, contains spring-flowering and fall-flowering races. The spring race develops a small shoot rapidly after the winter rains, goes on to flower from March to May, and dies in summer as its seeds ripen. In the fall race, on the other hand, a long taproot and dense leaf rosette develop during spring, the stem grows tall in summer, and the first flowers appear in August. The seasonal differences between the two races are maintained when the plants are grown side by side in a uniform garden.[31]

The main racial groupings in man are basically geographical in origin. The Indians are connected with America, the mongoloid peoples with east Asia, the Negroes with Africa, the white race with Europe, and so on. In the course of history, however, people have frequently migrated or been carried as slaves far from their homeland

[29] Levene and Dobzhansky, 1959.
[30] Schnakenbeck in 1931; see Mayr, 1942, 198–99.
[31] Clausen, 1951, 50.

and into the territories of other races. The immigrant race and the indigenous race often find themselves in separate social castes or classes. Social contacts and marriage unions are carried out predominantly within, and only clandestinely if at all between, the social races, which thus form separate though sympatric breeding groups.

The distinction between geographical and ecological races is relative. The environment changes along any geographical transect. Different geographical areas thus possess different environmental conditions. The races of a species inhabiting the different geographical areas therefore face the challenge of different ecological conditions. Conversely, although the separate habitats within any single territory are by definition ecologically different, they are also segregated spatially. Ecological races are at the same time microgeographical races. As Mayr states: "All geographical races are also ecological races, and all ecological races are also geographical races."[32]

Nevertheless, there is a real difference in the patterns of racial distribution in species composed of geographical races and those composed of ecological races. Consider the contrast between *Drosophila pseudoobscura*, with its races segregated into different large areas, and *Gilia achilleaefolia* with its intermingled races. Since the races of *Gilia achilleaefolia* are segregated more according to habitat than by geographical region, it is appropriate to recognize them as ecological rather than geographical races. Although geographical and ecological factors both enter into the racial differentiation in each case, the relative importance of the two factors differs in the two cases, and a segregation of races which is due primarily to geographical distance can be distinguished from racial segregation due primarily to habitat differences.

The same considerations apply to the distinctions between geographical races and the other types of non-geographical races—altitudinal, seasonal, biological, pollination, and social—which are special forms of ecological races. Spatial factors and ecological factors are combined in each case but do not always have the same relative effectiveness in determining the racial differentiation. The categories of altitudinal, biological, and social races are attempts to recognize the real situations in which elevation, food niche, or social stratum is more important than geography in channelizing interbreeding.

[32] Mayr, 1947, 280, also 1942, 193.

The distinction between ecological races and sympatric species is not always sharply drawn. We can say that ecologically segregated population systems are ecological races if they intergrade and sympatric species if they do not. But between the extremes of complete intergradation and non-interbreeding there are many intermediate degrees of isolation which do not fall clearly into either the category of ecological race or that of sympatric species. The partially interbreeding, partially isolated, sympatric population systems can, of course, be assigned arbitrarily to one or the other of the aforenamed categories, or (preferably, in my opinion) they can be designated as sympatric semispecies.[33] The assemblage of sympatric semispecies or of hybridizing sympatric species is then called a syngameon.[34]

Diplacus puniceus, longiflorus, and *calycinus* behave as interbreeding ecological races in some areas and as non-interbreeding neighboring sympatric species in others. These facts can best be summarized by referring to the three entities individually as sympatric semispecies and collectively as a syngameon.[35]

FROM RACIAL VARIATION TO SPECIES LIMITS

Races generally intergrade with one another. Racial variation is continuous or slightly discontinuous throughout the geographical area occupied by the population system. When we follow this more or less continuous intergradation, we come sooner or later to some major discontinuity in the variation pattern. Racial variation, in short, is circumscribed within limits. These limits are the boundary lines of species as that entity is defined in most higher animals and in many plants.

The population systems in the world, when considered as a whole, obviously do not form a single continuum; instead they are organized into several million separate continuums. Living organisms, in other words, form a huge discontinuum composed of many species. The discrete population systems can be seen on every hand, whether we survey the members of a genus or family, or the components of a biotic community. In the cat family, for example, we find lions, tigers,

[33] Sibley, 1954; Grant, 1957, 68.
[34] Grant, 1957, 67–68. According to Beaudry (1960, 237) Cuenot used the term syngameon in the same sense earlier, in 1951.
[35] Beeks, 1961, 1962.

leopards, jaguars, cheetas, bobcats, house cats, etc., all of which form discrete breeding units separated from one another by well-marked boundaries. Again a biotic community may consist of pine trees, oak trees, various sorts of shrubs, and herbs; worms, fungi, and bacteria in the soil; many species of insects feeding on the plants; insect-eating birds of various kinds; snakes, hawks, and small mammals preying on the birds; and so on. The discrete breeding units in nature are referred to by most modern evolutionists, following the older naturalists, as species.

As an example we may cite the thrushes of eastern North America as described by Mayr[36]:

The ornithologist unites ... all the smaller thrushes of this region in the genus *Hylocichla*. If we examine the variation within the genus in more detail, we find that it clusters very closely around five means, to which we apply the familiar names wood thrush (*Hylocichla mustelina*), veery (*H. fuscescens*), hermit thrush (*H. guttata*), gray-cheeked thrush (*H. minima*), and olive-backed thrush (*H. ustulata*). All five species are similar, but completely separated from one another by biological discontinuities. Every one of the five species is characterized not only morphologically, but also by numerous behavior and ecological traits. Two or three of them may nest in the same wood lot without any signs of intergradation; in fact, not a single hybrid seems to be known between these five common species.

POPULATION SYSTEMS IN THE COBWEBBY GILIAS

It may prove instructive to consider some of the types of population systems found in a single related group of organisms. The Cobwebby Gilias comprise a natural group of plants belonging to the genus Gilia (section Arachnion) of the family Polemoniaceae. The species to be discussed here occur in the mountains, valleys, and deserts of central and southern California. The physiographic features of this area are shown in Fig. 70. The plants with which we are concerned in the following discussion are all annual, diploid with the same number of chromosomes ($2N = 18$), sexual, and mainly cross-fertilizing, although some self-pollinating types exist. Despite numerous features in common, they exhibit a variety of patterns in their population systems.

Gilia latiflora grows in sandy places in the Mojave Desert and arid interior valleys of the South Coast Ranges. The location of the known colonies is shown by dots on the map in Fig. 71. It is safe to assume that many undiscovered and unmapped colonies of *Gilia latiflora* exist

[36] Mayr, 1942, 148, quoted from *Systematics and The Origin of Species*.

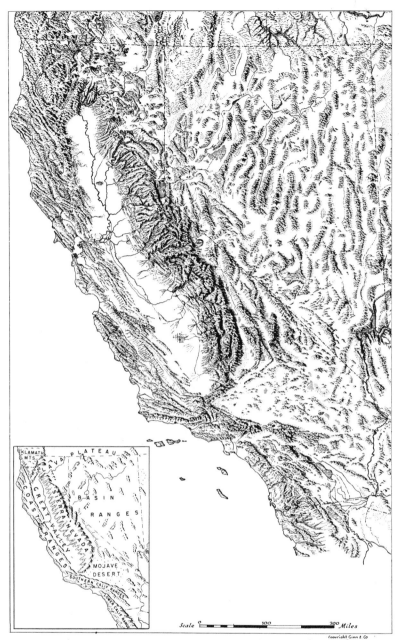

Fig. 70. Physiographic features of California and Nevada

Reproduced from E. Raisz, "Map of the Landforms of the United States," in W. W. Atwood, *Physiographic Provinces of North America* (Boston, Ginn and Company, 1940).

329

in nature, and therefore the total number of colonies in this species is certainly in the hundreds and probably in the thousands.

Well-differentiated geographical races can be distinguished in the different parts of the distribution area of *Gilia latiflora*. The colonies

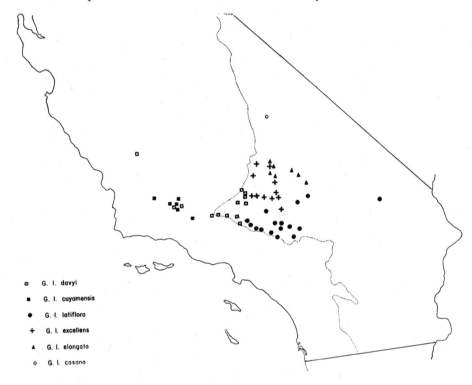

Fig. 71. Geographical and ecological races of Gilia latiflora

The dotted line marks the western boundary of the Mojave Desert.

Grant and Grant, 1956.

in Antelope Valley and the inner South Coast Range valleys form a race characterized by long purple flowers, and bearing the name *Gilia latiflora davyi* (see Fig. 71 and Plate IIa). Another race, *Gilia latiflora excellens*, with very long-tubed, yellow-throated corollas, occurs in the El Paso Mountains in the northern Mojave Desert (see Fig. 71 and Plate IIa). The populations in the drainage area of the Mojave River all have corollas with short tubes and yellow throats; these populations comprise the race known to taxonomy as *Gilia latiflora latiflora*. Complete intergradation occurs within *Gilia latiflora latiflora* from narrow-throated forms in the west to forms with very full corolla throats farther east, as shown in Plate IIa. The race *latiflora* intergrades with both *davyi* and *excellens*, and the latter with one another, where their areas merge.[37]

In addition to the continuously intergrading geographical races just mentioned, *Gilia latiflora* contains two pairs of ecological races. The races *excellens* and *elongata* occur in adjacent habitats in the El Paso Mountains of the northern Mojave Desert, where *elongata* grows on the low mountains and *excellens* grows on the sandy plains fingering into them. In intermediate habitats intergrading forms are found.[38] The second pair of ecological races consists of *davyi* and *cuyamensis* in Cuyama Valley in the South Coast Ranges (Fig. 71).

Far to the north in the Coso Mountains there is a disjunct and non-intergrading race known as *Gilia latiflora cosana* (Fig. 71).

Altogether six races considered worthy of taxonomic recognition are found in *Gilia latiflora*. For the most part the geographical areas occupied by the different races are adjacent and contiguous.[39] It is known from hybridization experiments that these races will cross freely with one another to produce vigorous and fertile hybrids.[40] Interbreeding can and does occur therefore between the races where they meet in nature. And since the areas inhabited by the different races merge more or less continuously into one another, the interbreeding between colonies and between races leads to a continuous series of intergrading forms.

In a related species, *Gilia leptantha*, the geographical areas occupied by the races are disjunct.[41] The southern Sierran race, *Gilia leptantha purpusii*, is isolated from *G. leptantha pinetorum* by a gap 30 miles wide at its narrowest point, and from *G. leptantha leptantha* in the San Bernardino Mountains by a gap 135 miles wide (see Plate IIb). The races of *Gilia leptantha* are known from artificial hybridizations to be capable of interbreeding freely.[42] Their opportunities of interbreeding in nature are greatly restricted, however, by the geographical gaps

---

[37] Grant and Grant, 1956. [38] Grant and Grant, 1956, 205–8.
[39] A. Grant and V. Grant, 1956. [40] V. Grant and A. Grant, 1960.
[41] Grant and Grant, 1956. [42] Grant and Grant, 1960.

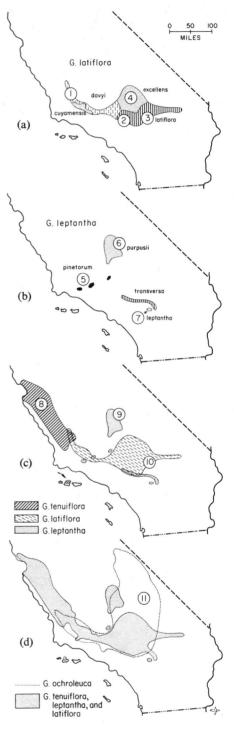

Plate II. Four types of population system, representing four stages of evolutionary divergence, in the Cobwebby Gilias

These plants occur in southern California, the physiographic features of which are shown in Fig. 70. The various kinds of Cobwebby Gilia have the geographical distribution patterns indicated in the maps on the left and the floral characteristics shown on the right. Flowers to same scale, approximately × 2.

(a) Continuously intergrading geographical races. *Gilia latiflora:* (1) subspecies *davyi*, (2) small-throated race of subspecies *latiflora* (3) wide-throated race of subspecies *latiflora*, (4) subspecies *excellens*. Regarding other types of races in *Gilia latiflora* see text and Fig. 71.

(b) Disjunct and distinct geographical races. *Gilia leptantha:* (5) subspecies *pinetorum*, (6) subspecies *purpusii*, (7) subspecies *leptantha*. The area but not the flower of subspecies *transversa* is shown.

(c) Largely allopatric but marginally sympatric semispecies. (8) *Gilia tenuiflora*, (9) *G. leptantha*, (10) *G. latiflora*. These semispecies hybridize in some areas but are non-interbreeding in other areas where they occur sympatrically.

(d) Sympatric species. The *Gilia latiflora-leptantha-tenuiflora* group and (11) *Gilia ochroleuca.*

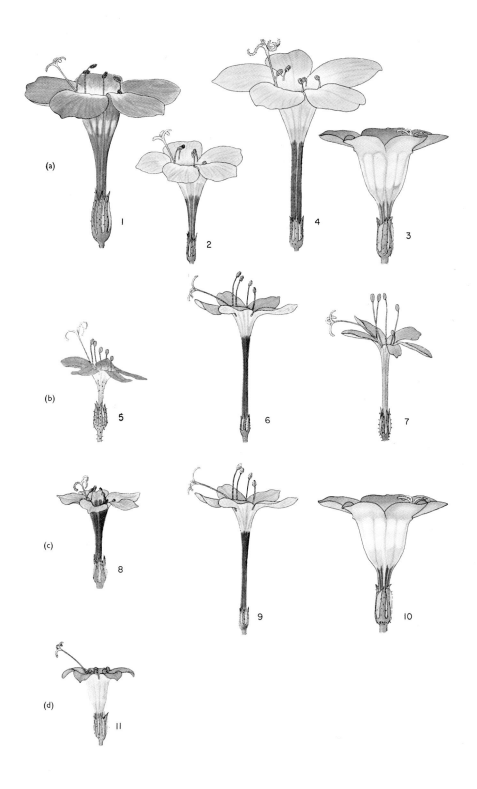

(a)

1

2

4

3

(b)

5

6

7

(c)

8

9

10

(d)

11

between their respective areas. Consequently these races tend to be discontinuous morphologically as well as geographically.

But the geographical barriers between the races of *Gilia leptantha* are not insurmountable. Seeds may occasionally be transported by birds or wind from one disjunct area to another. The migrants could then cross with the resident individuals. The effects of rare interbreeding between neighboring but isolated races can be seen in the partial intergradation between *purpusii* and *pinetorum*.

The extreme form of *purpusii* occurs along the upper Kern River in the southern Sierra Nevada. It is, as shown in Plate IIb, very different from the extreme form of *pinetorum* on Mount Pinos in the South Coast Ranges. But populations of *purpusii* in the Greenhorn Range in the southwestern Sierra Nevada approach *pinetorum* in their morphological characteristics; conversely, the easternmost population of *pinetorum* in the Tehachapi Mountains approaches *purpusii* morphologically. The phenotypic and genetic discontinuity between the two races narrows down where their areas approach each other and hence where migration and interbreeding are most likely to occur.

The different patterns of geographical distribution found in *Gilia latiflora* and *Gilia leptantha* are due to differences in the nature and distribution of the habitats occupied by the two species. *Gilia latiflora* is a plant of open sandy plains and slopes in desert or semidesert country. This habitat has a continuous distribution from the western Mojave Desert to the inner South Coast Range valleys. The distribution of *Gilia latiflora* is also continuous. *Gilia leptantha*, on the other hand, occurs chiefly in yellow pine forests, which in southern California are confined to the middle and higher elevations of separate mountain ranges. The habitats suitable for *Gilia leptantha* are disconnected and so are the populations. Consequently the geographical variation is continuous in *Gilia latiflora* but discontinuous in *G. leptantha*.

Gilia tenuiflora consists of four named geographical races, some of which are contiguous and others disjunct. There is continuous intergradation between a slender-flowered race in the northern part of the South Coast Ranges and a southern race with broader flowers (*Gilia tenuiflora tenuiflora* and *G. t. amplifaucalis*, respectively). Thus the corolla throat ranges from 3.0 to 3.8 mm wide in *tenuiflora* and from 3.8 to 5.0 mm wide in *amplifaucalis*; the corolla lobes are 2.5 to 4.2 mm wide in *tenuiflora* and 4.0 to 5.5 mm wide in *amplifaucalis*. On the other hand, a geographical and morphological discontinuity separates the southern mainland race (*amplifaucalis*) from a race on Santa Rosa Island (*hoffmanni*). The geographical gap can be seen in the map in Plate IIc. The morphological discontinuity is evident in the range of variation in flower size in the two races; the length of the corolla tube and throat varies in *amplifaucalis* from 7 to 11 mm, and from 13 to 14.5 mm in *hoffmanni*.[43]

The three population systems, *Gilia latiflora*, *leptantha*, and *tenuiflora*, are fairly closely related as shown by their possession of many common characters and their ability to form semifertile hybrids.[44] The distribution areas of the three species are different for the most part but have overlapping boundaries (see Plate IIc). Where these population systems coexist in the same territory they usually form separate breeding groups and retain their distinctive morphological features. This is not always the case however. *Gilia latiflora* and *G. leptantha* hybridize locally on the desert slopes of the San Bernardino Mountains. In the South Coast Ranges 100 miles to the northwest, the same two entities coexist without any signs of interbreeding.[45]

Gilia latiflora, *leptantha*, and *tenuiflora* interbreed and intergrade less freely than races but more freely than would be expected of unequivocal species. Being neither typical races nor good species, the named entities should be regarded individually as semispecies. The group as a whole, that is, *Gilia latiflora*, *leptantha*, and *tenuiflora*, is then a syngameon.[46]

The *Gilia latiflora-leptantha-tenuiflora* syngameon can be compared collectively with still another species of Cobwebby Gilia, *G. ochroleuca*, with respect to geographical distribution, morphological characters, and interbreeding. *Gilia ochroleuca* in various racial forms occurs in the Mojave Desert and the pinyon-juniper woodland of mountains west of the desert. As shown in Plate IId, the area of *Gilia ochroleuca* coincides with that of the *Gilia latiflora* group over thousands of square miles. In many localities populations of *Gilia ochroleuca* can be found growing intermixed with or adjacent to populations of *Gilia latiflora* or *G. leptantha*.

[43] Grant and Grant, 1956.   [44] Grant and Grant, 1960.
[45] Grant and Grant, 1956.   [46] Grant and Grant, 1956, 214; Grant, 1957, 63–64.

In the South Coast Ranges this sympatry is genetical but not ecological; it is, in other words, neighboring sympatry. There *Gilia ochroleuca* occurs in pinyon-juniper woodland, while *Gilia leptantha* grows in nearby yellow pine forests and *G. latiflora* in dry washes and valley bottoms in the same mountains. It is both ecological and genetical, it is biotic sympatry, in parts of the Mojave Desert where individuals of *Gilia ochroleuca* and *Gilia latiflora* grow side by side on the open sandy plains.

Gilia ochroleuca is distinguished from *Gilia latiflora, leptantha,* and *tenuiflora* by a number of morphological characters. In both population systems the flowers are borne in pairs on pedicels or stem branches; in *Gilia ochroleuca* the pedicels of a pair are equal or nearly equal in length, whereas in the *G. latiflora* syngameon the two pedicels are very unequal, one being long and the other short. The capsule is globular in *Gilia ochroleuca* and egg-shaped in the *Gilia latiflora* syngameon. The stamens are short in *Gilia ochroleuca* and long in the *Gilia latiflora* group of semispecies. The differences between the two population systems in the length of the stamens and color of the corolla throat can be seen in Plate II. These differences are consistent throughout the entire range of distribution of each group. The morphological discontinuity between *Gilia ochroleuca* and the *Gilia latiflora* group is clear-cut.[47]

It is known from artificial crossing experiments that *Gilia ochroleuca* and the *Gilia latiflora-leptantha-tenuiflora* syngameon are incapable of interbreeding.[48] The discreteness of *Gilia ochroleuca*, and the well defined morphological discontinuity between it and the *Gilia latiflora* syngameon, indicate that the two population systems do not interbreed even where they coexist in nature. There is nothing arbitrary about the decision that *Gilia ochroleuca* forms a distinct sympatric species in relation to *Gilia latiflora, leptantha,* and *tenuiflora.*

Gilia ochroleuca and *Gilia mexicana* are related but entirely allopatric population systems. *Gilia mexicana*, which ranges from southern Arizona and New Mexico to northern Mexico, is separated by a wide geographical gap from *Gilia ochroleuca* in southern California. There is also a prominent morphological discontinuity between the two

[47] Grant and Grant, 1956.   [48] Grant and Grant, 1960.

population systems.[49] From hybridization experiments we know that the two entities cross with difficulty to yield highly sterile F_1 hybrids.[50] *Gilia ochroleuca* and *G. mexicana* are thus a pair of allopatric species.

HISTORICAL AND MODERN APPROACHES TO THE SPECIES PROBLEM

The reality of the species as an organizational and reproductive unit in nature has been a subject of debate in biology for over a century. There have been in general two schools of thought on the species problem among naturalists and evolutionists. Some observers hold that species are objectively real units in nature; others contend that they are artificial categories created by man for purposes of classification. Let us briefly sketch the historical development of the two viewpoints: first the concept that real units exist to which we can assign the name species nonarbitrarily, and second the alternative concept of species as arbitrarily defined categories. Then let us search for a possible ground of common agreement.

It is noteworthy, to begin with, that the existence of discrete population systems in nature is recognized by animals. Various predatory vertebrate animals learn to discriminate between edible and poisonous or noxious species of insects by their form, odors, and appearance, as we saw in Chapter 5, and the higher animals generally discriminate between their own and foreign species for purposes of mating, as we shall see in Chapter 13. A sense of discrimination for the kinds of animals and plants is well developed in primitive man. Thus in New Guinea the native Papuans recognize 137 species of birds and have a distinct name for each. Ornithologists now recognize 138 species in the same region; there are two species of small greenish bush warblers for which the Papuans had only one name.[51]

The organization of population systems into discontinuous units was recognized again by the early naturalists, who applied the term species to the discrete breeding units. As John Ray put it in 1686: "No more certain criterion of a species exists than that it breeds true within its own limits."[52]

[49] Grant and Grant, 1956.   [50] Grant and Grant, 1960.   [51] Mayr, 1955a.
[52] Darlington, 1937.

Linnaeus in the *Critica Botanica* (1737) stated[53]:

All species reckon the origin of their stock in the first instance from the veritable hand of the Almighty Creator: for the Author of Nature, when He created species, imposed on his Creations an eternal law of reproduction and multiplication within the limits of their proper kinds. He did indeed in many instances allow them the power of sporting in their outward appearance, but never that of passing from one species to another.

This view Linnaeus stated again in other words in the *Philosophia Botanica* (1751) and the *Genera Plantarum* (1764). In the latter work he says: "We reckon as many species as there were diverse and constant forms created in the beginning."[54]

Many naturalists after Linnaeus restated the same idea in various forms. Buffon (1749), followed by Kant (1775), defined species as reproducible groups set apart from the rest of creation by sterility barriers.[55] Voigt (1817), Oken (1830), Lindley (1831), Gloger (1833, 1856), and Godron (1853) defined species as groups of individuals which interbreed with one another but not with other species.[56] Thus Gloger stated in 1856: "A species is what belongs together either by descent or for the sake of reproduction."[57]

Darwin in *The Variation of Animals and Plants under Domestication* (1868), in discussing the validity of ranking three fossil forms of cattle (Bos) as species, concluded[58]: "But what is of most importance for us, as showing that they deserve to be ranked as species, is that they co-existed in different parts of Europe during the same period, and yet kept distinct." From the context of Darwin's statement it is clear that he regarded his criterion of species, namely, distinctness under conditions of coexistence, as being widely accepted by the naturalists of that day. Darwin made similar statements in *The Origin of Species* (1859, 1872).

The concept of species as distinct breeding groups within a biotic community was given its first well thought out and consistent formulation by Karl Jordan in 1905.[59] Speaking of a single locality near Göttingen, Germany, Jordan writes[60]:

The three common Pieris of the gardens of Goettingen, the carab beetles of the Hainberg, the physopodes in the flowers of the Botanical Garden, the mice on the

[53] Quoted in translation by Ramsbottom, 1938.   [54] Ramsbottom, 1938.
[55] Buffon, 1749; Kant, 1775.   [56] Mayr, 1957a, 9; Lindley, 1831.
[57] Mayr, 1957a, 9.   [58] Darwin, 1868, ch. 3.   [59] Mayr, 1955c, 1957a, 1957b.
[60] Quoted in translation by Mayr, 1955c 53.

fields of the Weend, the bembids on the sand of the shore of the Leine, they all prove that the living inhabitants of a region are not a chaotic mass of intergrading groups of individuals, but that they are composed of a finite number of distinct units which are sharply delimited against each other and each of which forms a closed unit . . . . The units, of which the fauna of an area is composed, are separated from each other by gaps which at this point are not bridged by anything. This is a fact which can be tested by any observer. Indeed all faunistic activity begins with the searching out of these units. A list of the species that occur in a region is an enumeration of such independent units which with Linnaeus we call species.

The viewpoint of the naturalists was further developed in the early modern period of this century by such vertebrate systematists as Stresemann (1919), Rensch (1929), and Mayr (1940, 1942); by the botanist DuRietz (1930); and by the population geneticist Dobzhansky (1935).[61] Through these pathways the concept of species as isolated and objectively demarcated entities, the so-called biological species concept, has entered the mainstream of modern evolutionary thought. Most modern evolutionists, in other words, continue to use the term species in its traditional sense as referring to the discrete breeding units in nature, reserving the term race for the intergrading groups within a species.

The second school of thought can trace its beginnings to Darwin and Wallace, who argued that there is no essential difference, but only a difference of degree, between species and races, for these entities grade into one another. The dividing line between population systems at the racial and at the specific level can only be fixed arbitrarily, and the definition of these respective categories is consequently a subjective matter. As Darwin put it[62]:

Hence, in determining whether a form should be ranked as a species or a variety, the opinion of naturalists having sound judgment and wide experience seems the only guide to follow. We must, however, in many cases, decide by a majority of naturalists, for few well-marked and well-known varieties can be named which have not been ranked as species by at least some competent judges . . . .

Many years ago, when comparing, and seeing others compare, the birds from the separate islands of the Galapagos Archipelago, both one with another, and with those from the American mainland, I was much struck how entirely vague and arbitrary is the distinction between species and varieties . . . .

Certainly no clear line of demarcation has as yet been drawn between species and

[61] See Mayr, 1957a, 17–22, for literature references.
[62] Darwin, 1859, ch. 2, also 1872, ch. 2.

sub-species—that is, the forms which in the opinion of some naturalists come very near to, but do not quite arrive at the rank of species; or, again, between sub-species and well-marked varieties, or between lesser varieties and individual differences. These differences blend into each other in an insensible series . . . .

From these remarks it will be seen that I look at the term species as one arbitrarily given for the sake of convenience to a set of individuals closely resembling each other, and that it does not essentially differ from the term variety, which is given to less distinct and more fluctuating forms.

The viewpoint expressed by Darwin in *The Origin of Species* was adopted by his immediate followers in the nineteenth century. This same viewpoint, or a more extreme version of it, is held today by many taxonomists dealing with plants, insects, and other groups. Thus Bessey stated in 1908[63]:

Nature produces individuals and nothing more . . . . Species have no actual existence in nature. They are mental concepts and nothing more . . . . Species have been invented in order that we may refer to great numbers of individuals collectively.

Again in 1954 Burma states: "The basic unit of a biological taxonomic system is the individual." All taxonomic units are "arbitrarily erected, man-made constructs," and species in particular are "highly abstract fictions."[64] Davidson maintains[65]:

Species . . . are mental units rather than biological units. The biological units are the individuals and these functioning individuals are interrelated through their phylogenetic lineages.

Bessey, Burma, Davidson, Gates, Gregory, Mason,[66] and others put into printed words a concept of species which is held by a number of taxonomists specializing in plants, insects, protistans, and other groups of organisms. Their viewpoint, while related to that stated by Darwin in *The Origin of Species*, differs from the latter in one important respect. Darwin tacitly accepted the old naturalist's concept of discrete species, although he had to emphasize a different aspect of the problem in order to convince a generation of biologists brought up on Linnaeus and Cuvier of the truth of the evolution theory, whereas Bessey, Burma, Davidson, and others specifically reject the naturalist's species concept. It is necessary, therefore, to deal with Darwin's viewpoint and with that of Bessey, Burma, and Davidson separately. Let us

[63] Bessey; quoted by Mayr, 1957a. [64] Burma, 1954. [65] Davidson, 1954.
[66] Gregory, 1928; Mason, 1950; Gates, 1951.

consider the merits, or lack of merits, of the latter school of thought first.

The nominalistic species concept of Bessey *et al.* contains two tenets: first, individual organisms are objectively real units, and, second, species are not. As regards the first tenet, let us remember that in the many animals, plants, and protista that are dioecious or self-incompatible the single individual is not a self-reproducing unit; the minimum reproductive unit in such organisms is a population of two. And in the numerous organisms which, like Drosophila and man, exist as permanent heterozygotes, the smallest self-perpetuating unit is a population composed of many individuals. In social organisms the individual is to an even greater extent a subordinate member of a higher organization.

As regards the second tenet, the very fact that distinct groupings of organisms, and often the same groupings, are recognized by observers as different as wild animals, primitive humans, local naturalists, and evolutionary biologists shows us that the term species in its traditional usage applies to objective and real entities in nature. Those entities are the breeding groups which are separated from one another by the bridgeless gaps of Jordan. To the extent that such discrete breeding groups exist in the world, the delimitation of species is a phenomenon of nature and not an arbitrary procedure of man. To this same extent the species concept of the naturalists and evolutionists, from Ray to Jordan and Mayr, is not a preconceived notion imposed artificially on nature, but is on the contrary a theoretical generalization based on and derived from a very large number of observations of the organic world.

Why, we might ask, do some students deem it necessary to deny the objective reality of species while affirming the existence of individuals? Is it that the boundary lines of individuals are clear-cut while those of species are poorly demarcated? Is there in fact something wrong with the demarcation of species in nature that gives them an arbitrarily defined status not shared by individuals? Or is the difference in the point of view in regard to species and individuals due to differences in the methods of studying the two entities?

Now no one denies that the task of defining the boundary lines of species encounters real difficulties, some practical and some theoretical, in many particular cases. But let us note that our methods of studying species are calculated to draw attention to difficulties of delineation, while our methods of studying individuals are not. We feel obliged to classify into its species every individual that we find in nature and every specimen in our museum collections. Of course, some individuals in every group, and many individuals in some groups, do not fit into any system of natural species—notwithstanding the injunction in Article II of the Rules of Botanical Nomenclature that every plant belongs to a species—because, as Darwin pointed out, a continuous process of evolution takes place in our world.

If we attempted to apply the taxonomic method consistently and systematically to the component parts of the individual organism we might have difficulties in defining the boundary lines for this organizational unit too. Food, water, and air go into the animal body and excretory products go out into the environment: where precisely do you draw the boundary line to separate the material parts of the body from the material parts of the environment? Animal bodies unite in copulation and divide in reproduction: how again can you decide to which individual each and every cell belongs at every stage in the life cycle? It is fortunate for biology that no one has tried to answer these questions. For the attempt to stick annotation labels on every material particle and cell in the organism-environment complex would no doubt give rise to grave doubts in certain quarters about the concept of the individual. The reality of individual organisms as units of organization would then have to be debated for a century. Some thinkers would conclude that the individual organism is a figment of the imagination.

The cells within a plant body are poorly demarcated units of organization too. True, the cells are normally bounded by cell walls. But these walls are often broken by perforations and in the case of living parenchyma cells are traversed by protoplasmic strands or plasmadesmata which connect together the cells forming a tissue. The tapetal cells which serve to nourish the pollen mother cells in a stamen ultimately break down into a soupy mass, losing their cellular organization altogether. In some organisms like the slime molds (Myxomycetes) a cellular organization is never attained in the vegetative body; instead there is a gelatinous non-cellular mass of nuclei and cytoplasm called a plasmodium which moves and feeds in an amoeboid manner. In other words, cells are not found universally in living bodies, and where they do exist their boundaries may or may not be very definite, yet knowing all this we have not found it necessary to abandon the cell theory.

The species, considered as an inclusive, isolated, and discrete breeding group, is certainly not universal in nature, being absent in asexual groups for example, and where species are found, their discreteness varies greatly from case to case. The species as a unit of organization is probably no more and no less universal and well defined than the individual, cell, gene, atom, or any other unit with which we have to deal.

There is, to be sure, a sense in which the species *is* a universal category of classification. Although species, considered as a type of population system, are absent by definition in asexual and hence non-interbreeding organisms, it is a traditional practice to classify all organisms, sexual and asexual alike, into species. The term species is thus used in two quite different senses: as a *general* term to refer to the kinds of organisms, and in a *special* sense to designate the discrete and isolated population system. Some authors would restrict the usage of the word species to general formal classification, and give the evolutionist another technical term for the inclusive breeding group in nature.[67] Such proposals are not likely to find wide acceptance among evolutionary taxonomists acquainted with the historical background of their subject. In fact the dualistic usage of the term species is of long standing and will probably continue to exist.[68]

Perhaps the semantic impasse can be resolved most simply by adopting appropriate adjectives to distinguish the two usages of the word species. Thus the universal category of classification might be called the *taxonomic species*, and the inclusive breeding group the *biological species*. The adjectives, taxonomic versus biological, while open to some objections on purely etymological grounds, do seem to carry the right connotation. Leaving aside the semantic question, we are of course concerned here with the species as a biological unit.

Let us next compare Darwin's views on the species question as stated in *The Origin of Species* (and quoted earlier in this section) with those of K. Jordan (quoted above). It is interesting that the premodern

[67] I.e., Sonneborn, 1957; Gilmour, 1961, 42.   [68] Mayr, 1957b.

evolutionist Jordan and most modern evolutionists holding similar views are apparently poles apart from Darwin on a basic question like the nature of species. It is even more interesting that Darwin himself states one point of view regarding species in *The Origin of Species* (1859 and 1872) and another in *The Variation of Animals and Plants under Domestication* (1868). (See quotations previously given.) Indeed Darwin tacitly accepts the old naturalist's concept of species in *The Origin of Species*, as a careful reading shows, but *emphasizes* a very different concept in the same work. Herein is perhaps the key to the situation.

Each school of thought, the Darwinian and the Jordanian, is emphasizing one important aspect of the total picture. Interbreeding geographical races and non-interbreeding sympatric species undoubtedly exist in nature. But so do other types of population systems with intermediate geographical and breeding relationships. In the array of population systems in nature, in other words, we find good races, good species, *and* entities possessing some species-like properties combined with some race-like properties. The existence of the borderline cases does not entitle us to conclude that *all* species are arbitrarily defined entities; nor does it follow from the existence of true races and true species that all population systems will fall clearly into these two categories.

The recognition of just two main categories of population systems— the traditional categories of races and species—has hampered taxonomic thinking. The effort to classify the existing array of population systems into either races or species has had the result, whenever the numerous borderline cases came up for attention, of stretching both the race concept and the species concept to the point of terminological looseness. In view of this situation and on the basis of taxonomic experience with both birds and flowering plants, several students of the species problem have proposed to recognize a third category, the semispecies, for the borderline cases.[69]

A semispecies would be a population system which exhibits some species-like properties combined with some race-like properties. Thus semispecies might be more different morphologically and more

[69] Mayr, 1940, 267; Huxley, 1942, 407; Sibley, 1954; Grant, 1957, 68; Valentine and Löve, 1958, 159.

discontinuous in their variation pattern than is usual for races, without however possessing the barriers to interbreeding that are characteristic of true species. Or again, in areas of sympatric contacts, semispecies may interbreed and intergrade less freely than races but more freely than good species.

Races and species are distinguished by their breeding relationships, free interbreeding being characteristic of races and non-interbreeding typifying species. Where intermediate amounts of interbreeding take place between two or more population systems, however, so that referring to the entities in question either as races or as species can only be made by arbitrary decisions based on the breeding criterion, it is proposed to assign the entities to the intermediate category of semi-species. By this procedure the terms race and species can be reserved for those population systems which can be classified non-arbitrarily on the basis of their breeding relationships.

The classification of population systems is further clarified by the recognition that either races, species, or semispecies may be either allopatric or sympatric. This gives us the categories: contiguous geographical races, disjunct geographical races, and ecological races; allopatric or sympatric semispecies; and allopatric or sympatric species.

CLASSIFICATION OF POPULATION SYSTEMS

The array of population systems in nature can be classified on the basis of three criteria. These criteria, of which two have been discussed in this chapter, and the third will be discussed in more detail in the next chapter, all have to do in one way or another with the breeding relationships between population systems.

The first important consideration is whether two or more population systems do or do not interbreed, as inferred from intergradation or discontinuity in phenotypic characters. The second is whether they are allopatric or sympatric. Third, if two or more population systems do not interbreed, it is desirable to know whether or not the isolation is due to specific breeding barriers inherent in the organisms themselves, that is, to reproductive isolating mechanisms. (See Chapter 13 for definition of reproductive isolation.) On the basis of their relative phenotypic variation, relative geographical distribution, and mode of isolation we can classify population systems in the following way:

1. Population systems intergrading continuously in morphological or physiological characters, and hence judged to be interbreeding freely:
Allopatric—*contiguous geographical races*
Sympatric—*ecological races.*

2. Population systems intergrading discontinuously or partially, and judged to be interbreeding on a restricted scale:
Allopatric:
Differentiated morphologically or physiologically to a moderate degree—*disjunct geographical races*

Differentiated morphologically or physiologically to a considerable degree—*allopatric semispecies.*

Sympatric:
Not isolated reproductively—*ecological races*
Partially isolated reproductively—*sympatric semispecies.*

3. Population systems separated by a discontinuity in the pattern of morphological and physiological variations, and evidently not interbreeding:
Allopatric:
Not isolated reproductively—*allopatric semispecies*
Reproductively isolated—*allopatric species.*

Sympatric; reproductively isolated—*sympatric species.*

Related semispecies are, by definition, linked together by limited amounts of gene exchange into breeding groups of a higher order. An assemblage of semispecies is called a superspecies or a syngameon depending on the geographical relationships between the semispecies. A superspecies is a cluster of allopatric semispecies,[70] while a syngameon is a group of sympatric semispecies.[71]

The information which is necessary in order to classify population systems according to their breeding relations is attainable. Whether the information is in fact attained in any given case depends upon the taxonomic methods employed. A taxonomist who is strictly a museum

[70] Mayr, 1942, 169.   [71] Grant, 1957, 67–68.

worker may or may not have enough specimens at hand to distinguish between geographical races and sympatric species. A taxonomist who knows his organisms in both the museum and the field may be able to distinguish between geographical races and sympatric species, but could well be puzzled as to whether a geographically disjunct population is a geographical race or an allopatric species. The taxonomist or geneticist who studies the organisms in the experimental garden or laboratory but not in the field could answer the latter question; but he might not have the information needed for distinguishing between races and sympatric species. The worker who is at once a museum, field, and experimental taxonomist will, of course, be in the best position to classify population systems.

It may be objected that the taxonomist cannot always obtain the information required for a proper classification of population systems. This is particularly the case when the taxonomist is working in little known floras or faunas or with groups that cannot be studied experimentally. Yet given a few morphologically different specimens he may be called upon to furnish identifications. How is he to decide by inspecting a few specimens from a poorly known group whether these specimens belong to the same or to different races or to the same or different species?

Taxonomy like any other branch of biology has its practical as well as its theoretical functions and its exploratory as well as its explanatory phase. It so happens that the same system of naming and describing organisms is used for both practical and theoretical purposes, and for groups in their exploratory phase as well as for thoroughly studied groups. The practical routine task of identification and the exploratory task of preliminary classification often has to be carried out in spite of serious deficiencies in information. But what we are discussing here is the theoretical task of taxonomy to investigate the nature of population systems.

THIRTEEN

Isolating Mechanisms

Isolation, also, is an important element in the process of natural selection.
CHARLES DARWIN[1]

THE PROCESS which holds individuals together in populations, and populations together in races and species, is interbreeding, which leads to gene exchange. Populations and population systems are reproductive communities tied together by bonds of geneological relationship. Mating is not at random, however, between individuals belonging to different races and species. Barriers to interbreeding and to gene exchange exist within and between the larger population systems. These barriers are referred to as isolating mechanisms.[2]

In this chapter we shall first analyze isolation into its basic components. Then we shall survey the various isolating mechanisms in a synoptical fashion, illustrating each type with examples chosen mainly from plants. Finally we will consider the cooperative action of different isolating mechanisms.

THE COMPONENTS OF ISOLATION

The isolating mechanisms which reduce or prevent gene exchange between population systems are of many different kinds. In these varied isolating mechanisms we can distinguish, on the basis of the characteristics of the environment and of the organisms, three components. The three main factors that bring about isolation are spatial distance, the nature of the environment, and the reproductive characteristics of the organisms. The spatial, environmental, and organismic-reproductive components of isolation may be distinguished as follows.

Two or more population systems may be genotypically alike and fully capable of interbreeding, but may live so far apart that gene

---

[1] Darwin, 1959, ch. 4.   [2] Following Dobzhansky, 1937, 1951*a*, 180 ff.

exchange is reduced by the distance between them alone. This is spatial isolation.

The population systems may be adapted to different environments in the same territory, being genotypically different in respect to their ecological conditions, but not necessarily incapable of interbreeding. Gene exchange between the sympatric but ecologically differentiated populations is then limited by the availability of habitats suitable for the existence of the interpopulational hybrids. This is the environmental factor in isolation.

Finally, the organisms belonging to different population systems may differ genotypically in ways related to their ability to interbreed with one another. Thus individual members of the different populations may not be mutually attracted sexually, or they may have different mating seasons, or be intersterile. Where the barriers to gene exchange are due to the genotypically determined reproductive habits and capacities of the organisms themselves, we can speak of organismic, or reproductive, isolation.

The importance of the distinction between spatial and reproductive isolation has been pointed out by Dobzhansky.[3] The characteristics of the environment as a factor in isolation have been emphasized by a number of botanists, especially Kerner in the last century and Anderson in recent times.[4] I have modified and recombined the concepts of previous authors with as few changes in terminology as possible. Thus the botanical concept is introduced into our present context under the name environmental isolation. Dobzhansky's well-known term "reproductive isolation" is retained in essentially its original sense, except that ecological isolation is excluded. The latter is regarded here as a mixture of environmental and spatial components.

It will be noted that the division between spatial and reproductive isolation corresponds to the distinction between population systems which must live in different territories and those which could or do live together. Allopatric population systems which are isolated spatially and in no other way interbreed and intergrade if and when they come into contact. This is the behavior we have seen to be typical of

[3] Dobzhansky, 1951a, 180–81.

[4] Kerner, 1894–95, II, 587–88; Viosca, 1935; Anderson and Hubricht, 1938; Anderson, 1948, 1949; Epling, 1947a. Numerous later references are not cited here.

geographical races. Two or more biotically sympatric populations, on the other hand, must be reproductively isolated in order to maintain their separate genetic constitutions. Reproductive isolation is a characteristic of species.[5]

Environmental isolation is in an intermediate position. Population systems adapted to different habitats in the same area, and isolated by this environmental factor alone, may be either races or species depending on the characteristics of the environment. If the habitats intergrade, the populations will interbreed and intergrade as ecological races. But if the habitats are sharply segregated without intermediate zones, the interfertile population systems will not interbreed freely despite their potentiality for doing so, and may be regarded as species.

It not infrequently happens that two habitats will intergrade in one region but remain distinct in another area occupied by a given system of populations. One and the same population system then behaves as a pair of interbreeding races and as a pair of non-interbreeding species in different parts of its distribution area. Thus *Diplacus puniceus* and *D. longiflorus* intergrade along a broad front in southern California where the coastal plain meets the coastal mountain ranges; but in two areas farther north, where the respective ecological zones of the two population systems are well defined, *Diplacus puniceus* and *D. longiflorus* are neighboringly sympatric but non-interbreeding (Fig. 63 in Chapter 12).[6] These two entities are intergrading races in the southern and sympatric species in the northern part of their area of overlap.

It is useful for analytical purposes to distinguish between purely spatial, purely environmental, and purely reproductive factors in isolation. Any actual case of isolation, however, is likely to involve a mixture of these elements. We shall next consider briefly the possible paired combinations of the three isolating factors: (1) spatial and environmental, (2) environmental and reproductive, and (3) spatial and reproductive.

1. The spatial and environmental components of isolation are combined in actual situations. Insofar as the environment varies with latitude and longitude, different geographical areas differ, not only spatially, but also ecologically. Conversely, neighboring habitats, while

---

[5] Dobzhansky, 1951a, 180–81.   [6] Beeks, 1961, 1962.

different in their ecological conditions, are also separated by micro-geographical distances. A population of plants, insects, or small birds living in a meadow is not likely to mingle freely with a related population in an adjacent woodland, owing to spatial as well as ecological differences between the two habitats.

Nevertheless, on the one hand, we can recognize cases in which the spatial factor greatly outweighs the environmental factor, and on the other hand, cases in which the nature of the environment is the chief effective factor limiting gene exchange. Where the restriction of gene exchange is due primarily to the spatial remoteness of the populations, and only secondarily to the ecological differences between their respective areas, we can speak of geographical isolation. But where the isolation is due primarily to the segregation of different population systems into different habitats within the same geographical area, we have ecological isolation. If spatial and environmental isolation are abstractions, geographical and ecological isolation are their real counterparts.

In Chapter 12 we considered examples of geographically and ecologically (but not reproductively) isolated population systems in Drosophila, Diplacus, Gilia, and other organisms. Geographical isolation exists between the intergrading races of *Drosophila pseudoobscura*, of *Gilia latiflora*, and of *Gilia leptantha*. Ecological isolation separates the intergrading races of *Diplacus longiflorus-calycinus*, *Gilia achilleaefolia*, and *Pediculus humanus*.

The morphologically distinct but interfertile species of sycamore trees, *Platanus occidentalis* and *P. orientalis*, occur in widely separate geographical areas, namely, in the eastern United States and the eastern Mediterranean region, respectively. The climate is very different in the two areas. Although *Platanus occidentalis* and *P. orientalis* differ ecologically, the main factor preventing them from interbreeding under natural conditions is obviously the great distance between the two geographical areas. The effective barrier to gene exchange in this case is geographical and not ecological isolation.[7]

Ecological differences are the most important single factor preventing or reducing the interbreeding between several neighboringly sympatric,

[7] Stebbins, 1950, 199.

interfertile species of oaks in Texas. The following species are restricted to different soil types in the same region: *Quercus mohriana*, on limestone; *Quercus havardi*, on sand; *Quercus grisea*, on igneous outcrops; and *Quercus stellata*, on gravel, clay, or sandy clay. Along the contact zones between different soil types these species may hybridize locally to a limited extent. Elsewhere they remain distinct.[8] Since the environmental component is more important than the spatial component, the isolation is ecological.

2. Environmental barriers are usually combined with reproductive barriers in the case of species living in the same area. For example, two species of spiderwort, *Tradescantia canaliculata* and *T. subaspera typica*, are neighboringly sympatric throughout a large area in the Ozark region of the central United States. These two species are reproductively isolated in two ways: they bloom at different seasons, and their hybrids are partially sterile. They are also ecologically isolated, inasmuch as *T. canaliculata* grows in full sun and *T. subaspera* in rich soil in deep shade. Where cliffs are developed in the Ozark Mountains, the habitats of the two species come close together but remain distinct, and *T. canaliculata* can be found growing on the top and *T. subaspera* at the foot of the same cliff without interbreeding (Fig. 72a). Where erosion has broken the face of the cliff, however, intermediate ecological niches are formed. In such places hybridization on a limited scale between the two species leads to the production of some backcross types which inhabit the intermediate ecological zone (Fig. 72b).[9]

The two species of sage, *Salvia apiana* and *S. mellifera*, occur sympatrically over thousands of square miles in the coastal region of southern California. Although there are ecological differences between the species, *S. mellifera* being best developed in the foothills and *S. apiana* on the outwash plains, the two sages very commonly occur in contact. In such places they are reproductively isolated by at least three mechanisms. Their flowers bloom at different seasons and are normally pollinated by different kinds of insects as we shall see later. Furthermore, the F_1 hybrids are sterile, producing only a small proportion of viable seeds. The reproductive isolation between *S. apiana*

---

[8] C. H. Muller, 1952.   [9] Anderson and Hubricht, 1938.

and *S. mellifera* is not complete, however, and vigorous and semifertile
F$_2$ and backcross progeny can be obtained from artificial hybridizations. In nature the two species usually form distinct, non-interbreeding
population systems; but occasionally hybrid swarms are found. The
hybrids arise in habitats that have been disturbed by man or other
agents. Therefore it appears that the isolation of *Salvia apiana* and *S.
mellifera* in relatively undisturbed habitats is due to environmental
factors as well as to reproductive factors.[10]

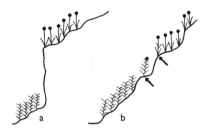

Fig. 72. Ecological relations between Tradescantia canaliculata (*above*)
and T. subaspera typica (*below*)

The former grows on rocky slopes in full sun, and the latter on rich soil in deep shade.
(a) Above and below a cliff the two species are distinct.
(b) In a ravine at the side of a cliff the species hybridize and produce backcross
    progeny (marked by arrows).
Anderson and Hubricht, 1938.

The environmental control of hybridization, which is well documented
by many examples in plants, is becoming recognized also in animals.
Two reproductively isolated species of towhees, the spotted towhee
(*Pipilo erythrophthalmus*) and collared towhee (*Pipilo ocai*) are neighboringly sympatric on the Mexican plateau. The spotted towhee
occurs primarily in oaks and brushy undergrowth, while the collared
towhee inhabits coniferous woods. No evidence of interbreeding
between these species can be found in parts of Oaxaca where the two
vegetation zones are well separated. In other areas in Mexico where
the clearing of forests by man has broken down the original zonation
of vegetation and opened up intermediate habitats, the two kinds of
birds hybridize and produce backcross progeny of various types. *Pipilo*

[10] Epling, 1947*b*; Anderson and Anderson, 1954; K. Grant and V. Grant,
unpublished.

erythrophthalmus and *P. ocai* comprise a pair of sympatric semispecies in some areas and a pair of sympatric species in other regions.[11]

3. The geographical and reproductive components of isolation are always combined in the case of allopatric species. In fact, all three components—geographical, environmental, and reproductive—are usually combined in such cases. An example of largely, though not entirely, allopatric semispecies is furnished by *Gilia latiflora, G. leptantha,* and *G. tenuiflora* (see Plate II in Chapter 12). *Gilia ochroleuca* and *Gilia mexicana* are wholly allopatric, reproductively isolated species (see Chapter 12).

CLASSIFICATION OF ISOLATING MECHANISMS

Let us summarize the preceding discussion and introduce the following topics by means of a classification of isolating mechanisms.

1. Gene exchange between population systems is reduced or prevented by the distance between them—*spatial isolation.*

   a. The populations live in different territories, viz., are allopatric— *geographical isolation.*

2. Gene exchange between population systems is limited by the availability of habitats suitable for the growth and subsistence of their hybrid progeny, that is, by environmental selection against the hybrids and hybrid derivatives—*environmental isolation.*

   b. The populations live in different habitats in the same region, being neighboringly sympatric—*ecological isolation.*

3. Gene exchange is restricted or blocked by genotypically controlled differences in the reproductive habits and fertility relationships of the individual organisms belonging to different population systems— *reproductive isolation.*

   I. The barrier to gene exchange lies outside the bodies of the organisms and is effective prior to fertilization. The interbreeding between individual members of different population systems is blocked by their external phenotypic characteristics, by the structural, physiological or behavioral differences between them—*external reproductive isolation.*

[11] Sibley, 1954.

c. The male and female reproductive organs, the flower parts or genitalia, of individual members of different population systems are not well coadapted structurally; therefore interspecific pollination or copulation is unsuccessful—*mechanical isolation.*

d. Psychological and behavioral factors oppose the union of male and female individuals or gametes belonging to different population systems—*ethological isolation.*

e. The period of sexual activity, the mating or flowering period, occurs in different seasons of the year or different times of day—*seasonal isolation.*

f. In organisms with external fertilization, the free-living male and female gametes produced by different population systems are not attracted chemically to one another—*gametic isolation.*

   II. The barrier to gene exchange operates inside the bodies of the organisms after the gametes or gametophytes come together. The block results from the unsuccessful interaction between the gametophytes, gametes, chromosomes, or genes of the different population systems—*internal reproductive isolation.*

g. In animals and plants with internal fertilization, cross insemination or pollination takes place between members of different population systems, but fails to lead to the production of hybrid progeny—*incompatibility.*

h. The hybrid is formed but is more or less inviable—*hybrid inviability.*

i. The F_1 hybrid grows to maturity but either its sex organs or its gametes are abortive and non-functional—*hybrid sterility.*

j. The hybrid produces F_2 or backcross progeny which, however, consist partly or entirely of inviable or sterile individuals—*hybrid breakdown.*

The classification of reproductive isolating mechanisms follows fairly closely that of Dobzhansky.[12] The species-separating mechanisms can be subdivided somewhat naturally into those that act externally to the individual organisms and prior to fertilization, and those that act within the individual organisms from the stage of insemination or

---

[12] Dobzhansky, 1951a, 181.

pollination onward.[13] The internal isolating mechanisms may operate in the parental generation (incompatibility), the F_1 generation (hybrid inviability or sterility), or in later generations (hybrid breakdown).

As Dobzhansky states[14]:

Considered physiologically, the agents which hinder or prevent the interbreeding of species have scarcely a common denominator. And yet, they have the same genetic effects: curtailment or stoppage of the gene exchange between populations.

MECHANICAL ISOLATION

Related species of plants frequently differ in their flower structures. It was postulated by several authors that these structural differences might sometimes operate to prevent cross-pollination between species. This prediction was made by Dobzhansky in 1937 for the orchids, Papilionaceae, and other plant families with complex floral mechanisms, by Epling in 1947 for Salvia, by Stebbins in 1950 for the milkweeds (Asclepiadaceae) and orchids including the genus Ophrys, Kullenberg in 1950 for Ophrys, and by Grant in 1949 for a wide variety of plant groups.[15]

The conditions for mechanical isolation in plants, and its relation to ethological isolation were summarized by Grant as follows[16]:

Interspecific pollination in a mixed population consisting of two intercompatible species of angiosperms, which are normally pollinated by animals rather than by wind or water, may be prevented in one of two ways. Either the floral mechanisms of the two species may differ in certain details so that the animals which pollinate one species are unable to enter the flowers of the other species, and, if they succeed in making their entrance, fail to touch the stigmas with pollen. Or it may happen that the pollinating animals themselves confine their visits to one kind of flower, which they recognize by its form and markings, and do not stray from species to species; this type of behavior has long been recognized in certain insects, particularly bees. . . . We may accordingly distinguish between barriers to interspecific pollination which arise from the floral mechanism itself, and those which owe their effectiveness to the habits of the pollinators. If the prevention of interspecific pollination comes about as a result of the structural contrivances of the flower, we have mechanical isolation . . . if cross-pollination between two species is mechanically possible, but does not occur owing to the constancy of the pollinating animals to one kind of flower, we may speak of ethological isolation. . . .

[13] Following Stebbins, 1950, 196, approximately. [14] Dobzhansky, 1951a, 180.
[15] Dobzhansky, 1937, 245; Epling, 1947b; Stebbins, 1950, 210–13; Grant, 1949.
[16] Grant, 1949, 82–83, quoted from *Evolution*.

Either mechanical or ethological isolation presupposes a certain level of complexity of both the flower and its agent of pollination. The structural complexity of the flower has the dual function of debarring all but certain types of animals access to the stores of nectar and pollen, and of rendering pollination impossible by those unwanted visitors that do find their way into the flower. The complexity of the animal vector of pollination consists of morphological adaptations for working the floral mechanism, sucking the nectar, and collecting the pollen, and, in some insects at least, also of specialized instincts of flower constancy, etc., which improve the efficiency of the worker. Flowering plants pollinated promiscuously by wind, water, or unspecialized insects will probably be incapable of developing barriers to interspecific pollination. We may expect the operation of floral isolating mechanisms to be confined, therefore, to those groups of angiosperms possessed of flowers sufficiently complex to insure their non-promiscuous pollination.

A survey of plants belonging to numerous genera and families with respect to their mode of pollination revealed an interesting correlation. In plant groups pollinated chiefly by specialized flower-visiting animals, such as birds, Lepidoptera, bees, and long-tongued flies, the related species of plants differ to a large extent in their floral characters. Thus between 37 percent and 54 percent of the taxonomic characters used to separate related species of nonpromiscuous angiosperms pertain to the corolla, stamens, and pistil. By contrast the promiscuously pollinated angiosperms are not differentiated into species predominantly on floral characters. Differences in floral mechanisms comprise only 15 percent of the taxonomic characters distinguishing related species of plants pollinated promiscuously by miscellaneous insects, and only 4 percent of the taxonomic characters in wind-pollinated angiosperms.[17]

The relatively high frequency of interspecific differences in the floral mechanism among plants with specialized methods of pollination suggests that mechanical and ethological isolating mechanisms could operate in such plant groups. And since complex floral mechanisms are found in many genera and families, mechanical and ethological isolating mechanisms could also be expected to occur widely throughout the angiosperms.[18]

In recent years a number of concrete cases of mechanical isolation in plants have been discovered and analyzed. A case was also described previously by Robertson in a forgotten paper of 1887.[19]

The two main species groups of columbines in western North

[17] Grant, 1949.   [18] Grant, 1949.   [19] Robertson, 1887; Holm, 1950.

America have very different floral mechanisms. *Aquilegia formosa* and its allies have nodding red and yellow flowers with short stout spurs (Fig. 73). The spurs which are 1 or 2 cm long contain nectar at the tips. Hummingbirds, hovering on the wing below the flowers, probe into the spurs for this nectar, the bird's bill being about the same length as the spur. In feeding, the head of the bill brushes the exserted central

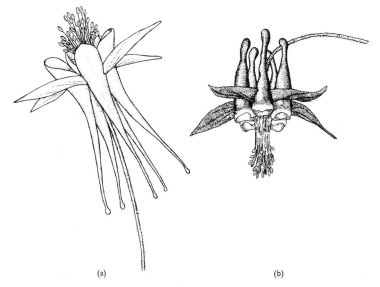

(a)   (b)

Fig. 73. Flowers of two species of Aquilegia

(a) *Aquilegia pubescens*   (b) *Aquilegia formosa truncata*
Life size. Grant, 1952*b*.

column of stamens and styles, and as the bird moves about within the plant population it carries pollen from one flower to the styles of another.[20]

The second group of western American columbines, consisting of the allopatric entities known as *Aquilegia chrysantha, longissima, pubescens,* etc., is characterized by erect pale yellow flowers with very long thin pendant spurs (Fig. 73). The usual pollinators of the *Aquilegia chrysantha* group are hawk moths (Sphingidae). While hovering above the flowers, these insects insert their long slender probosces into the spurs

[20] Grant, 1952*b*.

for nectar, and touch the stamens and styles with their heads. *Aquilegia pubescens* in the Sierra Nevada is visited and pollinated by the sphingid *Celerio lineata*; *Aquilegia chrysantha* in Arizona by *Celerio lineata* and *Phlegethontius sexta*; while *Phlegethontius sexta* and *P. quinquemaculatus* are the most probable but as yet unobserved pollinators of *Aquilegia longissima* in Texas and Mexico. The proboscis of the moth is in each case long enough to reach the nectar at the base of the floral spurs, as is evident from the following comparisons of spur lengths and proboscis lengths[21]:

Spur length (in cm)	Proboscis length (in cm)
A. pubescens, 3–4	C. lineata, 3–4.5
A. chrysantha, 4–7	C. lineata, 3–4.5
	P. sexta, 8.5–10
A. longissima, 9–13	P. sexta, 8.5–10
	P. quinquemaculatus, 10–12

Hummingbirds cannot obtain nectar from the extremely long-spurred forms of Aquilegia and do not try. Hawk moths have not been observed feeding on the short-spurred species, and if they did they would stand off so far from the stamens and styles, owing to the length of their proboscis, that they would pick up little or no pollen. The two groups of columbines are thus mechanically isolated. Although they are interfertile, representatives of the two species groups grow close together over large areas in the American southwest without interbreeding.[22]

The situation as regards the moderately long-spurred form, *Aquilegia pubescens*, is not so clear-cut. The nectar in the medium-long spurs of this species seems to be just within the reach of the hummingbird's bill and tongue. Hummingbirds may or may not feed on *A. pubescens* depending on the circumstances. It is significant that *A. pubescens* hybridizes freely with *A. formosa* in nature. The lowered effectiveness of the mechanical isolation barrier in this case is associated with an increased gene exchange between the two species.[23]

Mechanical isolation may be brought about by floral structures adapted to other combinations of pollinating animals. Three marginally sympatric species of Penstemon in southern California are pollinated

[21] Grant, 1952*b*, and unpublished. [22] Grant, 1952*b*.
[23] Grant, 1952*b*, and unpublished.

by hummingbirds, carpenter bees, and wasps, respectively: *P. centranthifolius* by Anna and Costa hummingbirds, *P. grinnellii* by Xylocopa, and *P. spectabilis* by Pseudomasaris.[24] The mechanical isolation of *Salvia mellifera* and *S. apiana* is due to floral mechanisms differentiated in relation to bees of very different size and body weight, *S. mellifera* being pollinated by medium-sized bees (Osmia, Apis, etc.) and *S. apiana* by large carpenter bees (Xylocopa).[25] Different species of the orchid Stanhopea are pollinated in Ecuador by different species of the same genus of bees, *Stanhopea tricornis* by *Eulaema meriana* ♂♂, and *S. bucephalus* by *Eulaema bomboides* ♂♂.[26]

Asclepias sullivantii, A. verticillata, and *A. longifolia* are pollinated by the same bumblebees, but different body parts of the same bee pick up pollen and contact the stigmatic chamber in the different species of flower.[27] Similarly, the two species of Pedicularis, *P. groenlandica* and *P. attollens*, in the Sierra Nevada of California are both pollinated mainly by the same kinds of bumblebees, *Bombus bifarius* and others; but the conformation of the flowers is such as to bring about pollination by the bee's head on *Pedicularis attollens* and by its abdominal venter on *P. groenlandica*.[28]

A fair number of cases are thus known now of plant species that grow together, bloom at the same time, and are more or less interfertile, yet do not hybridize in nature because of mechanical isolation. Besides the examples given above, such cases have been described in the orchid Ophrys in north Africa,[29] Mimulus (Scrophulariaceae) in California,[30] and inferred for the milkweed Sarcostemma in Mexico.[31] The mechanical isolation may be partial or complete. As expected, the mechanically isolated plants all possess complex floral mechanisms which are specialized in each species for a relatively restricted group of pollinating animals.

Mechanical isolation has been postulated also for the insects, in which related species generally differ in the structure of their genitalia. Apparently relatively few cases have been worked out in detail. Partial mechanical isolation exists between two species of butterfly in Europe,

[24] Straw, 1956.   [25] K. Grant and V. Grant, unpublished.
[26] Dodson and Frymire, 1961.   [27] Robertson, 1887; Holm, 1950, 498.
[28] Sprague, 1959, 1962.   [29] Stebbins and Ferlan, 1956.
[30] Nobs, 1954.   [31] Holm, 1950.

Erebia cassioides and *E. nivalis*. Structural relations between the sex organs interfere with intromission in the cross *nivalis* ♀ × *cassioides* ♂ (but not in the reciprocal cross), and the males break away from the females before copulation is completed.[32] In copulations within and between *Drosophila pseudoobscura* and its close relative *D. persimilis*, less sperm is delivered by the males to the females in interspecific crosses than in intraspecific matings. Intromission is apparently interfered with in some way in the interspecific copulations.[33] The shape of the penis is different in the two species.[34] In tsetse flies an accessory part of the male genitalia of one species pierces the abdomen of females of another species in hybrid matings, causing the death of the females, but if this spikelike organ is clipped off copulation is normal and hybrids are produced.[35]

ETHOLOGICAL ISOLATION

In the higher animals copulation is preceded by courtship during which the male and female individuals become mutually stimulated sexually. The mating urge is aroused by a variety of stimuli. The visible features of the animals including their color pattern and special adornments, their specific scents and mating calls, and their courtship performances all come into play.

Male and female animals of the same species are coadapted psychologically so that they respond sexually to the recognition signals and courtship behavior which are normal for their species. The visible appearance, odors, calls, and behavioral traits generally differ from one species of animal to another. The urge to mate is aroused only weakly or not at all in foreign combinations of males and females. The array of distinctive color marks and behavior patterns relating to courtship forms a very effective and important barrier to interspecific matings among the higher animals, as first pointed out by Wallace in 1878.[36]

For example, zebras are reluctant to mate with horses until they become accustomed to their different color patterns.[37] Attempts to induce copulation in the laboratory between closely related species of spiders (Agelenopsis) with similar appearances and courtships were

---

[32] Lorković, 1958. [33] Dobzhansky, 1947c. [34] Rizki, 1951.
[35] Vanderplank; see Patterson and Stone, 1952, 545.
[36] Wallace, 1878, 1889, 217–27. [37] Lull, 1947.

altogether unsuccessful.[38] The species of fish *Xiphophorus maculatus* and *X. helleri* hybridize rarely in aquaria, but no hybrid has been found among 7,000 specimens collected in nature from waters where they occur sympatrically. Several differences in the courtship patterns of the two species keep them isolated in nature.[39]

Mallard and pintail ducks can be crossed in captivity to yield vigorous and fertile F_1 and F_2 progeny. But natural hybrids of mallards and pintails are exceedingly rare, occurring in about 1 out of 100,000 birds, despite the fact that the two species breed in the same marshes, ponds, and creeks over the entire Northern Hemisphere. Where males and females of these two species of ducks meet in the wild they pay no attention to one another; they just don't like each other.[40]

Some birds have an engagement period before the first copulation. In the European yellow hammer (*Emberiza citrinella*) this engagement period lasts about six weeks during which time the birds may change partners several times until the final copulating pair is formed. Since the birds can identify members of their species by their characteristic appearance, behavior, and calls, the lengthy engagement period provides good opportunity for preventing hybridization with foreign species.[41] Wild hybrids are much less common in birds with preliminary pair formation than in birds lacking an engagement period.[42]

Mayr has pointed out that those groups of birds in which copulation is not preceded by an engagement period, such as hummingbirds (Trochilidae), grouse (Tetraonidae), and manakins (Pipridae), have elaborate courtship procedures associated with colorful breeding plumage. The distinctive colors and adornments of the male birds in such groups and their well-developed courtship behavior, which are well differentiated from species to species, enable the sexes of each species to recognize one another.[43]

In combinations of males and females belonging to different species, a reluctance to mate is commonly exhibited by both sexes. Or one sex in particular may reject the foreign mates. In some groups of animals, such as poeciliid fishes, the males discriminate.[44] Usually, however, it

[38] Gering, 1953.   [39] Clark, Aronson, and Gordon, 1954.
[40] Phillips; see Mayr, 1955*a*, 11–12.
[41] Diesselhorst, 1951; I am indebted to Mayr for this reference.
[42] Mayr, 1942, 255.   [43] Mayr, 1942, 255, 260.
[44] Haskins and Haskins, 1949.

is the female sex that exercises discrimination. This is the case *inter alia* in Drosophila.

Courtship in Drosophila involves visual, chemical, and mechanical stimuli. The flies recognize each other by sight, by smell—the receptors of which are located on the antennae—and by the perception of tapping motions and sounds. The recognition marks and behavior patterns differ perceptibly from species to species. Consequently a given fly can distinguish between sexes and between species by its various sensory perceptions of the characteristic recognition features and courtship performances of other flies. A Drosophila then responds sexually to the stimuli it receives from conspecific mates but does not respond to individuals of the opposite sex belonging to a foreign species.[45]

When males and females of *Drosophila pseudoobscura* and *D. persimilis* are intermixed in a laboratory cage, they copulate entirely or predominantly in intraspecific combinations. Thus in one experiment 90 percent of the females were inseminated by conspecific males while only 10 percent of the females were inseminated by foreign males.[46] The two closely related species, *D. pseudoobscura* and *D. persimilis*, are isolated ethologically.

The males and females of these two species can be placed together in various combinations in order to ascertain the proportion of conspecific and interspecific copulations under different conditions. For example, females of *D. pseudoobscura* can be grouped with males of both species, putting the choice of mates up to the female. Or the choice can be left to the males by combining the males of one species with the females of its own and the alien species. In experiments where the male flies have a multiple choice the ethological isolation is only partially effective. But where the females are given a choice of two kinds of males, they exhibit a very marked preference for the conspecific males. The strong ethological isolation between *Drosophila pseudoobscura* and *D. persimilis* is due primarily to the discrimination and choice of the females.[47]

The inhibitions of female Drosophilas against mating with foreign males can be broken down by environmental conditions that reduce

---

[45] Spieth, 1952.   [46] Streisinger, 1948.   [47] Merrell, 1954.

their sensory perceptions and powers of discrimination. Normal females of *Drosophila persimilis* always reject males of the distantly related *D. melanogaster* and nearly always reject males of the closely related *D. pseudoobscura*. However, the females of *D. persimilis* when under the influence of ether mate freely with either *D. melanogaster* or *D. pseudoobscura* males.[48] Females of *D. melanogaster* reject dark-colored or "ebony" males of the same species in the light but accept them in the dark.[49] The ethological isolation between *D. pseudoobscura* and *D. persimilis* disappears almost completely when the antennae of the female flies, which are the organs of scent, are removed.[50]

Bees, hawk moths, hummingbirds, and other animals which feed on complex flowers must learn how to work the floral mechanism—how to enter the flower and where to find the nectar and pollen—before they can gather food successfully. A flower-feeding animal, once it has learned how to work a given floral mechanism, can thereafter obtain more food in less time by continuing to visit other flowers of the same type. It can feed more efficiently by visiting different flowers of the same species repeatedly than by going randomly from one plant species to another. The selective advantage of flower-constant behavior, which has led to its development in the more highly specialized pollinating animals, is to be sought in the superior efficiency of those animals which get their food by repeated visitation to a single type of complex floral mechanism.[51]

The more specialized flower-visiting insects possess instincts of flower constancy impelling them to feed preferentially on one kind of flower for as long a time as they can successfully obtain food from it. They learn to recognize a species of flower by its distinctive form, color, and scent. The habits of flower constancy and the sensory perceptions for carrying out these habits are highly developed in most bees and hawk moths and are present in a somewhat lower degree in many butterflies and long-tongued flies. The behavior of flower-visiting birds is more capricious than that of insects. Yet hummingbirds can often be seen to visit systematically one species of flower at a time with only occasional experimental forays to other neighboring species.[52]

[48] Streisinger, 1948.   [49] Rendel, 1951.   [50] Mayr, 1950.
[51] Darwin, 1876, ch. 11; Grant, 1950*a*, 389, 392.
[52] Grant, 1949, 1950*a*, for review and literature references.

Related species of plants which are pollinated by specialized flower-visiting animals usually differ in their floral characters. Some of these floral differences between plant species operate as mechanical isolating mechanisms, as we saw in the preceding section. Other floral characters differentiating plant species have no evident mechanical function. For example, *Daphne alpina* is vanilla-scented; *D. striata*, lilac-scented; *D. philippi*, violet-scented; and *D. blagayana*, clove-scented.[53] Related species of plants frequently differ in the color pattern and shape of the petals. Floral characters such as these probably often serve as recognition features for flower-constant pollinators.[54]

The flower-constant behavior of pollinating insects and birds, whether due to instinct, choice, or both, sets up a system of assortative mating in the plant population. Cross-pollinations tend to be channelized within each plant species and reduced between species. Ethological isolation of plant species may arise as an incidental result of the flower-constant behavior of specialized pollinating animals.[55]

This incidental result is most likely to occur in situations where flower constancy is most advantageous for the pollinating animals. It is most likely to occur, therefore, between plant species with complex floral mechanisms, with well-marked distinguishing features on the flowers, and with extensive populations of individuals well stocked with floral food. Bees, hawk moths, hummingbirds, and other specialized flower-feeding animals will not go hungry in order to remain faithful to a certain plant species, nor are their flower-constant habits expressed fully in visitations to plant species with simple and open types of flowers. Interspecific pollinating visits are frequent, therefeor, and the ethological isolation is correspondingly weak or non-existent, where the floral mechanisms are simple and easily worked without a learning process, or where the numbers of individual plants and the supply of floral food are small.

The circumstances in sympatric populations of flowering plants may or may not be favorable for flower-constant behavior on the part of the pollinating animals. A wide range of conditions with respect to

[53] Kerner, 1894–95, II, 203.
[54] Grant, 1949, 1950a. I have since found that Wallace stated the same idea much earlier (1889, 318–19).
[55] Wallace, 1889, 318–19; Grant, 1949.

the randomness or assortativeness of pollination can be found in nature. Consequently ethological isolation in plants ranges from complete to partial to non-existent in different cases. We will consider here a few cases where ethological isolation exists.

Pedicularis groenlandica and *P. attollens*, which occur sympatrically in the Sierra Nevada of California, are both pollinated by bumblebees, mainly *Bombus bifarius*. Although the two species grow and bloom together in the same alpine meadows and are pollinated by the same species of bumblebees, no hybrids between them have ever been found. Part of the isolation is mechanical as noted in the preceding section. The mechanical isolation is reinforced by ethological isolation. Sprague observed mixed colonies of the two species at two sites in the Sierras during two field seasons. Some individual bees of *Bombus bifarius* worked the flowers of *Pedicularis groenlandica* exclusively; other individual bees of *Bombus bifarius* gathered food exclusively from the flowers of *P. attollens*. Since plants of *P. groenlandica* and of *P. attollens* were closely intermingled, growing within feet or inches of one another, the bees feeding on one Pedicularis species often had to pass over or avoid the other species in making their rounds. In no case was an individual bee observed to go from one species of Pedicularis to the other.[56]

Similar observations have been made in experimental gardens where related species or races of plants are grown intermixed and are pollinated selectively by honeybees (*Apis mellifica*). The cross-pollinations carried out by the bees were predominantly within each species or race in mixed plantings of Antirrhinum, Gilia, Brassica, Clarkia, and Papaver.[57] Assortative pollination of different species of Lamium by the bee Anthophora, of different floral morphological forms of the orchid *Dactylorchis fuchsii* by bumblebees, and of different color forms of *Lantana camara* by butterflies has also been reported.[58,59]

---

[56] Sprague, 1959, 1962.

[57] Mather, 1947, for Antirrhinum; Grant, 1949, for Gilia; Bateman, 1951, for Brassica; Lewis and Lewis, 1955, 249, for Clarkia; McNaughton and Harper, 1960*a*, for Papaver.

[58] Bennett in 1874 for Lamium; see Grant, 1950*a*. Heslop-Harrison, 1958, for Dactylorchis; Dronamraju, 1960, for Lantana.

[59] Review of earlier examples by Grant, 1950*a*. See also the interesting discussion by Heslop-Harrison, 1958.

Ethological isolation in flowering plants is a vicarious affair depending on the feeding habits of pollinating animals. This type of barrier is consequently much weaker in plants than in higher animals.

SEASONAL ISOLATION

The flowering season in most plants and the mating season in most animals other than man occurs only in certain periods of the year. Related species or races in both plants and animals frequently differ in their breeding season. The seasonal differences are often partial and overlapping, as where the peak of sexual activity is earlier by several weeks in one population system than in another, but their ranges overlap so that some members of each population system are in flower or in heat simultaneously. The seasonal isolation of the two population systems is then of course partial. It may also happen that two or more population systems differ, not only in their modal periods of sexual readiness, but also in their ranges. If the breeding seasons of the populations do not overlap at all, the seasonal isolation is complete.

In pine trees the female cones are receptive to pollen during a short period. This receptive period of the female cones lasts five days in *Pinus nigra* and *P. montana* and seven days in *P. sylvestris*. Pollen is shed from the male cones during a somewhat longer but still relatively short span of time.[60] The brief period of pollination in pines occurs in the spring time and is geared to the climatic conditions which are normal for each race and species.

The reproductive season in pines varies geographically within wide-ranging species, coming earlier in southern than in northern races and earlier again in low-altitude as compared with high-altitude races. Although pine pollen is blown long distances by the wind, therefore, the long-range dispersal of the pollen in either a latitudinal or altitudinal direction may not lead to wide cross-pollinations on an extensive scale. On the basis of the known duration of the receptive period of the cones in European pines, and the known rate of advance of the season in central Europe, the latitudinal distances that will bring about partial or complete seasonal isolation between pine trees can be estimated.[61]

In the case of pine populations growing closer than 100 km apart in a north-south direction, the wind-borne foreign pollen may arrive in

[60] Wettstein and Onno, 1948.   [61] Wettstein and Onno, 1948.

time to effect some cross-pollinations, but the great bulk of the seeds will result from pollination by native pollen. As between neighboring races of pines, therefore, a partial but pronounced seasonal isolation exists. Pollen grains blown 100 km or more from a parental tree in a latitudinal direction under central European conditions will have no genetical effects, and the seasonal isolation of pine populations at such distances is complete.[62]

Stebbins has described a case of seasonal isolation between related and interfertile species of pines which are adapted to different climates but overlap locally on the central California coast. The Monterey pine (*Pinus radiata*) which grows near the ocean in central and southern California, and is consequently adapted to a climate characterized by mild winters, sheds its pollen early in February. The Bishop pine (*P. muricata*) of the northern California coast and the Knobcone pine (*P. attenuata*) of the interior California mountains live in regions with cold winters and shed their pollen in April.[63]

The Bishop and Knobcone pines occur as sympatric neighbors of the Monterey pine at certain localities near Monterey Bay. Each species occurs in its characteristic habitat and is pollinated in its normal season, early February or April, in this area. Although the species concerned can be crossed artificially, and their hybrids are vigorous and fertile, little hybridization takes place between them where they meet in nature. Environmental isolation reinforced by seasonal isolation accounts for the fact that less than 1 percent of the trees in neighboring sympatric areas of *P. radiata* and *muricata* or *P. radiata* and *attenuata* are hybrids or hybrid derivatives of these species.[64]

The effectiveness of seasonal differences in the breeding period as a factor in isolation varies widely in different groups of plants and animals. In plants and animals which breed over a period of weeks or months, related species that differ at all in periodicity are likely to overlap broadly in their breeding seasons, so that the temporal isolation is relatively insignificant. This is the case in many sympatric species of land birds[65] and animal-pollinated flowering plants of temperate and arctic regions.

For example, *Salvia mellifera* blooms earlier than *S. apiana*, but the

[62] Wettstein and Onno, 1948.   [63] Stebbins, 1950, 209–10.
[64] Stebbins, 1950, 209–10.   [65] Mayr, 1942, 251.

two species are nevertheless in flower simultaneously during a period several weeks long. At the foot of the San Gabriel Mountains near Claremont, California, the first flowers of *Salvia mellifera* appear in early April, while the first flowers of *S. apiana* appear a week or two later. *Salvia mellifera* reaches full bloom in mid April and *S. apiana* in early May. The flowering season of *S. mellifera* ends in early middle May while *S. apiana* is still in good bloom. The flowering seasons of the two species are thus separate at the beginning and the end of spring, but overlap for about a month from mid April to mid May, and the seasonal isolation is accordingly slight.[66]

In organisms with short breeding seasons, on the other hand, we would expect to find more cases of complete or nearly complete seasonal isolation of related forms. Among animals, seasonally isolated populations with relatively brief reproductive periods are found fairly commonly in water birds, amphibians, fish, and invertebrates.[67] An example of seasonal isolation in the herring was mentioned in Chapter 12. In the plant kingdom sharply defined reproductive periods and the possibility of strong seasonal isolation are found in some wind-pollinated trees of the temperate zone, as exemplified by the pines, and in many animal-pollinated trees and epiphytes of tropical regions.

Many species of tropical plants have short-lived flowers. Different individuals of the same species all burst into bloom on the same day; but related species may flower on different days. Among several species of the orchid genus Dendrobium which bloom for one day in response to a temperature stimulus, *D. crumenatum* flowers on the ninth day after exposure to this stimulus, a closely related species flowers on the eighth day, and some other species flower on the tenth or eleventh day. Many tropical trees bloom only briefly at intervals of several years, which greatly increases the likelihood of non-simultaneity in the flowering times of different species.[68]

The length of the favorable growing season in any region is an important factor affecting the scope of seasonal isolation. The growing season reaches its maximum length in tropical districts with an equable ever-moist climate, where vegetative growth, flowering, and fruit formation can take place during any period of the year. In such

[66] K. Grant and V. Grant, unpublished.   [67] Mayr, 1942, 251–54.
[68] Holtum, 1953.

regions the opportunities for seasonal isolation of sympatric plant species are most numerous, and are apparently most exploited. Numerous species can bloom in different parts of the year, or in different years, with no overlap in flowering season.

A contrasting situation exists in hot deserts, mediterranean climates, alpine zones of temperate mountains, and arctic regions, where the period favorable for vegetative growth, flowering, and seed-ripening is restricted to a few months or even weeks in spring or summer. Simultaneity in flowering is enforced by the climatic conditions on most plant species growing together in the same desert, mediterranean, alpine, or arctic community. Such seasonal isolation as does occur is more often partial than complete, the species differing in their peaks of flowering but overlapping in their seasonal ranges. The example of *Salvia apiana* and *S. mellifera* mentioned above is typical.

Warm temperate and subtropical climates with a warm moist period six months or so long afford opportunities for seasonal isolation of sympatric plant species which are intermediate between those found, respectively, in equable tropical districts and in desert, alpine, or arctic zones. Several plant species can readily complete their flowering in different parts of a growing season six months long. In the Ozark region of southern Missouri, for example, there are five species belonging to the same section of the genus Phlox (section *Phlox*), each of which has a different flowering season. *Phlox bifida* blooms in early spring, *P. glaberrima* in late spring, *P. amplifolia* in early summer, *P. maculata* in late summer, and *P. paniculata* in late summer and autumn. Such a degree of differentiation in flowering period is not attained within any genus, still less within any section of a genus, of the Phlox family inhabiting western desert, coastal, or mountain regions with a brief favorable season.

Insofar as the factor of seasonal isolation alone is concerned, therefore, hybridization between sympatric plant species is evidently easier in areas with a compressed growing season than in areas with a prolonged warm moist period.

GAMETIC ISOLATION

In aquatic animals with free-living gametes, the sperm and eggs of distinct species may come into contact without fusing. The failure of fertilization in such cases may be due to some mutually incompatible biochemical reaction between the gametes of foreign species. When eggs of the sea urchin, *Strongylocentrotus franciscanus*, are placed in vessels of sea water containing spermatozoa of the same species at certain concentrations, between 73 percent and 100 percent of the eggs are fertilized. By contrast 0 percent to 1.5 percent of the eggs of a separate species of sea urchin, *S. purpuratus*, are fertilized by the same concentration of *S. franciscanus* sperm. In the reciprocal combination of *S. franciscanus* eggs and *S. purpuratus* sperm, fertilization is also inhibited.[69]

INCOMPATIBILITY

In higher terrestrial animals and plants with internal fertilization (and in aquatic types descended from them), interspecific copulation or pollination does not necessarily lead to the formation of hybrid individuals. The female individuals inseminated or cross-pollinated by foreign males frequently fail to give birth to live young, or to lay fertile eggs, or ripen sound seeds. Many developmental steps intervene between the deposit of sperm or pollen in the sex ducts of the female and the production of hybrid individuals by the mother. Blocks can and do occur at these various steps.

An incompatible reaction may block the functioning of sperm in a foreign vagina or of pollen tubes in a foreign style before the stage of fertilization. Or if fertilization occurs, the zygote or embryo may die at an early stage within the body of the mother. There are various prefertilization and postfertilization barriers[70] in the sequence of processes beginning with insemination or pollination, and culminating in birth, egg-laying, or seed-ripening. Since we frequently find different internal prefertilization and postfertilization barriers combined in the same hybrid cross, it is convenient to discuss them together.

The cross of the domestic sheep (*Ovis aries*) and the goat (*Capra hircus*) always fails. In the cross sheep ♀ × goat ♂ fertilization does not occur. The reciprocal cross goat ♀ × sheep ♂ often results in conception and implantation but the fetus is resorbed or aborted at an early stage.[71]

[69] Lillie in 1921; see Dobzhansky, 1951a, 190–91.
[70] Avery, Satina, and Rietsema, 1959, 235 *passim*.   [71] Rae, 1956.

In many interspecific crosses in Drosophila, for example in *D. arizonensis* × *D. buzzatii*, insemination is followed by a rapid secretion of a dense fluid from the vaginal wall into the vaginal cavity, and a swelling of the latter organ to three or four times its normal size. This so-called insemination reaction is provoked by some constituent of the semen other than the spermatozoa, and appears to be an immunological response of the vagina to a foreign material, perhaps a protein, in the semen of a distinct species. In extreme cases the inseminated females may die; or if they recover and mate again later with the same species of males the insemination reaction again takes place.[72]

The formation of a dense mass of material in the vagina effectively blocks off the reproductive tract. Live sperm may be present in the vagina for awhile after copulation, but either they die before reaching the eggs, or if fertilization occurs the fertile eggs cannot be laid. The net result is that hybrids are not produced. *Drosophila arizonensis* and *D. buzzatii*, for example, are weakly isolated ethologically and cross-mate fairly freely, yet never form hybrids owing to the insemination reaction.[73]

In the flowering plants the formation of viable seeds following cross-pollination depends on a long sequence of developmental processes. The pollen grains germinate on the stigma and the pollen tubes grow down the style and into the ovules. Thereupon one sperm nucleus discharged from the pollen tube fertilizes the egg nucleus inside the ovule to form a zygote, while a second sperm nucleus unites with another pair of female nuclei to form the so-called endosperm nucleus. The zygote next develops into an embryo; the endosperm nucleus develops into a nutritive endosperm tissue surrounding the embryo; and the ovule grows into a seed. Whereas these developmental processes usually proceed normally following intraspecific pollinations, they are liable to be upset in interspecific crosses, which consequently do not yield sound seeds.

Gilia tricolor is isolated from other species of Gilia (Polemoniaceae) by strong incompatibility barriers. Repeated attempts were made during four successive years to intercross 4 strains of this species reciprocally with 23 strains belonging to 11 other species in a total of

[72] Patterson and Stone, 1952, ch. 8.   [73] Patterson and Stone, 1952, ch. 8.

45 hybrid combinations. Several hundred flowers were artificially cross-pollinated in these hybridization experiments. Whereas different strains of *Gilia tricolor* cross freely with one another to yield vigorous hybrids, all attempts to cross *Gilia tricolor* with other species failed without exception.[74]

Cross-pollinations of *Gilia tricolor* with closely related species belonging to the same section of Leafy-Stemmed Gilias usually resulted in a reduced set of capsules containing numerous shriveled seeds. In certain hybrid combinations a few plump seeds were produced which, however, either developed into matroclinous non-hybrid progeny, or else did not germinate at all. For example, 72 flowers were pollinated in crosses between *Gilia tricolor* and the related *G. achilleaefolia*; 54 capsules matured on these flowers; they contained numerous shriveled seeds and a total of 2 plump seeds; the latter yielded 0 F_1 individuals. Again in crosses between *G. tricolor* and *G. capitata*, 42 flowers yielded 38 capsules and 0 plump seeds. The capsules would normally contain 10 to 30 sound seeds each following compatible intraspecific pollinations. In wide crosses between *Gilia tricolor* and distantly related species like *G. sinuata* the capsules did not even develop after cross-pollination.[75]

Gilia tricolor grows sympatrically and flowers simultaneously with various other species of Gilia in the California foothills. No doubt the pollinating bees and other insects, in lapses of flower constancy, often carry pollen from *Gilia tricolor* to its sympatric relatives or vice versa. Nevertheless, no hybrids or hybrid derivatives involving *Gilia tricolor* have ever been found in natural populations. Evidently the incompatibility block isolating *Gilia tricolor* from other species is as effective in nature as it is in the breeding plot.[76]

The barrier to crossing between *Gilia tricolor* and its congeners is absolute as far as we can determine from field observations and experiments. In the case of many plant species the incompatibility barriers are very strong but not insurmountable. An example of two species which can be crossed, but only with considerable difficulty, is provided by *Gilia latiflora* and *G. ochroleuca*. It will be recalled from the discussion in Chapter 12 that these two morphologically discrete species of Cobwebby Gilia occur sympatrically in the Mojave Desert and

[74] Grant, 1952a.   [75] Grant, 1952a, 1954b.   [76] Grant, 1952a.

South Coast Ranges of California without interbreeding (see also Plate II in Chapter 12).

It has not been possible to cross certain races of *Gilia ochroleuca* with *Gilia latiflora* at all. The western race known as *Gilia ochroleuca bizonata* can, however, be crossed with *Gilia latiflora latiflora* with some difficulty when the former is used as the female parent. Thus, 56 flowers of *Gilia ochroleuca bizonata* when pollinated by *G. latiflora* produced 18 capsules containing a total of 43 plump seeds. From these 43 seeds, 17 F_1 hybrid individuals were grown.[77]

The numerous diploid races and species of Cobwebby Gilia have been intercrossed in 133 hybrid combinations; representative diploid Cobwebby Gilias have also been crossed with diploid species belonging to related sections of the genus in 41 additional hybrid combinations. The strength of the incompatibility barriers in the different crosses can be measured in several ways. We can take (1) the average number of plump seeds formed per flower pollinated, (2) the percentage of the hybrid combinations attempted that yielded any hybrid offspring at all, and (3) the number of F_1 individuals obtained for each 10 flowers pollinated, as semi-independent measures of the crossability. The incompatibility barriers vary greatly in strength from cross to cross. Nevertheless, when the data are grouped according to the taxonomic relationships of the individuals crossed, some general correlations become evident. Table 15 shows that, with respect to three partially independent measures of crossability, the incompatibility barriers become consistently stronger with the remoteness of the systematic relationships.[78]

Whereas different individual plants from the same population can be crossed with the greatest of ease, weak barriers to crossing are found between geographical races of the same species (see Table 15). The incompatibility barrier is on the average quite strong between the different species of Cobwebby Gilia. The results described above for *Gilia ochroleuca* × *Gilia latiflora* are typical of interspecific crosses within the section.

Almost the only result of numerous attempts to cross Cobwebby Gilias with species belonging to the related sections of Leafy-Stemmed and Woodland Gilias was the production of abortive seeds. Two sound

[77] Grant and Grant, 1960.   [78] Grant and Grant, 1960.

seeds were obtained on one occasion, however, from the cross of a Cobwebby Gilia × Leafy-Stemmed Gilia, and two F_1 individuals were grown from these seeds. One natural hybrid of a Cobwebby Gilia × Woodland Gilia is known from a locality in the wild. The barriers to crossing between diploid species belonging to different sections of the genus Gilia are thus extremely strong but can be breached rarely.[79]

Table 15. Relative ease of crossing of diploid Cobwebby Gilias at different levels of divergence[a]

		LEVEL OF DIVERGENCE	SIZE OF SAMPLE
Type of cross	*Entities crossed*	*Number of hybrid combinations attempted*	*Number of flowers pollinated*
Inter-individual	Different individuals belonging to same population	4	116
Interracial	Different geographical races of same species	26	562
Interspecific	Different diploid species of Cobwebby Gilia	103	2,016
Intersectional	Diploid species of Cobwebby Gilia with diploid species of Leafy-stemmed or Woodland Gilia	41	528

[a] Grant and Grant, 1960, 451, also 439 and 474.

The incompatibility barrier in the Cobwebby Gilias is manifested at different stages of flowering and fruiting. A flower pollinated with foreign pollen may fail to set a capsule; the capsule may ripen but contain only or mainly shriveled seeds; a reduced number of plump seeds may develop; or numerous plump seeds may form in the capsule but fail to germinate. From the developmental standpoint, there is evidently not one incompatibility barrier but several.[80]

[79] Grant and Grant, 1960.   [80] Grant and Grant, 1960, 451.

A detailed embryological study of the incompatibility barrier between two species of Woodland Gilia, namely, G. *splendens* and G. *australis*, has shown that failure of crossing in this hybrid combination is due to a combination of causes. The pollen tubes in growing down the foreign style may fail to reach the ovules; where the male gametes reach the eggs they may fail to effect fertilization; the hybrid embryo if formed

Table 15. (Continued).

RESULTS

Average no. plump seeds per flower	Percent of hybrid combinations attempted that produced any hybrid offspring	Number of hybrid individuals per 10 flowers pollinated
17.8	100	22
15.2	73	12
3.7	43	3
.004	2	.038

may die at an early stage; and the endosperm tissue which surrounds and nourishes the embryo inside the developing ovule may disintegrate, causing the death of the embryo. The cross *Gilia splendens* ♀ × *G. australis* ♂ usually fails mainly because of the inability of the *australis* pollen tubes to reach the *splendens* ovules. The failure of the reciprocal cross, *G. australis* ♀ × *G. splendens* ♂, is due primarily to degeneration of the endosperm. Thus in the first case prefertilization barriers are operative, while in the second case, where these are less significant, postfertilization barriers come into play.[81]

The types of incompatibility barriers found between *Gilia splendens* and *G. australis* are in general like those established by many workers

[81] Latimer, 1958.

in a wide variety of angiospermous genera. Such crossing barriers have been found, for example, in Lilium, Hyacinthus, and Iris among monocotyledons; and in Aquilegia-Isopyrum, Brassica, Sisymbrium, Melilotus, Dianthus, Gossypium, Cucurbita, Solanum, Datura, Primula, and Galeopsis among dicotyledons.[82]

On the basis of the aforementioned and other studies we can say that there are two tissues in the female reproductive system of angiosperms which act as very important sieves in preventing the formation of hybrid seeds. The first of these is the stigma and style. The second is the endosperm.

Foreign pollen grains may fail to germinate on a stigma, as occurs in the cross *Datura meteloides* ♀ × *stramonium* ♂.[83] Or if they do germinate, the pollen tubes may burst while growing in the foreign style, as in *Iris tenax* ♀ × *tenuis* ♂.[84] Again, the foreign pollen tubes may grow too slowly down the style to reach the ovules; this is the case for example in *Gilia splendens* ♀ × *australis* ♂,[85] *Datura meteloides* ♀ × *stramonium* ♂, and *Iris tenax* ♀ × *tenuis* ♂. The pollen tubes of *Iris tenax* reach the *tenax* ovules in 30 hours, whereas the pollen tubes of *Iris tenuis* require 50 hours to reach the *tenax* ovules.[86] *Datura innoxia* and *D. quercifolia* pollen tubes grow at the average rate of 3.3 and 1.9 mm per hour, respectively, in their own styles.[87] Finally, even though the pollen tubes do reach the ovules the gametes may fail for some reason to unite. This result may be somewhat analogous to gametic isolation in organisms with external fertilization.

Several workers have observed that the production of hybrid seeds in incompatible crosses is enhanced by pollination with mixtures of foreign and domestic pollen. The cross A ♀ × B ♂ may yield few or no F_1s, but the cross A ♀ × (B + A) ♂ gives F_1s in greater abundance along with type A progeny. The stimulating effect of domestic pollen on interspecific crossing has been noted in crosses between corn (Zea)

[82] Lilium: Brock, 1954; Hyacinthus: Brock, 1955; Iris: Lenz, 1956, Smith and Clarkson, 1956; Aquilegia and Isopyrum: Skalinska, 1958; Brassica: Håkansson, 1956; Sisymbrium: Khoshoo and Sharma, 1959; Melilotus: Greenshields, 1954; Dianthus: Buell, 1953; Gossypium: Weaver, 1957, 1958; Cucurbita: Hayase, 1950; Solanum: Beamish, 1955; Datura: Avery, Satina, and Rietsema, 1959, especially ch. 14–15; Primula: Valentine, 1955; Galeopsis: Håkansson, 1952.

[83] Avery, Satina, and Rietsema, 1959, 239.   [84] Smith and Clarkson, 1956.

[85] Latimer, 1958.   [86] Smith and Clarkson, 1956.

[87] Avery, Satina, and Rietsema, 1959, 239–41.

and Tripsacum, wheat and rye, wheat and Agropyron, upland and Asiatic cotton, and between certain species of Clarkia.[88]

Perhaps these observations can be explained in the following way. The coordination of pollen and pistil in intraspecific crossings is probably controlled by hormones which are mutually adjusted within a species; the lack of coordination between pollen and pistil observed in interspecific hybridizations would then be due to disharmonious or inhibitory interactions between the reproductive hormones of dissimilar species. Where mixtures of domestic and foreign pollen placed together on a stigma promote the formation of hybrid seeds, it can be suggested that the domestic pollen releases hormonal stimuli which are favorable for the action of the foreign pollen.[89]

It by no means follows that mixtures of pollen will always or even generally have a stimulating effect on the fertilizing ability of the foreign pollen. From a hormonal standpoint the effect of dual pollination could well be just the opposite.[90] In point of fact it is the opposite in a number of actual cases, where the cross A ♀ × B ♂ produces some F_1 hybrids, but the cross A ♀ × (B + A) ♂ produces by contrast mainly or entirely progeny of type A and few or no F_1s. In other words, where pollen grains from the same population are placed together with pollen from a different race or species on a stigma, the former are likely to be prepotent over the latter and effect most or all of the fertilizations.[91]

If pollen from a distinct species be placed on the stigma of a castrated flower, and then after the interval of several hours, pollen from the same species be placed on the stigma, the effects of the former are wholly obliterated, excepting in some rare cases.[92]

A preponderance of non-hybrid individuals in the progeny derived from dual interracial or interspecific pollinations has been noted in corn, cotton, Streptocarpus, and Gilia.[93] Thus, in cotton, when a mixture of *Gossypium barbadense* and *G. hirsutum* pollen is applied to *barbadense* stigmas, the fertilizations are carried out predominantly by the *barbadense* pollen; and when the same mixture is applied to *hirsutum*

[88] Randolph, 1955, on corn; Kiss and Rajhathy, 1956, on wheat crosses; Anonymous, 1956, on cotton; H. Lewis, personal communication, regarding Clarkia.

[89] Maheshwari, 1957, 390.

[90] Lamprecht, 1954; Maheshwari, 1957; Arnold, 1958.

[91] Darwin, 1876, ch. 10.   [92] Darwin, 1876, ch. 10.

[93] Demerec, 1929, on corn; Darlington and Mather, 1949, 253, on Streptocarpus.

stigmas most of the fertilizations are performed by the *hirsutum* pollen.[94]

Weak but easily surmounted crossability barriers exist between two races of *Gilia capitata*, namely, the coastal sand dune race *chamissonis* and the interior foothill race *capitata*. These barriers are expressed in a statistical reduction in the numbers of capsules and seeds set following interracial pollinations as compared with the numbers produced by control pollinations within either race. When flowers of *capitata* are pollinated solely with *chamissonis* pollen, or *chamissonis* solely by *capitata* pollen, however, numerous capsules containing numerous sound seeds each are produced, and numerous F_1 hybrids can be grown.[95] But when these crosses are repeated using mixtures of *capitata* and *chamissonis* pollen, the results are very different. The cross *capitata* ♀ × (*capitata* + *chamissonis*) ♂, which was repeated in two different years, produced numerous *capitata* progeny but not one F_1 hybrid. The 106 progeny of the reciprocal cross ch ♀ × (ca + ch) ♂ consisted of 105 *chamissonis* individuals and only one F_1 hybrid.[96]

The prepotency of domestic over racially or specifically alien pollen is not difficult to account for. We have seen that pollen grains may germinate more poorly on a foreign stigma or the pollen tubes may grow more slowly in a foreign style than the native types of pollen. We also saw (in Chapter 10) that a strong competition between pollens in the style leads to a selective elimination of the less-efficient male gametophytes. Under these conditions the pollen produced by closely related individual plants will usually outcompete the pollen derived from more distantly related races or species when both types are deposited on the same stigmas. Now the natural pollinating agents—wind, insects, birds, etc.—when they cross-pollinate races or different species, normally do deposit a mixture of domestic and foreign pollen grains in a flower. The stigma and style of the flower then operate as a sieve which sorts out and eliminates most or all of the foreign male gametes before they reach the stage of fertilization.

A second important barrier to hybridization in the angiosperms lies in the endosperm.[97] In the flowering plants this tissue, which is the essential growth medium and nutritional supply of the young embryo,

[94] Kearney and Harrison, 1932.   [95] Grant, 1950*b*, 273–74.
[96] Grant, unpublished.   [97] Brink and Cooper, 1947; Brink, 1952.

is a product of fertilization of certain female nuclei by a sperm released from the pollen tube. In hybrid crosses, therefore, the endosperm as well as the embryo has a hybrid genetic constitution. Hybrid endosperm frequently exhibits abnormalities of growth and metabolism and may degenerate as a tissue. The deterioration of the endosperm is then soon followed by the death of the hybrid embryo. This sequence has been observed in interspecific crosses in the genera Lilium, Hyacinthus, Iris, Aquilegia-Isopyrum, Brassica, Sisymbrium, Gossypium, Datura, Gilia, Primula, and Galeopsis among others.[98] That the death of the embryo is caused by the failure of the endosperm has been demonstrated in certain cases in Datura, Solanum, Gossypium, Iris, and other groups by removing the hybrid embryo from the endosperm at an early stage and growing it on an artificial culture medium; the embryo may then develop into a vigorous hybrid plant.

The endosperm, like the style, thus acts as a sieve in the angiosperms to sort out many of the products of interspecific crossing. The orchids are notorious for the ease with which hybrids can be produced between different species, genera, and even subtribes.[99] It is probably significant in this connection that the orchids, unlike the great majority of angiosperms, lack endosperm in their seeds.[100]

A hybrid embryo in a seed must be mutually adjusted, not only with the endosperm, a tissue of hybrid origin, but also with the seed coat, a tissue formed by the mother plant. A famous case occurs in the cross *Linum perenne* ♀ × *L. austriacum* ♂, wherein the hybrid embryo grows to maturity but is incapable of sprouting through the seed coat. If the embryo is artificially dissected out of its seed, however, it can grow into a vigorous hybrid plant.[101]

HYBRID INVIABILITY

An interracial or interspecific hybrid contains two sets of genes derived from two genotypically very different population systems. The well-differentiated genotypes may happen to work well together in the heterozygous combination. It is to be expected in many cases, however, that the genes of different population systems will fail to interact

[98] References given in footnote 82.   [99] Lenz and Wimber, 1959.
[100] Brink, personal communication.
[101] Laibach in 1925; see Brink and Cooper, 1947, 522; and Stebbins, 1950, 215.

favorably, with the result that growth processes of either a specific or general nature are disturbed or blocked in the hybrids. Interracial and interspecific F_1s in both plants and animals are in fact frequently, though by no means always, less viable than the parental types. Furthermore, the hybrid inviability can manifest itself at any stage from the zygote to the mature organism, and can range in degree of expression from lethality to slight constitutional weaknesses.

Since sexual reproduction is a continuous process, and furthermore a process carried out by organisms living in an environment, hybrid inviability grades into other isolating mechanisms, particularly incompatibility, hybrid sterility, and environmental isolation. A hybrid zygote is formed but the embryo dies before birth or seed-ripening: is this incompatibility or hybrid inviability? The hybrid grows to maturity but its sex organs do not develop properly: is this hybrid inviability or hybrid sterility? The hybrid succumbs in the natural environments but can be raised under artificial conditions and with the aid of special growth substances: is this hybrid inviability or environmental isolation? If it seems important to do so, we can of course define what we mean by hybrid inviability arbitrarily.

The viability exhibited by an organism is relative to the environmental conditions under which it lives. Environmental isolation implies that F_1 hybrids or their progeny are inviable in any natural habitat available to them; they may of course be quite successful in certain artificial environments. We encounter, however, cases of hybrids which are constitutionally weak under a wide range of normal environmental conditions, and these cases are best categorized as hybrid inviability. The constitutional defects that result in poor development of, say, leaves or bones, may not differ fundamentally from the defects resulting in abortion of the sex organs, but it is convenient to treat the latter under the heading of hybrid sterility. In higher animals and seed plants which retain the embryo within the body of the mother, it is not easy to say whether the failure of hybridization in any given case is due to incompatibility or to hybrid inviability at an early embryonic stage.

In the flowering plants, as we have seen, the collapse of the endosperm tissue is a frequent cause of the death of the hybrid embryo, which proves viable enough when grown by itself on an artificial medium.

This is the case, among other examples, in the incompatible cross between two species of cotton, Gossypium hirsutum ♀ × G. arboreum ♂. Here, however, it is known that the hybrid endosperm is also capable of normal growth in the absence of a hybrid zygote or embryo. The embryo and the endosperm of Gossypium hirsutum × arboreum, though viable separately, have an antagonistic interaction in the developing ovule. The failure of hybridization apparently begins with the embryo. The diffusion of some substance from the embryo into the endosperm starts the breakdown of that tissue, which in turn causes the starvation of the embryo.[102] It is a moot point whether this is incompatibility or hybrid inviability. However we wish to treat the matter, the fact remains that the cross between Gossypium hirsutum and G. arboreum usually fails.

Lethal or semilethal hybrids have been found in many groups of plants and animals: Drosophila, platyfish, toads, wheat, Crepis, Nicotiana, Gossypium, Clarkia, and Gilia.[103] In toads, for example, the F_1 of Bufo valliceps ♀ × Bufo americanus or related species dies in the embryonic stage of gastrulation, while in the reciprocal cross of Bufo americanus group ♀ × B. valliceps ♂, a certain proportion of the F_1s are inviable in the late larval stage but other F_1 individuals are viable.[104]

The lethality or semilethality of certain hybrids has been traced to the complementary action of particular alleles of one or more genes in most of the genera listed above.[105] Thus in the American species of cotton, hybrids between certain strains of Gossypium hirsutum and G. barbadense have inrolled leaves, corky stems, and bushy growth form, and are more or less sterile.[106] This abnormal syndrome of features is due to the complementary action of two alleles, or more probably two pseudoalleles, of the so-called "corky" gene, which is represented in G. hirsutum by ck^x and in G. barbadense by ck^y.[107]

Hybrids of Gossypium arboreum and G. hirsutum can be produced by the use of certain crossing techniques. The F_1 generation derived from particular strains of these two parental species then consists of normal vigorous plants and red-leaved dwarfs in definite ratios. The red dwarf phenotype is engendered by the complementary action of two

[102] Weaver, 1957.   [103] Gerstel, 1954; also Stebbins, 1958.   [104] Volpe, 1959.
[105] Gerstel, 1954.   [106] Stephens, 1946.   [107] Stephens, 1950.

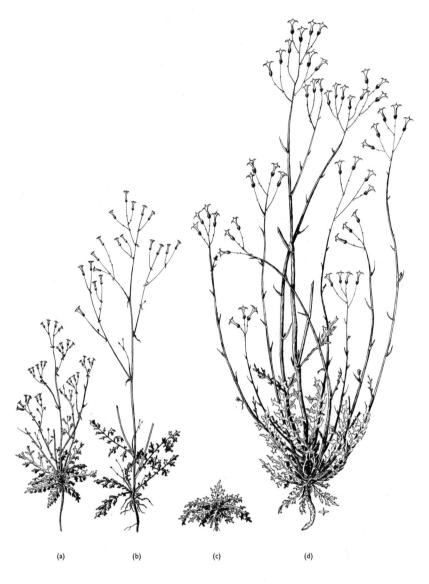

Fig. 74. *The normal and dwarf types of* F$_1$ *hybrids between* Gilia australis
and G. splendens

(a) *Gilia australis*   (b) Normal viable F$_1$
(c) Semilethal F$_1$ (d) *Gilia splendens*

lethal factors, Rl_a carried in one strain only of *G. arboreum*, and Rl_b carried in most strains of *G. hirsutum*. The two lethal factors could either be alleles of the same gene or separate complementary genes. In either case they have complementary interactions in the hybrid. The behavior of the red dwarf hybrids is such as to suggest that the nature of the complementary interaction is to change a physiological equilibrium in the plants, and more particularly to induce the production of an excessive amount of some growth substance.[108]

A similar syndrome of features is found in progeny of the cross *Gilia australis* × *G. splendens*.[109] Among the hybrids are dwarf plants with a somewhat bushy form and red leaves which usually fail to produce flowering stems. The semilethal red dwarfs produce new buds and leaves but make no net gains in growth owing to the rapid breakdown of the older parts; they appear to use up their food materials as fast as they synthesize them. It is easy to imagine that a foreign combination of genes in the hybrids might upset a metabolic balance between constructive and oxidative processes to produce the observed results (Fig. 74).

The red dwarf condition must be caused by disharmonies between specific genes rather than between the whole genotypes of *Gilia australis* and *G. splendens*, since the F_1 generation includes normal plants as well as semilethal dwarfs (Fig. 74). The genic basis for this situation is probably fairly simple, because both the F_1 and F_2 generations exhibit clear-cut segregation into normal plants and red dwarfs in definite ratios. That the semilethal condition is an effect of the complementary action of genes or alleles carried separately in *Gilia australis* and in *G. splendens* is shown by the fact that progenies of either species alone, whether inbred or crossbred, do not contain any red dwarfs. The segregation ratios obtained so far are not decisive as to whether the complementary effect is produced by homologous alleles of a single gene, having unfavorable joint actions, or by a set of complementary genes.[110]

Remarkably similar anomalies of growth have been found also in first-generation hybrids between *Papaver dubium* and *P. rhoeas*. Here again the plants are dwarfed and bushy with malformed leaves and often fail to flower.[111]

[108] Gerstel, 1954.   [109] Grant and Grant, 1954; Latimer, 1958.
[110] Grant and Grant, 1954; Latimer, 1958.   [111] McNaughton and Harper, 1960*b*.

Fertile individuals belonging to different population systems, when crossed with one another, frequently give rise to sterile hybrids. Thus the mule, the F_1 hybrid between the horse and donkey, has been noted since ancient times not only for its physical endurance and surefootedness, but also for its inability to have offspring.

We saw in an earlier section that hybrids can be produced with some difficulty between the sympatric desert species, *Gilia ochroleuca bizonata* and *Gilia latiflora*. These F_1 hybrids are viable but highly sterile. They produce anthers containing mainly abortive pollen grains, and ovules which do not develop into seeds even under the most favorable conditions of pollination. In 1953 three hybrid individuals which were pollinated with one another by bees for several weeks produced a total of three seeds, which could not be germinated. In 1957 another F_1 generation was grown and was treated in three different ways in an effort to obtain later generation progeny. Four hybrid plants were set out in a field with plants of the *Gilia latiflora* parent, which produces abundant good pollen, and were open pollinated by bees for ten weeks. Other hybrid individuals were used as females in artificial backcrosses with *Gilia ochroleuca bizonata* in the greenhouse. And eight other hybrid plants were allowed to bloom together and become self-pollinated or intercrossed over a ten-week period. No seeds were formed by any of the hybrid plants in 1957 under any of these conditions.[112]

Hybrids may exhibit any degree of reduction in fertility, some hybrid combinations being fully fertile, others semisterile, still others highly sterile, and some being for all practical purposes completely sterile. Interracial hybrids are usually fertile or semisterile. The degree of sterility of interspecific hybrids varies greatly according to the cross and the group of organisms. The mule is highly sterile and has been observed to give birth to colts in only a few rare instances, but the F_1 hybrid of the mallard and pintail ducks on the other hand is apparently quite fertile. The F_1 hybrid of *Gilia ochroleuca* × *G. latiflora* is highly sterile, but the hybrids between species of Aquilegia are highly fertile.

The sterility of hybrids encompasses a diversity of phenomena. These phenomena can be reduced to some order by considering

[112] Grant and Grant, 1960, 460.

the stage in the reproductive process at which the block occurs and the genetic basis of the block. It is useful in this regard to combine the separate classifications of sterility phenomena proposed by Federley, Müntzing, and Dobzhansky.[113] As a result of an unfavorable combination of genes, the sex organs may fail to develop properly in a hybrid, leading to what we may call diplogenic sterility. Or if the sex organs develop, the formation or functioning of the gametes may be abnormal; this is gametic sterility. Gametic sterility can be caused either by a disharmonious gene combination in the hybrid, or by structural differences between the chromosomes that interfere with normal pairing and separation of these bodies at meiosis; accordingly, two main types of gametic sterility can be distinguished, namely, genic sterility and chromosomal sterility. Of course the various stages and causes may be combined in actual cases.[114]

The sterility of many animal and some plant hybrids is of the diplogenic type. The abortive development of sex organs in the mule is probably genic in origin. The two individual plants in the F_1 generation derived from a rare successful cross between a Leafy-Stemmed Gilia and a Cobwebby Gilia were fairly normal in their vegetative parts and flowered but failed to develop normal anthers. Abortive development of the stamens is a frequent occurrence in plant hybrids.

Genically determined gametic sterility is found in many animal hybrids. The cross between *Drosophila pseudoobscura* and *D. persimilis* yields fertile F_1 females but sterile F_1 males. Backcrossing the hybrid females to either parental species again gives sterile male progeny. An elegant analysis of the nature and causes of this sterility barrier has been made by Dobzhansky.[115]

In the hybrid males of this cross the external sex organs are normal but the testes are usually abortive and the process of sperm formation is disturbed in various ways. The stage of meiosis in the testes does not follow a normal course; the chromosomes do not all pair in the first division, and the second division is not completed. The products of meiosis, the spermatids, then degenerate. There is cytogenetic evidence

[113] See Dobzhansky, 1951a, 214; also Stebbins, 1958.
[114] Good reviews of hybrid sterility are given by Dobzhansky, 1951a, ch. 8; Stebbins, 1958.
[115] See Dobzhansky, 1951a, 220–27, for summary.

that the disturbed meiosis and gamete formation is not due to differences in the structural arrangements of the chromosomes between the parental species, but is caused instead by particular genes. These sterility genes are distributed on all the main chromosomes in the complement, that is, on the X, II, III, and IV chromosomes. Each of these chromosomes carries at least two such genes and probably many more. Their complementary effect in the interspecific hybrid is to upset the formation of functional sperm.[116]

This effect is most pronounced and the sterility is greatest in backcross progeny which have numerous autosomal chromosomes (II, III, and IV) derived from one parental species together with an X chromosome from the other species. A combination of X chromosomes and autosomes from the same species in a backcross male produces semifertility. This indicates that the balance between different sterility genes on the X and autosomal chromosomes is an important factor determining the degree of sterility. The sterility genes on the different autosomal chromosomes have additive effects. Another factor influencing the sterility barrier is the allelic form of the various sterility genes. Races of *D. pseudoobscura* from Mexico and of *D. persimilis* from the Pacific northwest produce hybrids with very aberrant spermatogenesis, while the races of *D. pseudoobscura* from the northwest and *D. persimilis* from the Sierra Nevada produce hybrids which, though sterile, have a more normal meiosis.[117]

The sterility of many plant hybrids, particularly in herbaceous genera such as Nicotiana, Gossypium, grasses, Viola, Clarkia, Gilia, etc., is chromosomal in origin and gametophytic in stage of expression.[118] The parental types of plants differ in the structural arrangement of their chromosomes, that is, their chromosomes are differentiated with respect to translocations, inversions, and other rearrangements. A proportion of the products of meiosis in a structural hybrid lacks a balanced chromosome set and is consequently inviable. Either the structurally differentiated chromosomes do not pair and separate properly in the first meiotic division, or, if the chromosomes separate to the poles in the usual way, they form daughter nuclei carrying deficiencies and duplications for particular chromosome segments.

[116] Dobzhansky, 1951a, 220–27.   [117] Dobzhansky, 1951a, 220–27.
[118] Reviews in Stebbins, 1950, 218–27, and 1958, 174–78.

Many or most of the daughter nuclei, and the gametophytes and gametes destined to develop from them, then lack the internal balance in their chromosome sets needed for functioning in pollination and fertilization. (A more detailed explanation of the mechanism of chromosomal sterility is given in Chapter 16.)

Chromosomal sterility may be illustrated by the hybrid of *Gilia ochroleuca* × *G. latiflora* which has 18 chromosomes. Pollen mother cells in metaphase of meiosis always have some reduction and usually a marked reduction in chromosome pairing. Whereas each parental species regularly forms 9 chromosome pairs (or bivalents) at meiosis, the maximum pairing ever seen in their hybrid was 8 bivalents and 2 unpaired chromosomes, and this was seen in only 1 out of 73 pollen mother cells analyzed (see Fig. 75a). By far the majority of the pollen mother cells had between 3 and 6 bivalents (Fig. 75b), and many had few or no bivalents (Fig. 75c, d). Configurations characteristic of translocations and inversions, such as chromosome chains, morphologically dissimilar pairs, and chromosome bridges, were seen in a number of cells. The distribution of the paired and unpaired chromosomes to the poles was irregular, many chromosomes being left out of the daughter nuclei (Fig. 75e). All these irregularities of meiosis resulted in the production of mainly—98 percent or more—inviable pollen. Presumably a similar abortion of female gametes took place in the ovules. We have seen that despite repeated attempts to obtain F_2 or backcross progeny from *Gilia ochroleuca* × *latiflora*, this hybrid has so far proven completely sterile under experimental conditions.[119]

HYBRID BREAKDOWN

The disharmonious genic and chromosomal combinations which produce hybrid inviability and hybrid sterility in the F_1 generation can be formed by the sexual process also in F_2 and later generations. If the F_1 hybrids have any progeny, these may well include many inviable and sterile individuals. We have seen in previous sections that the hybrid inviability of *Gilia australis* × *splendens* reappears in the F_2 generation, and that the male sterility of *Drosophila pseudoobscura* × *persimilis* recurs again in the B_1 generation.

Many unfavorable gene recombinations do not arise until the F_2

[119] Grant and Grant, 1960.

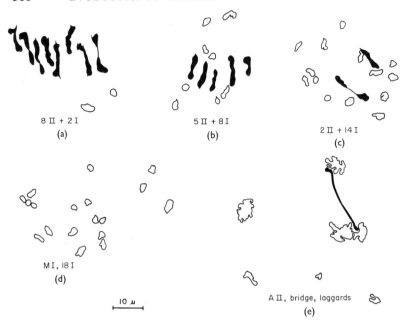

8 II + 2 I
(a)

5 II + 8 I
(b)

2 II + 14 I
(c)

M I, 18 I
(d)

10 μ

A II, bridge, laggards
(e)

Fig. 75. Meiosis in different pollen mother cells of the F₁ *hybrid of* Gilia ochroleuca × G. latiflora

Paired chromosomes (bivalents) are drawn in black, unpaired chromsomes (univalents) in outline. The total number of chromosomes is 2N = 18.
(a) Metaphase with 8 bivalents and 2 univalents.
(b) Metaphase with 5 bivalents and 8 univalents.
(c) Metaphase with 2 bivalents and 14 univalents.
(d) Metaphase with all 18 chromosomes unpaired.
(e) Anaphase of second division with lagging chromosomes and a chromosome bridge.
Grant and Grant, 1960.

generation or later. An F_1 hybrid may then be vigorous or fertile but produce weak or sterile progeny. The two diploid species of Zausch-neria, *Z. cana* and *Z. septentrionalis* (Onagraceae), produce a vigorous and semifertile hybrid, but the F_2 generation consists mainly of dwarfish, weak, rust-susceptible plants. In one hybrid combination, 2,133 F_2 plants were grown, not one of which was viable, while in another cross only 6 out of 361 F_2 individuals could be grown successfully.[120]

The F_1 hybrid between two species of tarweeds, *Layia gaillardioides*

[120] Clausen, Keck, and Hiesey, 1940, 247–49.

and *L. hieracioides* (Compositae), is vigorous and semifertile, but 80 percent of the F_2 individuals are weak and many of these are dwarfish or lethal.[121] Similarly, many F_2 progenies derived from interspecific crosses in Gilia include a fair to large proportion of weak, dwarfed, or lethal types. In the F_2 of the semifertile hybrid *Gilia leptantha* × *G. latiflora*, for example, about half of the individuals are as vigorous as the parents and the other half are weak or dwarfed.[122] In Layia, Gilia, and other plants an interspecific F_2 generation often exhibits continuous variation from fully vigorous individuals through plants slightly smaller or weaker than normal to inviable types.

COOPERATION BETWEEN DIFFERENT ISOLATING MECHANISMS

Any given type of isolating mechanism is likely to reduce the gene flow between species without cutting it off entirely. A combination of several incomplete isolating mechanisms may, however, bring about an essentially complete separation of the species. Most species are generally isolated, not by single mechanisms, but by combinations of different mechanisms working in cooperation.

Gene exchange between the sympatric species, *Drosophila pseudoobscura* and *D. persimilis*, is prevented by at least five and perhaps six isolating mechanisms. (1) Ecological isolation: *D. persimilis* occurs in cooler and moister habitats than *D. pseudoobscura*; also the food preferences of the two species are somewhat different. (2) Mechanical isolation (?): The shape of the penis is different in the two species; furthermore less sperm is transferred in interspecific copulations than in intraspecific ones; but whether these two facts can be interpreted as mechanical isolation remains to be determined. (3) Ethological isolation: The females of either species accept their own males more readily than the foreign males. (4) Seasonal isolation: *D. persimilis* is more active in the morning and *D. pseudoobscura* in the evening. (5) Hybrid sterility: The F_1 males are sterile. (6) Hybrid breakdown: The backcross progeny are weak in both sexes and sterile in the male sex.[123]

No one of these isolating mechanisms is sufficient in itself to prevent interspecific hybridization entirely; yet taken together the isolation of

---

[121] Clausen, 1951, 109–11.   [122] Grant and Grant, 1960, 472–74.
[123] Dobzhansky, 1951b, 1955c.

the two species is complete. Although countless millions of flies belonging to these two species live and breed sympatrically throughout a wide area in western North America, hybridization between them is extremely rare if not non-existent.[124] Only two female flies out of many thousands inspected have been found to be inseminated by males of the opposite species, while no F_1 hybrids have ever been collected in the wild.[125] If hybrid individuals should ever arise, the internal barriers of hybrid sterility and hybrid breakdown would come into play, and further block the gene flow from one species to the other.

Two species of Leafy-Stemmed Gilia, G. *millefoliata* and G. *capitata chamissonis*, grow together on the coastal strand of central California. They are isolated in five ways. (1) Ecological isolation: G. *capitata chamissonis* occurs on sand-dunes and G. *millefoliata* on flats. (2) Mechanical and ethological isolation: G. *capitata chamissonis* is large-flowered and bee-pollinated, while G. *millefoliata* is small-flowered and self-pollinating; there are, furthermore, characteristic differences in the color pattern and odors of the flowers that would encourage species-constant visitations by the insects. (3) Seasonal isolation: G. *millefoliata* blooms earlier than G. *capitata*. (4) Incompatibility: Hybrids are very difficult to produce by artificial crosses in the experimental garden. (5) Hybrid sterility: The F_1s when they can be obtained are chromosomally sterile to a high degree, producing only about 1 percent of good pollen grains and no F_2 seeds.[126] It is not surprising, therefore, that in a number of localities where Gilia *millefoliata* and G. *capitata chamissonis* occur sympatrically, no trace of hybridization between them can be found.

Related species or semispecies of plants often hybridize in nature. The natural hybridization occurs between plant species or semispecies that are weakly isolated reproductively as well as between species with strong reproductive isolating mechanisms. Conversely, species with weak reproductive isolation may maintain themselves fully distinct in parts of their sympatric range. Earlier in this chapter we mentioned cases in Tradescantia, Salvia, and Aquilegia. Sterility barriers and other mechanisms of reproductive isolation between *Tradescantia canaliculata* and *T. subaspera*, or between *Salvia apiana* and *S. mellifera*,

[124] Dobzhansky, 1951b.   [125] Dobzhansky, personal communication.
[126] Grant, 1954b.

do not keep these species from hybridizing in certain localities. Notwithstanding the interfertility between the *Aquilegia formosa* group and the *A. chrysantha* group, and their known potentiality for hybridizing, these species remain quite distinct over large areas of sympatric overlap.

The crucial factor determining the presence or absence of interspecific hybridization, in these and many other plant groups that have been carefully investigated from this standpoint, is the availability of environmental niches suitable for the growth of the hybrid types. Where the habitats occupied by the respective species remain distinct in nature, hybridization does not take place in spite of weak reproductive barriers; but where the old habitats are disturbed and new open habitats are created, either by natural processes or human activities, even highly intersterile species may be found to hybridize.[127] Environmental isolation exercises an ultimate censoring effect on hybridization in many groups of plants.

Environmental isolation may prove to be an important ultimate controller of hybridization also in many animal groups, particularly among the smaller animals that are fairly closely tied to specific habitats. In the Mexican towhees the breakdown of ecological isolation is correlated with hybridization, as we have already seen. Many species of animals which rarely or never hybridize in nature will interbreed freely in captivity when the environmental situation is greatly altered. This is the case, for example, with mallard and pintail ducks.

Do these considerations mean that environmental isolation is more important than reproductive isolation in preventing interbreeding between species? Not necessarily. The various mechanisms of reproductive isolation between sympatric species of ducks, towhees, columbines, and sages are probably as strong as they need to be to prevent hybridization in the normal environments occupied by those species. But radical changes in the environment may permit events of gene exchange to take place which the existing reproductive isolating mechanisms would be adequate to block in the ancestral environment. Environmental isolation, in other words, does not necessarily outweigh reproductive isolation in importance, but the environment may sometimes hold the balance of power between the processes promoting reproductive isolation and those promoting hybridization.

[127] Epling, 1947a; Anderson, 1948, 1953.

FOURTEEN

Ecological Relations Between Species

THE REPRODUCTIVE PHASE of sympatric species or semispecies is characterized by breeding relations—by interbreeding or its avoidance—as discussed in the preceding chapter. In their non-reproductive or secular phase the sympatric population systems interact ecologically in various ways. The ecological relations between sympatric species range from direct competition to mutual dependency, as we shall see in this chapter.

INTERSPECIFIC COMPETITION

Any population of organisms must obtain from its environment certain factors necessary for its maintenance. If it is an animal, fungus, or protistan it requires food; if a green plant, it needs sunlight, water, carbon dioxide, and various minerals; and in any case it must have a certain amount of space, or *Lebensraum*, in which to sink its roots and spread its branches, or gather food, or seek cover, or find nesting sites. The resources of food, raw materials, and space are limited, either within a given region or over our planet as a whole, whereas the ability of organisms to multiply in numbers is unlimited; and under these circumstances certain environmental factors may become limiting whenever the growth of populations gives rise to demands on the environment which, in the aggregate, exceed the available supply of resources.

Now in an area or world populated by a single species of organism, the competition would take place between individuals and local populations, and insofar as these units differed genetically in their ability to stay alive and reproduce in their environment, this competition would lead to changes in the genetic composition of the one species involved. But the world is manifestly not populated by a single species; it is populated instead by some 4.5 million species at the present time; and each geographical area has its biota composed of hundreds or thousands of species. In an area inhabited by diverse species, therefore, a competition for the means of life may arise, not only between the individuals and local populations of one species, but also between the different species.

Interspecific competition presupposes a number of conditions. It presupposes in the first place that two or more species live in the same territory; secondly, that these species require some of the same environmental resources, that they eat the same kinds of food, nest in the same sites, tap the same supplies of soil moisture, etc. Thirdly, it is a precondition for interspecific competition that the environmental resources utilized in common by two or more species are available in limiting amounts.

With regard to the second condition above, it should be noted that closely related species are generally similar in their ecological requirements and in their adaptations for meeting these requirements. Distantly related species, being more dissimilar in their genotypes, usually differ ecologically as well as in other characteristics. However, in cases of evolutionary convergence two or more species derived from distantly related ancestors may develop very similar modes of life. Widely distant species, finally, belonging to different major groups—classes, phyla, kingdoms—make their living in very different ways.

For example, the related species, coyotes and wolves, *Canis latrans* and the *Canis lupus* group, are similar ecologically insofar as they both live on fresh meat which they obtain by running down other animals in pairs or small packs. Other species of mammals related to the canines may eat the same kind of food but hunt for it by different methods, as do the cats; and still more distantly related mammals may live on such non-canine types of food as grass, flying insects, nectar, marine plankton, etc. However, the Tasmanian wolf, Thylacinus, a marsupial mammal which is only remotely related to the placental family Canidae, nevertheless lives by a doglike method of hunting. Canis and Thylacinus are convergent in their ecological requirements and adaptations.

The mode of nutrition and the ecological demands of Canis, finally, are utterly different from those found in still more distantly related forms of life, as for instance worms, trees, or fungi.

Assume that the first and third conditions for interspecific competition are fulfilled; that is, two or more species live together in the same territory, and require some resources that are present in limited amounts in this territory. Then from the foregoing considerations we may expect that the degree of the competition between these species will vary widely according to their ecological similarities. In general, the competition will be keenest between closely related species or between such remotely related species as have converged in evolution; it will be less severe between more distantly related species that have not undergone convergence; and between members of different major groups the competition will be negligible.

These differences in the relative strength of interspecific competition were appreciated by Darwin, who discussed the problem in *The Origin of Species*[1]:

As species of the same genus have usually, though by no means invariably, some similarity in habits and constitution, and always in structure, the struggle will generally be more severe between species of the same genus, when they come into competition with each other, than between species of distinct genera.... The dependency of one organic being on another, as of a parasite on its prey, lies generally between beings remote in the scale of nature.... We can dimly see why the competition should be most severe between allied forms, which fill nearly the same place in the economy of nature....

In terms of our concrete examples, coyotes and wolves might enter into competition for rabbits as a source of food. Following the introduction of the dingo (Canis) into Australia, probably by aboriginal man in recent prehistoric times, a direct competition arose between this wild hunting dog and the indigenous Tasmanian wolf (Thylacinus), both of which have the same general food habits. A reduced though definite competition may take place between Canis and other carnivorous mammals, such as cats or bears. But competition is not a factor in the ecological interactions between Canis and the remotely related and ecologically dissimilar forms belonging to other major groups.

The closeness of the phylogenetic relationship between different species is correlated with the degree of the competition between them only insofar as phylogenetic relationship is in general correlated with similarity of ecological requirements. It is the latter factor which really

[1] Darwin, 1859, ch. 3; order of sentences changed in quotation.

determines the directness of competition between species. Two species which live in exactly the same places and utilize the same range of food types and/or raw materials may compete directly for several or many resources. Two species with totally different ecological requirements will not compete at all. And between these extremes lies the vast number of cases where two species overlap in some of their ecological requirements but differ in others. Thus two species of birds, A and B, might eat some kinds of seeds in common, but if A also relies on other larger seeds which are not taken by B, while B in turn feeds on smaller seeds than A is able to handle, the competition for food between A and B is only partial.[2]

Let us next consider the third condition stated above for the development of interspecific competition, the condition, namely, that some factor in the environment which is necessary for the different species is present in limiting and restrictive amounts. That factor might be space, food, or raw materials. Indeed in a densely inhabited area it might be some particular aspect of a single factor. Thus two species of birds might find enough insects in an area to support their respective populations and have ample space for food-gathering, but be limited by the number of suitable nesting sites available. Or a given area might afford an abundance of sunlight, water, and most minerals for plant life, but have an insufficient quantity of some particular mineral, say, boron, which is required by two or more plant species and which consequently becomes the critical factor limiting their population growth. The limiting factor may come into play at one particular time in the year or even at intervals in a longer cycle. Needless to say, a pair of species may be limited by a combination of factors.

Ecologists find, as a broad generalization applying to many biotic communities, that interspecific competition is strong among the producers, decomposers, and carnivores, but relatively weak among the herbivorous animals. Carnivores are usually limited in numbers by the supply of food animals, for which they compete. Plants, as primary producers of organic matter, may compete for space, moisture, light, minerals, and so forth. Fungi and other organisms of decomposition produce antibiotic substances with which they depopulate their competitors. But herbivorous animals in undisturbed natural communities

[2] Mayr, 1948, 212.

seem to be limited not so much by the available supply of plant food as by their predators, and therefore they do not compete so strongly with one another for common food resources.[3]

It is by no means a simple task to specify the quantitative level at which an environmental resource becomes limiting. Consider the requirements of different plants for different raw materials. Some such materials, like boron, are required only in minute amounts, while other materials, like water, are needed in copious quantities; and, again, the amount of water that would suffice for the growth of a desert plant would be far below the minimal requirements of a rain-forest plant. Trees make far heavier demands on the moisture and minerals in the soil than do annual herbs. A large grazing animal requires a much larger territory than a small herbivorous mammal. Much more food is needed to support a population of carnivorous mammals than is needed by a population of carnivorous beetles.

Therefore, even if it were possible to specify in physical units of measurement the minimal amounts of boron, water, space, or meat that are required by each species of organism, these absolute measurements would have little relevance to the competition between species with somewhat different requirements. In the last analysis, a quantity of some environmental resource can be said to be limiting for organisms only if and when that quantity is small in relation to the requirements of the organism. The quantitative level of any environmental resource, which sets the stage for interspecific competition, is relative to the quantitative demands made by the species in question.

The species of organisms with a large biomass, that is, with a large total amount of living substance, represented in the form of many small individuals or few large ones, will have the greatest requirements for various resources and will thus tend to exhaust rapidly the available supplies of those resources. By contrast the species with a small biomass, the species of small and uncommon organisms, will make relatively slight demands on their environment. It would seem to follow as a general rule that a state of interspecific competition would be reached more immediately by the larger and more numerous forms of life, and would be more delayed in the case of the smaller forms

---

[3] Hairston, Smith, and Slobodkin, 1960.

which have modest requirements. A competition for food would be expected to develop more rapidly between different species of carnivorous mammals than between species of carnivorous beetles. The common species of trees probably compete more readily with one another for water and minerals than do the different common species of annual herbs in the same area.

The degree of competition between species is thus affected by two sets of factors, the ecological similarities between those species and the extent of their ecological demands. If two species require many of the same resources, they will compete directly; and if they require some of those resources in large amounts relative to the supply in the environment, they will enter into competition rapidly. But if the species utilize mainly different resources or utilize the same resources in relatively small amounts, the competition will be indirect and delayed.

INTERSPECIFIC SELECTION

Given a state of direct and immediate interspecific competition, what is its outcome? Before attempting to answer this question let us recall that all species differ genetically by definition. It is reasonable to suppose that these genetic differences control the ecological characteristics as well as the other features of the species. Hereditary ecological differences have in fact been found between species in every case that has been analyzed. However, the magnitude of the interspecific ecological differences and the types of ecological characters involved vary from one pair of species to another. Consequently, although any two species certainly differ genetically, and almost certainly differ ecologically, they may *or may not* differ in their genetically determined ability to get and utilize the environmental resource that happens to be the subject of interspecific competition in any given case. Now one effect of interspecific competition is independent of the ecological differences between the species, but another effect is not.

Consider two isolated areas, say two islands, each of which is populated with canine animals belonging to two ecologically similar species, A and B. Assume that the supply of some resource on each island will support 100 canine animals only. The limiting factor on one island is the number of rabbits, rodents, and other game animals, and on the other is the supply of drinking water. Assume further that the area is in either case actually filled to capacity with 50 individuals of species A and 50 more of species B.

The first effect of the competition between A and B on either island will be a restriction on population growth over and above the existing total of 100 animals. The tendency of the two species to increase in numbers is checked by an elimination of individuals which keeps the total population down to 100 individuals.

The second and evolutionarily interesting effect of this competition will depend on the relative efficacy of the two species in getting their share of the limiting resource. We will consider separately the competition between A and B for water on the one island and for food on the other.

If the genetic differences between A and B do not directly or indirectly affect their ability to get water, and if water is the only resource for which they compete, the restriction on the natural increase of the animals will be random with respect to their species affinities. That is, the numbers of A and of B will fluctuate from year to year and from generation to generation around an average frequency of 50 A:50 B within the limits of chance. The competitive elimination may reach considerable proportions but is non-selective.

The genetic differences between A and B are quite likely to affect in some way their efficacy in hunting. Let us assume that one of the species, say A, is a more successful hunter than the other, owing to greater speed, more cunning, better concealing coloration, or some other trait. Then on the island where food is the limiting factor, the numbers of species A will gradually rise in proportion to the numbers of B. The eventual outcome of this trend, if it continues for a long enough time, will be the complete replacement of B by A. The composition of the species mixture on this island changes, though its total size remains constant, from the initial 50 A:50 B, to 51 A:49 B, to 98 A:2 B, and perhaps finally to 100 A:0 B. There is, in short, a systematic change in the relative frequency of the two species during time.

Let us note that on each hypothetical island there is a constant restriction on the population growth of both species. Where the two species do not differ adaptively *with respect to the factor which they are competing for*, the limitation of population growth operates randomly on both species alike. But where one species does possess some inherent advantage over the other in getting the resource for which both species are competing, the competitive elimination has a non-random or selective component, which is expressed in a systematic change in the relative abundance of the two species.

SPECIES REPLACEMENT

The question of what actually happens when two species requiring some of the same resources live together has been approached in different ways. One approach used by naturalists since the time of Darwin is to carry out observations in natural communities. The great advantage, indeed the indispensability, of this approach lies in the fact that the problem before us is the interactions between species in nature. But the corresponding disadvantage of the naturalist's approach stems from the infinite variety and complexity of the ecological relations between species in any natural community. As Darwin noted[4]:

We can dimly see why the competition should be most severe between allied forms, which fill nearly the same place in the economy of nature; but probably in no one case could we precisely say why one species has been victorious over another in the great battle of life.... It is good thus to try in our imagination to give any form some advantage over another. Probably in no single instance should we know what to do so as to succeed. It will convince us of our ignorance on the mutual relations of all organic beings; a conviction as necessary as it seems difficult to acquire.

In recent years the complex problem of interspecific competition and its effects has been lifted out of nature and brought into the laboratory. Mathematical models dealing with the situation of interspecific competition were proposed independently by Lotka (1925, 1932) in the United States and Volterra (1926, 1931) in Italy.[5] Experiments with mixed laboratory populations of Paramecium and yeast were carried out by Gause (1934, 1935) to test the validity of these models.[6] Since the 1940s experiments have been performed with competing species of grain-eating insects, flour beetles, Drosophila, etc., by such workers as Crombie, T. Park, Merrell, and others.[7] We will briefly review some of the results of these experiments here.

[4] Darwin, 1859, ch. 3.
[5] See Allee, Emerson, Park, Park, and Schmidt, 1949, for references.
[6] Gause, 1934.
[7] For reviews see Andrewartha and Birch, 1954, ch. 10; and Slobodkin, 1961, chs. 7, 8, 11.

In the experiments with Paramecium Gause introduced 20 individual protozoans into a tube containing water, salts, and fresh bacteria (*Bacillus pyocyaneus*) for food. The temperature, volume, chemical composition, and bacterial supply of the medium were kept constant, and the waste products of the Paramecium were removed at regular frequent intervals. The number of individual Paramecia in a culture was estimated daily from counts of small samples of the medium.[8]

Where a single species of Paramecium, either *P. aurelia* or *P. caudatum*, was introduced into a tube at the beginning of an experiment, it multiplied rapidly during the first 4 to 8 days, from the initial 20 individuals to several hundred or more, and then ceased to increase further in numbers during subsequent days. A population of Paramecia grew up to the limits set by a constant food supply and expanded no more thereafter. The ceiling on population growth happened to be higher for *P. aurelia* than for *P. caudatum* under the conditions of the experiments.[9]

Gause next placed *P. aurelia* and *P. caudatum* together in the same tubes, keeping the environmental conditions for the mixed cultures the same as those for the single-species cultures, and thus forced the two species to compete for the same limited food supply under constant experimental conditions. *Paramecium aurelia* increased in numbers more rapidly than *P. caudatum* during the first week, and eventually replaced *P. caudatum* completely. The species *P. aurelia* exhibited a selective advantage over the species *P. caudatum* under the environmental conditions of the experiments. The selective advantage of *P. aurelia* over *P. caudatum* consisted partly of a greater insensitiveness of the former species to the excretions of the food bacterium, *Bacillus pyocyaneus*.[10]

In another series of experiments a different strain of bacillus which does not excrete a harmful product was furnished as food. In these experiments *P. caudatum* survived while *P. aurelia* died out.[11] The relative adaptive advantages of the two Paramecium species were thus reversed when the environmental conditions were altered in particular ways.

[8] Gause, 1934; see also Andrewartha and Birch, 1954, 351–55.
[9] Gause, 1934; Andrewartha and Birch, 1954, 351–55.
[10] Gause, 1934; Andrewartha and Birch, 1954, 424–26.
[11] Gause, 1934; Andrewartha and Birch, 1954, 424–26.

Park reared two species of flour beetles, *Tribolium confusum* and *T. castaneum*, on limited quantities of flour in tubes kept at constant temperatures and humidities. These beetles produce a new generation every 2 or 3 months. The usual result of competition for food between the two species after about 10 or 20 generations was the replacement of one species by the other. The identity of the victorious species varied according to the environmental conditions under which the competition took place. Thus in a series of replicated competition experiments at one temperature, *T. confusum* proved superior and usually survived at 30 percent humidity, while at 70 percent humidity *T. castaneum* usually survived. The superiority of *T. castaneum* over *T. confusum* at 70 percent humidity disappeared, however, when a sporozoan parasite Adelina was present, for *T. castaneum* usually survived only when this parasite was absent, whereas *T. confusum* generally emerged victorious when the parasite was present.[12]

The theoretical conclusion deduced by Volterra and Lotka and verified experimentally by Gause, Crombie, Park, and others can be summarized in the words of Hutchinson[13]: "Under constant conditions two species utilizing, and limited by, a common resource cannot coexist [indefinitely] in a limited system." We may add the further condition that the two species must differ genetically in their ability to get or utilize the resource for which they are competing.

Our next task is to attempt to relate the foregoing rule to the situation in nature. Undoubtedly many situations arise in nature in which one or more of the above conditions are not fulfilled. In such cases there is no reason to expect interspecific selection to run the full course to complete replacement. We will consider some of the factors permitting coexistence between ecologically similar species later.

In many other natural situations all the conditions necessary for a selective elimination of one species from a competing pair may be present. Two or more species: (1) enter into direct and immediate competition; (2) the competition has to be fought out within a circumscribed area; (3) one species is inherently superior to the other(s) in this particular competition and under the prevailing environmental conditions; and (4) those environmental conditions remain relatively

[12] T. Park in 1948 and 1954; see Andrewartha and Birch, 1954, 426–29.
[13] Hutchinson, 1957, 417.

constant during a period of time long enough to carry the process of interspecific selection to completion.

The struggle for existence between species in nature, like that between individuals within a species, is very real, and the adaptations of species for surviving in the competition are many and varied. There are the innumerable adaptations for getting and utilizing an important resource in short supply. In addition there are various devices for harming competitor species: the weapons of many animals, the antibiotics produced by some fungi and used against their competitors, and the toxic substances produced by some plants which kill the seedlings of other plant species growing near them. In the competition between *Paramecium aurelia* and *P. caudatum* neither species apparently exerted any direct interference on the other, but in the competition between the two species of Tribolium, on the other hand, the presence of crowds of *T. confusum* had a direct inhibiting effect on the fecundity of *T. castaneum*.[14]

In the case of the doglike mammals in Australia, the dingo and the Tasmanian wolf, the process of interspecific selection ran to complete replacement over very large areas. The Tasmanian wolf, which was formerly widespread in Australia, became restricted after the introduction of the dingo, and has disappeared from most of its former range, hanging on at present only in the mountains of Tasmania.[15] Darwin cited the expansion of one species of swallow and the concomitant decline of another swallow in the United States; the decrease of the song thrush in Scotland resulting from the increase of the missel thrush; the replacement of a large-bodied species of cockroach by the small Asiatic cockroach in Russia; and so on.[16]

The replacement of archaic and inferior species by more modern and efficient forms has occurred all through earth history. The dinosaurs became extinct and their place was taken over by the mammals; the marsupial mammals were replaced long ago on most continents by the placental mammals; this replacement of some marsupial species by their superior placental counterparts started later in the isolated continent of Australia and is still going on there; and now over the whole

[14] Gause, Park; see Andrewartha and Birch, 1954, 424, 429.
[15] See Sanderson, 1955, 21, 196.
[16] Darwin, 1859, ch. 3.

world man is reducing the other larger mammals, placental and marsupial alike, to relics preserved in zoos and national parks.

As a result of the intense competition and selection between species, the dominant forms of life at any one stage in earth history represent the surviving descendants of only a small minority of the species which existed in some previous era. A rich diversity of vertebrate animals existed at the beginning of the Mesozoic era. Of the presumably many thousands of vertebrate species living at the beginning of the Mesozoic, only about two dozen have left any descendant lines that are still alive, and only about eight of the earlier species have given rise to descendants that are successful in the Recent fauna. Some 98 percent of the living families of vertebrates can trace their ancestry back to approximately eight species out of all the thousands existing in earlier periods.[17]

In the broad scheme of evolution, species play a role somewhat analogous to that of mutations within a species. Mayr has pointed out that a new species, like a new mutation, is a trial change wrought in its phyletic line. The majority of such trial changes will prove unsuccessful in competition with existing forms. But a small minority of these changes at either the genic or the specific level may represent improvements in the machinery of life. These will be preserved by natural selection and passed down to the descendants in later generations.[18]

THE COEXISTENCE OF ECOLOGICALLY SIMILAR SPECIES

It is evident from the most casual observation that ecologically similar species do coexist rather frequently in nature. In such cases one or more of the conditions leading to the replacement of one species by another are presumably wanting so that both species manage to survive. In analyzing the conditions which permit two ecologically similar species to coexist, either temporarily or permanently, it will be helpful to begin with a series of hypothetical examples; and for this purpose we may reconsider our pair of canine species living on the same island, and eating the same type of food, letting their conditions vary one at a time.

[17] Wright, 1956.   [18] Mayr, 1955b, 47.

I. The populations of the two canine species A and B can have the same food requirements, yet not compete for food.

1. This would be the case if the food supply on their island is sufficient for 100 animals, whereas the foundation populations of both A and B consist of only a few individuals, and neither species has yet multiplied in numbers up to the carrying capacity of the land. To be sure, the growth of populations is relatively rapid in most organisms, but it does require time, and during that time the two potentially competitive populations will not actually compete.

2. It would also be the case if the numbers of individuals in populations A and B are kept in check, not by the available supply of food which they use in common, but by some other factor which preys upon them both, such as larger predators, parasites, or disease organisms. If the immunity of A and B to predation or disease is about the same, the two species will of course continue to coexist in spite of the fact that they consume the same food resources.

II. Let us suppose that species A and B do compete directly and immediately for food.

3. Suppose further that A is superior to B in this competition. Then it is a matter of time until B becomes extinct. But the extinction of B does require time, and in the meantime the two species coexist. If the selective advantage of A over B is slight, the time lag between the beginning and the end of competitive coexistence will be long.

4. The adaptive superiority of A over B is manifested in one environment (E_1), whereas under other environmental conditions (E_2) the relative selective values of A and B are reversed and B is superior to A. It is possible that during the time when A is replacing B the environment may change from E_1 to E_2, reversing the trend and giving B a new lease on life. In fact the environment may fluctuate cyclically, from E_1 to E_2 and back to E_1 again, and if the period of the environmental cycle is shorter than the time required for complete replacement of either competing species, both A and B can continue to coexist indefinitely.

5. The numbers of individuals in each species are relatively small, and although A is superior to B in the food competition, some chance accident unrelated to the adaptive superiority of A might befall many individuals of that species. A random component in the competitive

elimination of the canine animals comes to the rescue of B, keeping both species in the race. Of course chance could equally well cause the extinction of either species as preserve them both in a state of co-existence.

6. Now let us suppose that A and B are adaptively equal with respect to the competition for food. Perhaps all the varied characteristics of A and of B concerned with food-getting are the same. Or, more plausibly, it might be that A possesses an adaptive advantage in one character, but B has a compensating advantage in some other character, as where A is a faster runner but B has a better sense of smell. In this case the interspecific competition ends in a stalemate.

7. The two related species A and B are likely to be similar but different in their ecological requirements. The direct competition between them stems from their ecological similarities, whereas their ecological differences provide an avenue of partial escape from that competition. Thus A and B might eat some kinds of food in common, for which they compete, but in addition A is better fitted than B for running down large rabbits, while B is better able to subsist on small field mice and berries. To the extent that the environment provides a variety of food materials, and to the extent that A and B can live in somewhat different facies or food niches in their heterogeneous environment, they can continue to exist in equilibrium.

8. If B is losing in competition with A, it will be to its advantage to escape from the island during the season of greatest scarcity, if it can, and retreat to some other area where the competition is not so severe, returning to the island when the food supply is more abundant again. Periodical migratory movements correlated with unfavorable seasons would of course be difficult in our hypothetical case of island-dwelling canines, but are otherwise common in the mammals as well as in many other animal groups.

9. We must bear in mind the possibility that even though one species, say B, becomes extinct on our island it (or some other similar species C) might be reintroduced from an outside source, starting the process of interspecific selection all over again. If the island lies within the radius of occasional dispersal of canine animals from some adjacent land area, the repeated immigration of new individuals of B would tend to counteract the selective elimination of the same forms on the island.

Some of the foregoing possibilities can be illustrated by concrete examples.

We have seen that the herbivorous animals in undisturbed natural communities are seldom limited by the supply of food. Herbivores are most likely to be limited in numbers by their predators. Among herbivores, therefore, we find examples of coexistence of species without evidence of competition or the avoidance of competition for food resources.[19]

A case studied in detail by Ross involves a group of six related species of leafhoppers, *Erythroneura lawsoni* and its relatives (Jassidae, Hemiptera), which feed and breed on the leaves of sycamore (*Platanus occidentalis*) in the eastern United States. Several species of the *Erythroneura lawsoni* group can be found occupying the same rather narrow food niche at the same time without giving any indications of interspecific competition. If interspecific competition were a factor affecting the relative abundance of the species of sycamore leafhoppers, one would expect trees with small leafhopper populations, in which competition would be at a minimum, to have the greatest average diversity of species; and trees with the largest leafhopper populations and hence the strongest interspecific competition would be expected to have the least diversity of species. But this is not what Ross found. Sycamore trees with small leafhopper populations usually harbor only the most common species, *E. lawsoni* and *E. arta*, whereas trees with large numbers of individual leafhoppers almost always support all six species. Ross concludes that interspecific competition is not the chief factor governing the density of populations or limiting the coexistence of species in leafhoppers and in many other insect groups.[20]

Savile has pointed out that in poor habitats in the arctic region where the plant cover is not closed, as on bleak islands or clay or gravel plains, there is not much competition between plants, either individuals or species. In such places individuals belonging to related species intermingle in the same ecological situations. As many as seven species of *Saxifraga* can be found growing together in what appears to be the same niche on Ellef Ringnes Island in the Canadian arctic archipelago.[21]

The time required for the replacement of one species by a superior competitor may be relatively long, particularly if the adaptive advantage

[19] Hairston, Smith, and Slobodkin, 1960.   [20] Ross, 1957.   [21] Savile, 1960.

of the selectively favored species is slight. Andrewartha and Birch have pointed out that in the laboratory experiments carried out with competing species of beetles (Tribolium, Rhizopertha, etc.), in which new generations arise every two or three months, four years or more elapsed in several cases before complete replacement ended the state of coexistence.[22]

Related species wherever studied have been found to differ in their ecological characteristics. It is not safe to assume, however, that ecological differences are synonymous with adaptive differences under a given set of environmental conditions. One species may possess one selective advantage, while another species is favored in a different way, so that a balanced coexistence of both species results.[23]

Furthermore, even where the competing species do differ adaptively, the adaptive differences are not *necessarily* related, directly or indirectly, to their competition in each and every case. It is conceivable, for instance, that two species of trees standing side by side in a dense rain forest, and drawing on the same supplies of light, air, water, and minerals, might be equally well fitted physiologically to extract those resources from their environment and use them in food manufacture, for species of plants belonging to different families and orders are often very similar in their basic physiological processes, and the interspecific competition could thus end in a stalemate.

As we saw in the preceding section, a certain species of Paramecium, beetle, or other organism may well be adaptively superior to a competitor species under one set of environmental conditions (E_1), but inferior to the same competitor under other conditions (E_2). If, therefore, the environment fluctuates from E_1 to E_2 and back in less time than is required for complete replacement to take place, both competing species may persist indefinitely.[24]

Drosophila melanogaster and *D. funebris* eat the same food, but in laboratory experiments one species multiplied faster than the other on fresh food, while the other species multiplied faster on stale food. When the food was changed at regular frequent intervals in the experiments both species persisted.[25]

Different species of green algae coexist in small ponds. It was found

---

[22] Andrewartha and Birch, 1954, 432.   [23] Skellam; see Hutchinson, 1957.
[24] Hutchinson, 1957.   [25] Merrell in 1951; see Andrewartha and Birch, 1954, 433.

in one study that the algae *Haematococcus pluvialis* (Volvocales) is favored when the pond dries up, while Chlamydomonas, Scenedesmus, and Chlorella have the advantage when there is standing water in the pond. If, as sometimes happens in nature, a pond alternately dries up and refills on a frequent cycle, both Haematococcus and Chlamydomonas, Scenedesmus, and Chlorella coexist indefinitely.[26]

Or again, where one set of environmental conditions (E_1) favors one species (A) in a competing pair, and other conditions (E_2) favor the competitor species B, if the different environmental conditions E_1 and E_2 themselves coexist simultaneously in a habitat or territory, the species A and B can persist sympatrically too. This is another way of saying that two species which cannot live together indefinitely in a homogeneous environment may be able to coexist permanently in a heterogeneous environment.

Gause found that if two species of Paramecium are forced to compete in a homogeneous culture medium, one species always dies out. But if the laboratory environment is diversified, the interspecific competition does not inevitably run to complete replacement. Thus where the food supply is a suspension of yeast, *Paramecium bursaria* will feed mainly on the bottom and *P. caudatum* in the upper layers, and the two species exist together in a balanced equilibrium.

It is evident that a heterogeneous environment provides a basis for continual coexistence only insofar as the different niches present can be occupied by the different species. The same vertically stratified suspension of yeast which permits *Paramecium bursaria* and *P. caudatum* to coexist does not lead to stable mixtures of *P. aurelia* and *P. caudatum*; for the latter two species both feed in the same upper layers, and in competition experiments between them one species always replaces the other eventually.[27]

Natural environments are always heterogeneous to some extent, and are frequently heterogeneous in ways that can be taken advantage of by different species with similar ecological requirements. By avoiding complete direct competition in their diversified environment, the ecologically similar (but non-identical) species may coexist permanently.

Lack showed that the ecological relationships between related

[26] Proctor; see Hutchinson, 1957.
[27] Gause in 1935; see Andrewartha and Birch, 1954, 431.

sympatric species of birds are usually such as to reduce the direct competition between them. In northern Europe there are three species of crossbills (Loxia) which pry the nuts out of the seed cones of coniferous trees with their specially adapted beaks. The two-barred crossbill (*Loxia leucoptera*) has a small beak, the common crossbill (*L. curvirostra*) has a medium-sized bill, and the parrot crossbill (*L. pytyopsittacus*) is heavy-beaked. The first species feeds chiefly on the soft cone of larch, the second on spruce cones, and the third on the hard cones of pine. Observations indicate that when the common crossbill does try to get seeds out of pine cones it is less competent at this task than the parrot crossbill.[28]

Different species of finches (Geospiza and relatives) on the same islands in the Galapagos group are specialized for feeding on different foods in different niches. For example, on Tower Island, *Geospiza magnirostris* eats seeds on the ground, *G. conirostris* feeds on cactus, and Certhidea picks small insects off leaves (see Fig. 90 in Chapter 18).[29] In the *Geospiza magnirostris* species group, *G. magnirostris* proper has a huge beak, *G. fortis* a medium-large beak, and *G. fuliginosa* a small beak. These three species all feed on the same general types of food, in the same places, by similar methods. But there are also some significant differences in their diets. Thus *G. magnirostris* and *G. fortis* (but not *G. fuliginosa*) eat the large hard fruits of manzanilla (Hippomane). The small-beaked *G. fuliginosa*, on the other hand (but not *G. magnirostris*), utilizes small grass seeds as a staple food. The diets of the three species, though similar, are differentiated in a statistical way, at least during some seasons of the year.[30]

Four species of seals inhabiting the Ross Sea in the Antarctic all differ from one another in their main food preferences. The main elements in the diet of each species are as follows. The crabeater seal (*Lobodon carcinophagus*) eats euphausiid crustaceans exclusively; the Weddell seal (*Leptonychotes weddelli*) feeds mainly on fish; the leopard seal (*Stenorhinchus leptonyx*) eats penguins and seals mainly but also takes fish; and the Ross seal (*Ommatophoca rossi*) eats cephalopods.[31]

The three species of buttercup, *Ranunculus bulbosus*, *repens*, and *acris*, exist sympatrically in old meadows in England, where however

[28] Lack, 1947, 65–66.   [29] Lack, 1947, 66.   [30] Lack, 1947, 61–62.
[31] Wilson in 1907; Lack, 1947, 141.

each species occupies a slightly different zone on the uneven ground surface. *Ranunculus bulbosus* characteristically occurs on small rises or ridges, *R. repens* in furrows, and *R. acris* at intermediate levels between high ground and low ground in the same meadows. Harper found that plants of the three species transplanted to a series of pots in which the water table ranged from flooded to well-drained, simulating the different conditions in an uneven meadow, succeeded equally well under the various conditions. But when seeds of the three species were sown in pots with water tables of varying depth, the establishment of the seedlings was not uniform. The seedlings of *R. bulbosus* became established best on well-drained soil, while *R. repens* established best in waterlogged soil. In most cases where two species were sown together only one survived to reach flowering. "This suggests," as Harper points out, "that conditions for seedling establishment play a major role in determining the habitat occupied by the mature plants."[32]

Many other examples of ecological differentiation between sympatric species have been worked out in various groups of birds and mammals, Drosophila and other insects, sponges, Paramecium, and plants.[33]

Two sympatric species which require some resources in common may not be able to avoid direct competition, even though they differ somewhat in their ecological requirements, if the total supply of necessary resources becomes scanty at some critical season of the year. If the species cannot avoid a restrictive competition while remaining in the same territory, one or both of them may still be able to escape from the home territory temporarily during the season of scarcity. In animals with means of locomotion the escape frequently takes the form of annual migrations[34]; in plants the escape may consist of a period of annual dormancy. Two species of ducks which breed in the same ponds in the far north but spend the rest of the year in different tropical areas, and two species of annual plants which grow and bloom together in the spring but pass through the dry season as dormant seeds, clearly

---

[32] Harper, 1958.

[33] See *inter alia*, Lack, 1947, for birds and mammals; Baldwin, 1953, 374–75, for Hawaiian honeycreepers; Cooper and Dobzhansky, 1956, for California Drosophila; Dobzhansky and da Cunha, 1955, for Brazilian Drosophila; da Cunha, Shehata, and Oliveira, 1957, for Brazilian Drosophila; Hartman, 1957, for sponges; Hairston, 1958, for Paramecium, and Harper *et al.*, 1961, for plants.

[34] Lack, 1947, 135–36.

face the struggle for existence in their mutual territory—the competitive struggle with one another as well as the struggle against the physical elements—during only a part of each year. The periodical escape from competitive coexistence is one way by which both species may survive to compete again.

ON THE LIMITED VALIDITY OF THE PRINCIPLE OF ECOLOGICAL DIFFERENTIATION IN SYMPATRIC SPECIES

It has been found in many cases that species which appear to be living together successfully and permanently in nature are differentiated ecologically in some way that reduces the direct competition between them. Several examples of such ecological differences between sympatric species of crossbills, finches, seals, and buttercups were mentioned in the preceding section.

Some authors contend that ecological differences designed to reduce competition are not only widespread, but are universal, in sympatric species. It is held to be axiomatic that "two species with similar ecology cannot live in the same region."[35] This conclusion is regarded as a corollary of the principle of competitive exclusion in a limited homogeneous environment as stated by Volterra, Lotka, and Gause. The axiom that species with similar ecologies cannot coexist then leads to the further deduction that wherever two species are in fact living together they must be ecologically differentiated in some significant respect and are probably occupying separate niches.[36]

The conclusion that species must be ecologically different in some way in order to coexist is *apparently* supported by the evidence of field observations, which almost invariably reveal ecological differences between related sympatric species. Or, if the mode of ecological differentiation between the sympatric species has not been found in a particular case, it can nevertheless be assumed to be there, awaiting a more refined investigation, since the species could not both persist without it.

In opposition to the foregoing idea, several students have argued that we should not accept as axiomatic the statement that two species

---

[35] Lack, 1947, 62; Harper *et al.* 1961, 224, for similar statement. This view is often referred to as Gause's law.

[36] Lack, 1947, 62.

cannot coexist in the same niche.[37] Ross's case of sympatric but non-competitive species of sycamore leafhoppers, Savile's case of arctic saxifrages, and others like them should be accepted at face value. As Hutchinson aptly remarks, the principle that two sympatric species must be occupying different ecological niches is a good empirical generalization, which is true except when it isn't.[38]

From the preceding discussion in this chapter it should be evident that species may be found living together for a variety of reasons besides differentiation in their ecological requirements. We can recognize that the principle of the ecological differentiation of sympatric species is of very great and widespread importance in nature without accepting it as universally valid.

SELECTION FOR ECOLOGICAL DIVERGENCE BETWEEN SYMPATRIC SPECIES

One of the important conditions for a permanent coexistence of species is a differentiation between them with respect to their requirements for the chief limiting resources in the environment. This condition of ecological differentiation is fulfilled in many cases of species coexistence, as we have seen. We may next inquire how the widespread and common condition of ecological differentiation has come about. It can be shown that ecological divergences between species will be promoted by interspecific selection.

Assume that two species of animals, A and B, can both seek a common type of food in the same range of environmental niches, E_1, E_2, and E_3. Assume further that A is better adapted than B for feeding in E_1 and, conversely, that B is superior in E_3. In an area occupied by either A or B alone, where interspecific competition does not come into play, the single species present will feed throughout the full range of niches. But in an area inhabited by both species the final result of interspecific selection will be the complete replacement of B by A in E_1 and of A by B in E_3. In short, species A, which is potentially capable of feeding in a wide range of niches, is actually restricted to E_1 and E_2, and species B likewise is restricted by the competition to particular niches, in this case to E_2 and E_3.

[37] Andrewartha and Birch, 1954; Ross, 1957; Hutchinson, 1957; Savile, 1960.
[38] Hutchinson, 1957, 417.

Each species is likely to contain genetic variations with different adaptive values in the different environmental niches. Thus species A can be supposed to consist of the genotypes A_1, A_2, and A_3, which possess the highest fitness, respectively, in niches E_1, E_2, and E_3. A similar array of genotypes (B_1, B_2, B_3) with the same relative order of fitness in the three niches (E_1 to E_3) can be supposed to exist in species B. But A_1 is superior to any genotype of B in E_1, and B_3 is superior to any form of species A in E_3. The process of interspecific selection leading to niche restrictions of the two species in the area of sympatric coexistence now entails a selective elimination of genotypes in each species. The rise to dominance of species A in niche E_1 is accompanied by a decline in the relative frequency of genotype B_1 within species B. And, conversely, when species A loses its hold on niche E_3 it may also lose its genotype A_3. In the process of becoming more strongly differentiated ecologically, each species has also become more narrowly specialized genetically.

In the hypothetical case just examined the symbols E_1, E_2, and E_3 are assumed to represent different feeding niches. These could be visualized as different altitudinal zones for mountain-inhabiting birds or different feeding levels for Paramecia. The symbols could equally well stand for different types of food, thus seeds of different sizes for birds[39] or different species of yeasts eaten by Drosophilas,[40] with the same general results, namely, divergent nutritional specializations under certain conditions of interspecific selection.

The white-eye, *Zosterops palpebrosa*, breeds from sea level to the high mountains in Burma, where there is no other species closely related to it. But in Malaya and Borneo where a related species of Zosterops occupies the middle and higher zones, *Z. palpebrosa* is restricted to the lowlands. And in Java, Bali, and Flores where there are two other related species, one in the higher regions and one on the coast, *Z. palpebrosa* is restricted to the middle altitudes.[41] Lack has described similar cases in the Galapagos finches. A given species of finch frequently exhibits a wider ecological amplitude on islands where particular competitor species are absent than on islands where it faces interspecific competition.[42]

[39] Lack, 1947, 135; Mayr, 1948, 212.   [40] Dobzhansky and da Cunha, 1955.
[41] Stresemann in 1939; Lack, 1947, 30.   [42] Lack, 1947.

These relations can be seen again in the Cobwebby Gilias of southern California. *Gilia diegensis* occurs in a variety of ecological zones from the coastal plain and interior valleys up to pine forests in the mountains of San Diego County where no closely related species are found. But, farther north, where *Gilia leptantha* inhabits the pine zone and *G. latiflora* the lower valleys, *G. diegenis* is restricted to the higher valleys. These examples provide good indirect evidence of the role of interspecific competition and selection in promoting divergent ecological specializations among related sympatric species.

THE INFLUENCE OF ECOLOGICAL DEMANDS ON THE CLOSENESS OF COEXISTENCE BETWEEN ECOLOGICALLY SIMILAR SPECIES

There are great differences between the species belonging to different major groups in the extent of their requirements for space, food, and raw materials. The ecological demands of species are correlated with the size and abundance of their individuals. Common species of trees require far more space and raw materials than common species of herbs. The larger animals require much more space and food than the smaller forms.

Differences are also found in the closeness of coexistence of ecologically similar species. We saw in Chapter 12 that related species may be allopatric, neighboringly sympatric, or biotically sympatric; in other words, they may live in separate territories, in different habitats within the same territory, or in different niches in the same habitat. Further examples of the relative degree of ecological differentiation between species have been examined in this chapter, and we have seen that species may coexist in the same niche in some instances. The term coexistence covers a multitude of sins, for species may live together in various degrees of intimacy!

It will now be shown that the two phenomena are causally connected. The extent of the ecological demands of two or more related species is an important factor determining the closeness of coexistence between them, for the available supply of environmental resources will in general be used up more rapidly by the larger than by the smaller plants and animals. The interspecific competition will accordingly be keener on the average and will lead to stronger selective pressures in the larger forms than in the small-sized organisms. The interspecific selection is likely to lead to complete replacement in the case of organisms with very large requirements, to a marked ecological differentiation in organisms with moderately large demands, but may not exert pronounced effects in organisms with slight demands.

Ross, who found six related species of leafhoppers all feeding on sycamore leaves without any apparent signs of interspecific competition, attributes the absence of competition to the small size of these insects. The small size of the leafhoppers makes it possible for them to maintain themselves without depleting the food supply, even within a restricted niche, for their total food requirements are small. It is difficult, as he notes, to imagine six species of elephants maintaining populations in a single territory if restricted to sycamores for food—or even if unrestricted as to food source. In general, "the number of species which could occupy the same niche would be inversely proportional to the food requirements, hence usually to the absolute size, of the organism."[43]

The fauna of any area generally includes a large number of species of small animals but only a few species of large animals. Plant-eating insects are present in greater numbers of species than herbivorous mammals; carnivorous insects greatly exceed carnivorous vertebrates in species diversity. The explanation is partly that the larger animals require more space and food than the smaller ones.[44]

The largest mammals with the heaviest demands for food and space, as exemplified by the big cats, bears, and elephants, form predominantly allopatric species. Among very small animals and protistans with relatively slight demands, as exemplified by Drosophila, mosquitoes, and Paramecium, biotically sympatric assemblages of 6 to 14 related species have been described. And between these extremes are the numerous medium-sized animals with intermediate ecological requirements and intermediate degree of sympatry.

The relation between ecological demands and ecological differentiation is well illustrated by a comparison of the feeding habits of adult and larval forms in the Lepidoptera. The caterpillars feed heavily on plant materials. It is well known that different species of Lepidoptera living in the same area are generally specialized for feeding on different

[43] Ross, 1957.
[44] Hutchinson and MacArthur, 1959.

kinds of host plants in the caterpillar stage. Some species of caterpillars feed exclusively on cruciferous plants, others on solanaceous plants, or Aristolochia, or pines, etc. The adult butterflies and moths, which can subsist on moderate amounts of nectar, are not nearly as specific in their food niches as their larval forms. In temperate regions, different species of adult Lepidoptera commonly obtain nectar from the same wide range of flower species in an area, while seeking out particular plant species on which to lay their eggs and rear the next generation of caterpillars.

The adult hawk moths (Sphingidae), which are heavy-bodied, metabolically active, and long-lived, are somewhat intermediate in their food requirements between lepidopterous caterpillars and adult butterflies generally. Correlated with the intermediate extent of their demands for food is an incipient degree of divergence and specialization in the feeding habits of adult hawk moths in regions where a number of species coexist. In the American southwest two of the most common flower-feeding hawk moths, Celerio and Phlegethontius, differ markedly in proboscis length, Celerio being moderately long-tongued (3 to 4.5 cm) and Phlegethontius very long-tongued (8.5 to 12 cm). In this area the different species of adult hawk moths feed partly though not exclusively on different species of flowers: Celerio on a wide range of flowers with short or moderate tubes, and Phlegethontius on these and also on long-tubed flowers in such genera as Mirabilis, Aquilegia, Oenothera, and Datura.

In the California flora there are characteristic differences in the amount and degree of sympatry attained by species of annuals as compared with species of large perennial plants. The annual herbs tend to form flocks of biotically sympatric species. Among the perennial herbs, shrubs, and trees in the same flora, by contrast, related species usually grow in neighboring habitats or ecological zones.

The different species of oaks and pines, for example, generally occupy different ecological zones from coast to interior and from valley floors to high mountains, overlapping marginally in range. The same type of ecological segregation is found in such perennial herbaceous genera as Calochortus, Iris, and Polemonium. Thus on a transect across northern California, *Polemonium carneum* occurs in the cool moist coastal region; *P. caeruleum* in the coniferous forest; *P. californicum* in the upper part of the coniferous forest or in the subalpine zone; *P. pulcherrimum* in the alpine zone above treeline; and *P. eximium* on the high peaks.

The sympatric species flocks in the California flora occur exclusively among the annual plants in such genera as Clarkia, Mentzelia, and Gilia. In Gilia one frequently finds two or three species intermingled in the same habitat, and it is not unusual to find six or seven related species growing together in the same spot of open ground. There is some ecological differentiation between these biotically sympatric species, as can be shown by detailed studies, but it is obviously not so great as to prevent close coexistence.

Within the category of California annual plants a further distinction can be seen between the large annuals and the diminutive ones in the closeness of coexistence. The distinction is well illustrated by the Cobwebby Gilias. The large species, *Gilia latiflora, leptantha, ochroleuca,* and others, frequently coexist in pairs and rarely in threes. The species of diminutive annuals in the same group, on the other hand, quite commonly form sympatric flocks in which three, four, or five species are intermingled in the same or similar niches.

The differences between perennial plants, large annuals, and diminutive annuals in the number and closeness of sympatric contacts between species are correlated with the relative ecological demands of the species belonging to the several life-form classes. These demands are of course greatest in the large perennial plants, are much smaller in the annuals, and reach their minimal proportions in the diminutive annuals.

The ecological requirements of different species of organisms form a spectrum ranging from very great to slight. There is also a spectrum in the frequency and closeness of the sympatric contacts between related species. The two spectra are correlated. The largest animals and plants tend to form predominantly allopatric species. In many groups of vertebrate animals and perennial seed plants with intermediate ecological requirements, related species coexist chiefly as ecologically isolated neighbors in the same territory. Flocks of biotically sympatric species living in similar or even perhaps identical niches are found mainly among the smallest types of plants, animals, and protistans. These correlations can be explained as a result of the more intense interspecific competition and selection in large organisms than in small forms.

THE COEXISTENCE OF ECOLOGICALLY DISSIMILAR SPECIES

Sympatric species with widely different ecological requirements do not compete and may instead be dependent upon one another in various ways, as animals are dependent upon green plants and predators upon prey. The dependencies may become mutual in some instances. A mutually dependent association exists between the various species of flowering plants in an area and the various flower-feeding and flower-pollinating insects and birds. An example of a much closer cooperation between species is furnished by the lichens with their mutually dependent algal and fungal members.

Ecological differences between sympatric species are frequently favorable to the species concerned, like a division of labor in human affairs. Darwin remarked that "the greatest amount of life can be supported on each spot by great diversification or divergence in the structure and constitution of its inhabitants."[45]

SELECTION IN THE BIOTIC COMMUNITY

A complex web of ecological relationships, some competitive but many others interdependent, links together the diverse species comprising a biotic community. The biotic community has a life of its own; it grows, maintains itself, extends itself under favorable conditions, and may even evolve.

Selection is an important factor determining the composition of a biotic community, the types of ecological relations between its component species, and hence its basic characteristics. The selective processes involved operate partly at the individual level but also partly at the higher levels of species and community, as we shall now see.[46]

One of the conditions affecting the survival of a biotic community is its stability. The stability of a community increases with the number of connections in the food web. Complexity favors stability. To illustrate this point we may compare the extremely different types of communities developed in the tropics and in the arctic.[47]

[45] Darwin, 1868, introduction.   [46] See Allee *et al.*, 1949, ch. 35.
[47] MacArthur, 1955; Hutchinson, 1959.

The tropical forest communities consist of large numbers of inter-dependent species and are stable. The numerous species of trees support numerous kinds of insects, which in turn provide food for numerous species of birds, and these serve as prey for many kinds of predatory animals. Any given species of leaf-eating insect is likely to be able to feed on two or more alternative species of plants; an insectivorous bird can utilize various alternative species of insects for food; a predatory animal can prey upon different species of birds. Fluctuations in the abundance of any single species do not bring about large sympathetic fluctuations in the numbers of some other species which uses the former as food, because the latter, whenever one kind of food source is scarce, can turn to an alternative food source. An elaborate system of checks and balances preserves stability within a complex community.[48]

The situation is otherwise in the relatively undiversified communities of the arctic and boreal zones, where species are linked together in simple food chains. If one kind of prey, say a rabbit species, becomes decimated by disease, its predators, having few or no alternative food animals, are likely to become decimated by starvation. After the disease has run its course, and while the numbers of predators remain small, the rabbit population may grow to epidemic proportions, until it is checked when the predator population also expands again. The populations belonging to the simple food chains are thus subject to violent cyclical oscillations. The community lacks the diversification that would cushion it against such fluctuations. Its lack of complexity is correlated with instability.[49]

Since diversified food webs with many checks and balances contribute to the stability of a biotic community, and since the stability of a community is a factor in its survival, the diversified communities will persist longer and will be more successful in nature than communities composed of few species. Although each species in a biotic community may fend for itself alone, integrated aggregates of these species tend to develop, because such aggregates are more stable and are preserved longer by selection acting at the community level. In short, the more stable communities outcompete and outlast the less stable ones.[50]

[48] MacArthur, 1955; Hutchinson, 1959.   [49] MacArthur, 1955; Hutchinson, 1959.
[50] MacArthur, 1955; Hutchinson, 1959.

It may be that the arctic biota is too young historically to have acquired yet the maximum diversity and stability that are possible under an arctic climate.[51] It has been suggested on the basis of some evidence that arctic marine communities are now in the process of evolution toward greater stability.[52] The geologically ancient forests of the tropical zone have presumably undergone selection for diversification and stability in the past.[53]

A rich diversity of animal life is supported in the tropics by a great diversity of plant life. The diversity of plants in the tropical rain forest has been commented on by many observers. To quote a classical writer, Humboldt[54]:

> In the temperate zone, particularly in Europe and Northern Asia, forests may be named from particular genera of trees which grow together as social plants, and form separate woods. . . . Such uniformity of association is unknown in tropical forests. The excessive variety of their rich sylvan flora renders it vain to ask, of what do the primeval forests consist. Numberless families of plants are here crowded together; and even in small spaces plants of the same species are rarely associated.

Three one-hectare plots (100 meters square) at two localities in Brazil were found to contain 60, 79, and 87 species of trees. It was estimated that less than half of the species present in the community were encountered in this survey, many rare and some common species being present in adjoining areas but absent from the plots sampled.[55] Comparable figures are obtained in other parts of the tropical forest belt.[56]

What is the cause of this species diversity of trees in the rain forest?[57] One of the explanations often given, that the tropical climate is favorable for plant growth,[58] will account for the large total amount of plant

[51] Hutchinson, 1959.   [52] Dunbar, 1960.   [53] Fischer, 1960.
[54] Humboldt, 1950, 194–95.   [55] Black, Dobzhansky, and Pavan, 1950.
[56] Richards, 1952, 230.
[57] A valuable discussion of the general problem of species diversity, animal and plant alike, in the tropics is given by Fischer, 1960. Another point of view on this problem has been presented recently by Klopfer and MacArthur, 1961. Recommending to the reader the task of consulting these papers in the original, I will set forth here and in the next section a line of thought which I have held for many years, and which runs parallel in places to Fischer's discussion. The following brief remarks, needless to say, can provide only an oversimplified analysis of one aspect of a very complex problem. It is the aspect, however, with which we are concerned in the present section.
[58] I.e., Belt (1874) and Richards (1952, 229) to cite just one traditional and one recent author.

material produced by trees, but does not tell us why this tree life is aggregated into many rather than few species. A diversity of ecological niches in the territory occupied by the forest will account for some of the species diversity, but does not provide a satisfactory explanation for the coexistence of those species which are not segregated ecologically in any observable way.

Similarly, the geological antiquity of the tropical zone, untouched by the ice ages that depauperated the temperate floras,[59] can only account in part for the relative floristic richness of the rain forest. The historical factor seems best fitted to explain why a large number of ecologically diversified species are preserved extant. It does not tell us why mixtures of ecologically similar species of trees have been preserved; indeed the long time available for interspecific competition might have been expected to have reduced the number of competitor species within each ecological niche to one or a few. The question remains, then, why many species of trees with no apparent ecological differentiation grow in mixed stands in the tropical forests.

The moist warm tropical climate is favorable for the growth, not only of plants, but also of plant diseases and plant pests. Each disease organism is likely to be specialized for attacking certain susceptible species of plants but not other resistant species. A dense crowding of one or a few species of trees in relatively pure stands in the forests presents a favorable situation for the development of epidemics, in which the disease can spread from one susceptible tree to a neighboring susceptible tree. It is because of plant diseases that the Brazilian rubber tree, *Hevea brasiliensis*, cannot be grown in plantations in the area where it is native. A dispersion of the individuals of each species in the forest, on the other hand, is a response to and protection against epidemics.[60]

Given dense stands of any species of tree in the rain forest, its ranks would be thinned in time by disease, and the vacant places would be occupied by other resistant species. This process continued over long periods would lead eventually to the presence of many ecologically similar species each of which is dispersed sparsely throughout the forest. This is the condition which actually exists. The mixed associations of tree species which occupy similar edaphic and climatic niches

[59] Richards, 1952, 230.   [60] Also Savile, 1960.

in a tropical forest may be explained as a result in part at least of selection for communities composed of disease-avoiding species.

THE INTERACTION BETWEEN SELECTION BY PHYSICAL FACTORS AND SELECTION BY BIOTIC FACTORS

An organism must be adapted to its total environment in order to survive. The total environment is highly complex and embraces many aspects, which for convenience can be resolved into three main groups. The organism must be fitted to the physical factors of moisture, temperature, light, chemical nature of the substratum, etc.; to the biotic factors posed by other species of organisms in the same community; and to other individuals of its own population, which we can designate as social factors.

The environmental agencies that exert selective pressures on the organism are consequently also threefold. Natural selection may have reference to the physical factors of the environment, to the biotic environment, and to the presence of conspecific individuals in the social environment. While we can recognize these three classes of environmental selection for purposes of analysis, there is probably always an interaction between them in practice, and evolutionary changes are rarely if ever determined purely by one set of factors alone. We shall consider here with the help of some simple examples the interaction between selection by the physical factors and selection by the biotic factors of the environment.

Desert plants are adapted to withstand drought—a physical adaptation. But the shortage of plant life in deserts enhances its desirability for herbivorous desert animals. Corresponding to this condition, desert plants are outstanding in their development of spines protecting the leaves and stems from being devoured by animals. Biotic adaptations are a corollary here of physical adaptations.

In the general course of evolution of land life, adaptations to withstand drought and extreme temperatures have been a primary evolutionary goal. The selective factors most directly involved in the colonization of the land environment have been the physical factors, for the first forms of land life had to develop the adaptations for meeting the problems of moisture and temperature on land before anything else. Where the problem of unfavorable moisture and temperature conditions was solved most easily and hence solved earliest in the history of life, namely, in the warm moist tropics, there is now the richest flora and fauna. The strong competition between ecologically similar species, and the relations between host and parasite species, have led here to numerous biotic specializations. Biotic adaptations have been superimposed on the physical adaptations in the tropics.

The desert and polar organisms are the present-day heirs to the old challenge of adapting to unfavorable climatic conditions. Progressive evolution guided by physical selection is still going on in desert and polar regions. Biotic selection is not without its important effects here, as exemplified by the spines on desert leaves and stems, but owing to the relative poorness of the desert and artic biota in species, this biotic selection is auxiliary to physical selection, and is slight in comparison with the role of biotic factors in the tropics. Biotic adaptation is in a sense a luxury which can be supported only after the physical environment is brought under control by the appropriate physical adaptations.

But even in the emergence of primitive land life and later of desert and polar life, evolution has not been guided by physical selection alone. Crowded conditions in the sea stimulated the first amphibians and vascular plants to set foot on land. Many desert plants can trace their ancestry to tropical groups, and overcrowded conditions in the tropics may thus have stimulated the exploration of marginal arid environments. Looking at the causes of evolutionary changes in relation to physical factors, we find biotic factors occupying a prominent place in the background.

Biotic factors, then, have often initiated the exploration of new and unfavorable physical environments. Selection by physical factors enabled the organisms to actually conquer the new environment. As the conquest of the environment is completed by different organisms, evolving in parallel ways, biotic factors increasingly come into the picture again, and set the stage for a new wave of colonizations of another new environment.

FIFTEEN

Evolutionary Divergence

> Those forms which possess in some considerable degree the character of species, but which are so closely similar to some other forms, or are so closely linked to them by intermediate gradations, that naturalists do not like to rank them as distinct species, are in several respects the most important for us. CHARLES DARWIN[1]

THE PROCESS OF DIVERGENCE

IN *The Origin of Species* Darwin called attention to the fact that "organic beings have been found to resemble each other in descending degrees, so that they can be classed in groups under groups." Thus all the members of one order have some features in common; the members of each separate family in the order have more features in common; the members of each genus within a family are still more alike; and still closer resemblances are exhibited by the members of the same species within a genus.[2] The series continues on down from the species level through races to local breeding populations, as we saw in Chapter 12.

"The grand fact of the natural subordination of organic beings in groups under groups," which is a fact of nature and not an artifact of man-made classifications, can only be explained, according to Darwin, as the result of a process of evolutionary divergence. For if an ancestral form breaks up into subgroups which diverge in character through time, their modified descendants will be seen "to resemble each other in descending degrees, so that they can be classed in groups under groups." The genera belonging to the same family resemble one another because they have inherited something in common from their ancestral stock; the members of a single species are still more alike because their geneological relationship is closer.[3] Darwin used a diagram similar (and ancestral) to our Fig. 7 in Chapter 2 to make his explanation

---

[1] Darwin, 1859, ch. 4.   [2] Darwin, 1872, ch. 14.   [3] Darwin, 1872, ch. 14.

clear. Referring to the latter figure, we can see that the pattern of taxonomic relationships (Fig. 7, top) is best interpreted as a result, viewed at one time level, of a past history of evolutionary branchings (Fig. 7, bottom).

In Part 3 of this book we considered evolution as a change in the frequencies of alleles or genotypes within a population. The population was assumed to pass as a whole from one genetic composition to

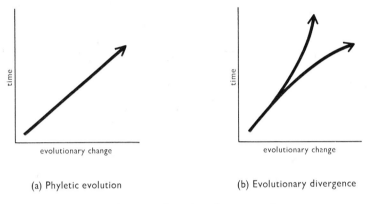

(a) Phyletic evolution   (b) Evolutionary divergence

Fig. 76. Two modes of evolutionary change

another during the passage of time. But a population may become different in another sense; it may split and become genetically different in its different branches.

The direction of evolutionary change can be viewed either as a progressive movement of one population from one state to another, or as a divergence between two or more populations. The evolutionary motion is linear when measured against time for a single population, but is relative when measured against the trends in other related populations. Darwin referred to the two evolutionary modes as "descent with modifications" and "the origin of species." Modern evolutionists often speak of phyletic evolution as contrasted with evolutionary divergence (Fig. 76).

The transformation of populations of the British moth, *Biston betularia,* from gray colored to black during a fifty-year period in the nineteenth century provides a case of phyletic evolution (see Chapter 9). The Illinois corn-selection experiment, also described in Chapter 9, furnishes an example of evolutionary divergence. It will be recalled that one original population of corn plants with ears borne midway on the stems split in the course of fifty generations into two derivative populations, one with ears borne high and the other with ears borne low on the stems.

Considering the evolution of organisms on earth in its broadest aspects, divergence is the process which has given rise to the approximately 4.5 million contemporaneous species and the innumerable races within these species, while phyletic change is the process involved in the progressive transformation of each race or species from its ancestral form.

It is true that paleontologists in dealing with specimens from different time levels on the same phyletic line may have to assign one species name to an ancestral form and one or more other species names to various morphologically different descendant forms. The ancestral species has given rise in the course of time to the descendant species. But successional species are stages in phyletic evolution. They are represented by points located at different places along the length of one arrow in Fig. 76a or b. We are concerned here with the development of gaps between diverging lines of evolution, as represented by the branching of the arrow in Fig. 76b. In Fig. 77 Population A' and Species A represent two different successional species, whereas Species A and Species B are separate contemporaneous species.

By what criteria do we measure relative divergence? As phyletic lines diverge, they become increasingly different in their morphological and physiological characters, and in the genotypes determining these characters. They develop different adaptive properties which are based on different allele combinations. Furthermore, as the two lines become more different in their genetic constitutions, the degree of isolation between them also increases (Fig. 77). Isolation is a condition which comes about as a by-product of divergence and which, to complete the cause-effect cycle, promotes further divergence. Finally, the ecological relationships between diverging phyletic lines undergo changes if and when these lines pass from a state of non-competitive allopatry to a state of competitive coexistence.

If the process of evolutionary divergence continues over a long period of time, the diverging phyletic lines will become progressively more different in genetic constitution and ecological requirements and more strongly isolated. By these criteria two colonies may diverge in time to the stage of separate geographical races. With further divergence the races may become separate semispecies and eventually distinct sympatric species (Fig. 77).

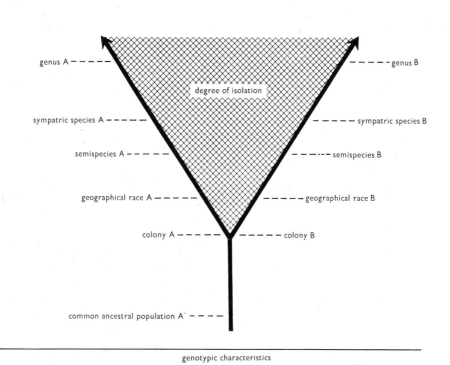

Fig. 77. Stages of primary divergence

In the Illinois corn-selection experiment a single ancestral population of corn with medium-high ears gave rise in the course of fifty generations of differential selection to one derivative line with tall stems and high ears and another with short stems and low ears. Now the low-ear plants flower earlier in the season than the high-ear plants. The two derivative populations not only differ in several gene-controlled morphological and physiological characters, therefore, but are in addition seasonally isolated.[4] If the reproductively isolated corn populations were to grow together in the same fields, so that a struggle for existence between them would arise, they would be exposed to a stimulus to diverge still further in their ecological requirements.

THE GEOGRAPHICAL THEORY OF SPECIATION

The geographical theory of species formation, anticipated by Darwin, advanced by M. Wagner in 1889, developed by K. Jordan and D. S. Jordan in 1905, and further developed and advocated in the modern period by Rensch (1929, 1933), Dobzhansky (1937, 1951), Mayr (1942, 1947), and others,[5] holds that the stage of geographical races is a normal and even necessary preliminary to the stage of sympatric species in cross-fertilizing organisms. Spatial isolation of populations, in other words, is a usual and perhaps an essential condition for the development of reproductive isolation between them.

The chief theoretical argument in favor of this theory is that the interbreeding of individuals within a population would usually swamp out in its initial stages any evolutionary divergence developing in a local area; whereas two or more spatially isolated populations may accumulate many genetic differences, including those which make for environmental and reproductive isolation.[6] Later in this chapter we will examine instances of incipient reproductive isolation between races.

The above argument has a special cogency as regards the origin of internal reproductive isolating mechanisms. There are two aspects to incompatibility, hybrid inviability and hybrid sterility between two population systems, A and B. Different individuals of A must be able to intercross freely with one another to produce vigorous and fertile progeny, and the individuals of B must likewise be interfertile, but reproduction is blocked in the hybrid combination A × B. The genes which bring about internal isolation of the heterogamic combination A × B or its heterozygous progeny do not affect the homogamic combinations or homozygous types A × A or B × B.[7]

Single isolation-inducing mutations arising within a population would produce their effects first as infertile or inviable individuals in

[4] Bonnett, 1954.   [5] See Dobzhansky, 1951a, 204–5, for references.
[6] Dobzhansky, 1937, 1951a. Mayr, 1942, 1947.
[7] Dobzhansky, 1937, 255–56, 1951a, 202–4.

that population. Such individuals and the mutations they carry would be selected against. As Dobzhansky states[8]:

> In sexual and cross-fertilizing species, a great difficulty is encountered in the establishment of any reproductive isolating mechanism in a single mutational step. Since mutants appear in populations at first as heterozygotes, inviable and sterile heterozygotes are eliminated, regardless of how well adapted might be the corresponding homozygotes. This consideration is fatal to Goldschmidt's (1940) theory of evolution by "systemic" mutations. These mutations are supposed to induce isolation of the newly emerged species from its ancestor. Even if the inviability or sterility of heterozygotes be supposed to be incomplete, these heterozygotes will be discriminated against by natural selection . . . .
>
> It is probable that the formation of isolating mechanisms entails not single mutational steps but the building up of systems of complementary genes. Assume that a population has the genetic constitution aabb, where a and b are single genes or groups of genes, and that this population is broken up into two allopatric, geographically isolated parts. In one part, a mutates to A and a local race AAbb is formed. In the other part, b mutates to B, giving rise to a race aaBB. Since individuals of the constitutions aabb, Aabb, and AAbb interbreed freely, there is no difficulty in establishing in the population the gene A. The same is true for the gene or genes B, since aaBB, aabB, and aaBB interbreed freely. But the cross AAbb × aaBB is difficult or impossible, because the interaction of A and B produces one of the reproductive isolating mechanisms. If the carriers of the genotypes AAbb and aaBB surmount the extrinsic barriers separating them, they are now able to become sympatric, since interbreeding is no longer possible.

Thus workable isolating mechanisms can be produced most effectively by systems of complementary genes. And the separate components of a set of complementary factors for reproductive isolation can be established in the different populations most readily during a period of geographical isolation.[9] Dobzhansky's hypothesis, considered above in relation to internal barriers, applies in certain instances also to the development of seasonal, ecological, and other external isolating mechanisms.[10]

So far as the taxonomist is able to infer the course of evolutionary divergence from the comparative distribution patterns of existing population systems, the theoretical expectation that geographical isolation should precede reproductive isolation, and that sympatric species develop from geographical races, is confirmed by the evidence of minor systematics for many groups of birds, mammals, fishes, insects, flowering

[8] Dobzhansky, 1951a, 203–4, quoted from *Genetics and the Origin of Species.*
[9] Dobzhansky, 1951a, 204.   [10] Dobzhansky, 1937, 255.

plants, and other organisms. The taxonomic arguments for the geographical theory of species formation have been well summarized by
Mayr.[11]

Every stage of divergence expected on the basis of this theory, from
geographical races through various intermediate conditions to sympatric species, is found in nature, and often within a single related
group of organisms. Thus a complete series of stages in divergence is
found within the Cobwebby Gilias of southern California (see Chapter
12). (1) Continuous geographical variation is exhibited by the contiguous races *latiflora*, *davyi*, and *excellens* of *Gilia latiflora* (Plate IIa).
(2) Disjunct geographical races, exhibiting discontinuous geographical
variation, are found in *Gilia leptantha* (Plate IIb). (3) Ecological races
are exemplified by *davyi* and *cuyamensis*, and *excellens* and *elongata*,
in *Gilia latiflora* (Fig. 71). (4) Largely allopatric but marginally
sympatric semispecies are found in the *Gilia latiflora-leptantha-tenuiflora* syngameon (Plate IIc). (5) As an example of allopatric species we
have *Gilia ochroleuca* and *Gilia mexicana*. (6) As sympatric species we
can cite *Gilia ochroleuca* in relation to *Gilia latiflora*, *leptantha*, and
tenuiflora (Plate IId).

Mayr reports that about 12 percent of the population systems of
continental North American birds and 28 percent of those in the
Solomon Island fauna have reached a semispecific stage of divergence
more advanced than the stage of geographical races without however
attaining the status of sympatric species, so that in formal taxonomy
these bird populations are sometimes treated as subspecies and sometimes as (allopatric) species.[12]

Many cases are known in both animals and plants of intergrading
series of geographical races, the terminal and most extreme members
of which overlap in distribution without interbreeding.[13] The great
titmouse (*Parus major*) forms a chain of continuously intergrading
races across Eurasia. The race *major* with a yellow belly ranges through
Europe and Siberia to the Amur region in northeast Asia; the race
major also extends southeastward through Turkey to Iran where it
intergrades with a white-bellied race belonging to the *minor* group of
races; race *minor* then ranges through India and China to the Amur

---

[11] Mayr, 1942, 154–85. [12] Mayr, 1942, 165.
[13] Mayr, 1942, 180–85, for review of animal examples.

region, where it exists sympatrically with *major*.[14] Such overlapping rings of races have been found in warblers, gulls, kingfishers, mice (Peromyscus), salamanders, and butterflies among animals.[15] A case in the angiosperm genus Diplacus in southern California involving *D. puniceus* and *D. longiflorus*, which intergrade in some mountain areas but coexist without interbreeding in others, was mentioned in Chapter 12. The population system in such cases exhibits simultaneously the properties of geographical races and of sympatric species in different parts of its distribution area.

Lack has pointed out that the process of species formation in island-inhabiting birds has been active only on island archipelagos, like the Galapagos Islands and the Hawaiian Islands, where opportunities exist for geographical isolation on different islands; whereas the bird populations on solitary oceanic islands, while frequently diverging markedly from their mainland relatives, have never given rise to numerous new endemic species *in situ*.[16]

Some cases of sympatric species flocks, as for example the over 300 species of shrimps in Lake Baikal and the 174 endemic species of cichlid fishes in Lake Nyasa, which are isolated water bodies, at first seemed to require an extensive multiplication of species under conditions of sympatry, but on analysis the interpretation of sympatric species formation is found to be gratuitous. For the lake in question is a part of a vast present or past network of lakes and rivers, which could provide opportunities for geographical isolation of diverging populations. The new species with the reproductive isolating mechanisms they acquired during a period of spatial separation could then colonize the lake. Such colonizations continued over a long time would result in the accumulation of large flocks of sympatric species in the lake. And if changing environmental conditions led to the extinction of these species in the tributary water bodies, they would remain as endemics in the lake.[17]

THE CAUSES OF DIVERGENCE

The primary evolutionary forces—mutation, gene flow, selection, and drift—which bring about phyletic changes within populations will

[14] Rensch in 1933; see Cain, 1954, 54–58, with map.
[15] Mayr, 1942, 180–85; R. Stebbins, 1957.
[16] Lack, 1947, 150–58.   [17] Mayr, 1942, 1947.

account for the evolutionary divergences between them as well. Divergence develops when these forces act differentially in different branches of one population or in different related populations derived from a common ancestral stock.

If the primary evolutionary forces continue to act differentially in separate phyletic lines, the divergence will become progressively greater with the passage of time. The separate phyletic lines may then pass through the stages of races and species and eventually become different genera (Fig. 77) or reach still higher taxonomic levels of differentiation.

RACE FORMATION

The variation-producing and variation-sorting forces were dealt with in Part 3 and will only be mentioned here briefly as they relate to race formation.

The mutation process could theoretically be a predominant factor in racial differentiation in small isolated populations. If the mutation rate is large in relation to population size, mutation pressure will predominate over either drift or selection. The relation between mutation and drift was summarized briefly in Chapter 11; in general the allele frequencies in a population will be controlled by mutation when $u > 1/2N$. And if the mutation rate is as high as $1/1,000$ of the population size, the mutation pressure can outweigh selection.[18] It follows from these considerations that, in a series of separate small populations in which the mutation process can control allele frequencies directly, racial differences might be produced by the different types of mutations arising in the separate populations.

It will be recalled from Chapter 8 that, while the mutation process is the ultimate source of genetic variability, the chief immediate source in most populations of higher plants and animals is gene flow and recombination. It is easy to see that if gene flow occurs in one branch of the population system of a species but not in some other branch, the genes producing racial traits would tend to spread preferentially throughout the former populations. Widespread gene flow within a species, on the other hand, tends to unify the species and obliterate racial differences.

Rensch has compared the number of well-marked geographical races in the migratory and non-migratory species of smaller birds of the

[18] Ludwig; see Rensch, 1959, 9.

palearctic region. He found that the sedentary birds have an average of 7.2 races per species (for a sample of 81 species), whereas the migratory birds have only 3.2 races per species (average for 173 species).[19] Mayr has called attention to an interesting correlation between breeding habits and race formation in water fowl. In geese, which form semi-permanent families and nest in colonies, and which consequently become inbred, there are many well-marked races in each species, some of which are treated as species. But in the ducks, by contrast, which choose their mates in a more random fashion, from year to year, few or no races are developed even in species with very wide circumboreal distribution areas.[20]

Interspecific hybridization is an important source of variation in many groups of higher plants. Two species or semispecies that overlap on the margins of their respective distribution areas often hybridize on a limited scale in the zone of overlap. The variations engendered by local hybridization may be fixed in the populations by selection and/or drift. If the variations of hybrid origin do not spread throughout the entire population system of the species, but remain in one geographical region, as frequently happens, a racial differentiation will be promoted by the hybridization. Furthermore, each species or semispecies will be represented in the geographical area where it contacts the other species or semispecies by races which vary in the direction of the other entity. There is, in short, a convergence in the racial characteristics of the two species or semispecies in and around their area of overlap.

Thus the race *davyi* of *Gilia latiflora* approaches *Gilia tenuiflora* both morphologically and geographically (see Plate IIa and c). In all probability *davyi* has diverged racially from the ancestral stock of *Gilia latiflora* in the direction of *G. tenuiflora* as a result of an influx of genes from *G. tenuiflora* into *G. latiflora* during some past period of hybridization. This pattern of geographical variation involving racial convergence in the area of overlap between two species or semispecies is a very common one in higher plants. In many cases this pattern of racial variation can be associated with evidences of past or present hybridization.

Interspecific hybridization is a potential source of the genetic variations that can be utilized in race formation in the animals, protista, and

[19] Rensch, 1959, 23.   [20] Mayr, 1942, 241–42.

fungi as well as in plants. In any sexual organism the variability in the species can be increased by natural hybridization. Whether in fact the natural hybridization does or does not occur on a scale sufficient to raise the level of variability significantly is another question. In many plant groups it does. In many groups of animals, such as Drosophila among the insects and many mammals and birds, it does not. In other animal groups, the fishes for example, natural hybridization occurs fairly frequently and could be a source of racial variations. As regards many other groups of animals, particularly the lower forms, and the fungi and protista, the importance of hybridization as a source of variation remains to be determined.

The great role of natural selection in the formation of races can be inferred from the observation that racial characteristics are often adaptive. The adaptiveness of the racial characters in many plants and animals is demonstrated by two sets of correlations. First, the different races of a species have morphological and physiological characters that are related to the distinctive features of the environment in their respective areas. Second, the same general patterns of racial variation frequently recur in a parallel form in separate species inhabiting the same range of environments.

In birds and mammals, for instance, the geographical races inhabiting the areas with a warmer climate are generally smaller and the races occurring in the cooler areas are larger in body size. The size of a warm-blooded body is a factor in its thermodynamic efficiency. A small warm body, owing to a large surface in relation to the volume, gives off heat relatively rapidly into the air, whereas a large body with a smaller surface-volume ratio conserves its body heat to a larger extent. The conservation of body heat is adaptive in cold climates and the races of a species living in such environments tend to have large bodies. In warm areas, by contrast, the prevention of heat loss from the body does not have such a high adaptive value, in fact mechanisms increasing the irradiation of heat may be required for proper temperature regulation, and the races living in the warmer parts of the species area generally have the smaller body size.

Now if the racial variations in body size had the trend described above in just one or a few species, those variations might be ascribed to chance. But they occur in species after species among the birds and mammals. So general are these trends, indeed, that they are referred to as Bergmann's rule after the zoologist who described them in 1847. Only 16 percent of the passerine birds of the palearctic region do not conform to Bergmann's rule.[21] Therefore this pattern of racial differentiation cannot be a matter of chance. Since selection is the main antichance factor known in evolution, the development of racial variation conforming to Bergmann's rule is explained most satisfactorily by the hypothesis that it is a result of natural selection.

Many other characters in animals exhibit trends of geographical variation within species which are evidently adaptive, considered on their ecological merits alone, and which furthermore recur in parallel forms in species after species. Thus Allen's rule states that ears, tails, limbs, and other protruding body parts in birds and mammals are relatively shorter in the races living in the cooler areas and longer in the races inhabiting the warmer parts of the distribution area. Protruding body parts give off heat, which is likely to be disadvantageous for a warm-blooded animal living in a cold environment, but is not so disadvantageous and may even be advantageous in a warm climate. The adaptive trends of racial variation with respect to the length of protruding body parts have probably been established independently in many different species of warm-blooded vertebrate animals by the common factor of selection.[22]

Similarly, in plants the races of a species frequently exhibit characters that are correlated with the nature of the environment in their respective areas. *Hieracium umbellatum* is represented by two ecological races on the windy west coast of Sweden, by bushy plants with broad leaves and an expanded inflorescence on sea cliffs, and by prostrate plants with narrow leaves and contracted inflorescences on sand dunes; and as the rocky cliffs and shifting sand dunes alternate along this coast, so too does *Hieracium umbellatum* give rise alternately to bushy cliff-inhabiting races and stringy dune-inhabiting types.[23]

Turesson showed that the genetically determined traits fitting a plant race for a certain range of environmental conditions are paralleled by

[21] Rensch in 1939; see Mayr, 1942, 90.
[22] For excellent brief summaries of the numerous ecological rules in animals, see Mayr, 1942, 88–94; Rensch, 1960.
[23] Turesson, 1922.

the phenotypic modifications induced by the same environmental factors. For example, in *Atriplex litorale* the race on the wind-swept sandy west coast of Sweden is low and spreading, while the race on the sheltered south coast is tall and erect. When individual plants of the erect southern race were transplanted to sand and exposed to wind they became prostrate by phenotypic modification. There is a parallelism between the characters which arise as phenotypic responses in individual plants upon exposure to particular environmental conditions and the hereditary characters of the race which lives under the same environmental conditions. Turesson argues that if the phenotypic responses are adaptive, as is generally agreed to be the case (see Chapter 6), then the parallel racial characters must be adaptive too. Such racial characters can be looked upon as "genotypical responses of the plant species to its habitat" produced by selection.[24]

Different species of plants exposed to the same habitat conditions develop parallel races with similar characteristics. This parallelism between the racial variations of different species, which is like that mentioned earlier in animals, can be illustrated by the races belonging to a series of unrelated species which inhabit the island of Oeland off the southeast coast of Sweden. The races in question grow on a limestone substratum which has a low water content and, since the atmospheric humidity on the island is also low, a dominant feature of the habitat is drought. The species with island races belong to different families and genera, namely, *Artemisia campestris* and *Leontodon autumnalis* (Compositae), *Rumex acetosella* (Polygonaceae), *Silene maritima* (Caryophyllaceae), and *Allium schoenoprasum* (Amaryllidaceae). The island races of these species are all distinguished from the related mainland races by either a prostrate or dwarfish habit, and by small, narrow, depressed, or thick-walled leaves. The common direction of divergence of the island races is toward shoot and leaf characters that conserve moisture. There can hardly be any doubt that the common factor involved in the parallel differentiation of drought-resisting races in the various unrelated species of plants inhabiting the Oeland limestone is selection.[25]

Racial differentiation may be caused, finally, by genetic drift. The observations and interpretations now subsumed under the term drift

[24] Turesson, 1922.   [25] Turesson, 1922.

were made by a number of naturalists and taxonomists since Gulick in 1870.[26] The theoretical basis of these interpretations was first set forth by the population geneticist Wright in 1931 (see Chapter 11).

The condition which enables drift to operate most effectively in relation to the other evolutionary forces is a relatively small number of breeding individuals. This condition may be brought about in various ways. A given population may be small and isolated during many successive generations in some cases. In other cases a population may be large for many generations but subject to periodical reductions to a small size. Or, again, new populations are frequently founded by one or a few colonizing individuals emigrating from a large parental population, and as the immigrants carry a non-random sample of the existing genetic variations, the new population descended from them is likely to diverge racially from the old population.[27] In each case the small size of the breeding population, whether recurrent or periodical, promotes a racial divergence in respect to genetic composition and phenotypic characters which is partly, or in extreme cases perhaps wholly, independent of selection.

Naturalists have observed in many groups of plants and animals that species with a colonial population structure tend to be differentiated into numerous local races. The relation between isolation of colonies and haphazard racial variation was perhaps first pointed out explicitly by Gulick in 1870 with regard to land snails on Pacific islands.[28] On Hawaii, Tahiti, and other islands in the Pacific the valleys which are inhabited by the snails are separated by sharp ridges, and the snail populations are accordingly isolated spatially from one another. Associated with this colonial subdivision of the species is a marked differentiation of the snails into local races. The distribution of the racial characteristics does not correspond to any known geographical or ecological gradients, but is irregular or haphazard in many instances, suggesting that the chance fixation of sporadic variations in the different semi-isolated populations has played an important role in race formation. This situation was described in the early years of this century for snails of the genus Achatinella on Hawaii (by Gulick in 1905) and of

---

[26] Mayr, 1942, 234; 1959, 3; and personal communication.
[27] Mayr, 1942, 237.   [28] Mayr, personal communication.

the genus Partula on Tahiti and Moorea (by Crampton in 1916 and 1932).[29]

Recently described cases of marked colony-to-colony variation in the snail Cepaea in France, in man in Greenland, and in the plants Gilia and Clarkia in California were mentioned in Chapter 11. It is not necessary to assume that selection has no effects in the formation of local races in these groups, indeed selection may well have important effects, but the random component in the racial variation, more specifically the irregular distribution of racial characteristics in relation to environmental gradients, suggests that the factor of drift is also involved.

Naturalists have likewise long recognized that the peripheral populations of a species, and especially the insular populations belonging to predominantly mainland species, are likely to exhibit aberrant characteristics.[30] An example is provided by the kingfishers of the *Tanysiptera galatea* group on New Guinea. The mainland races of these birds are very much alike from one end of New Guinea to the other, whereas the populations on several surrounding islands are strikingly different from the mainland populations and from each other. This is evident from the taxonomic division of *Tanysiptera galatea* into three poorly defined races on the mainland and five very distinct races on the surrounding islands. A sixth insular form, *Tanysiptera hydrocharis*, has reached the stage of a separate species (Fig. 78). Although the island populations are numerically large today, they have probably descended from small numbers of colonizing individuals which migrated from the mainland.[31]

The amount of divergence in the kingfisher populations of the New Guinea region is not correlated with environmental differences. New Guinea is 1,500 miles long and 200 to 500 miles wide, and embraces a wide range of environmental conditions from constantly moist rain forest at one end to seasonally dry monsoon forest at the other. Yet the kingfishers are similar throughout this vast and environmentally diverse territory. The islands inhabited by the kingfishers, on the other hand, are 100 miles or less from New Guinea and lie in the same climatic and vegetational zones as the adjacent parts of the mainland.

[29] Dobzhansky, 1951a, 170.   [30] Mayr, 1942, 38, 152–53, 234–38.
[31] Mayr, 1942, 152–53, 236.

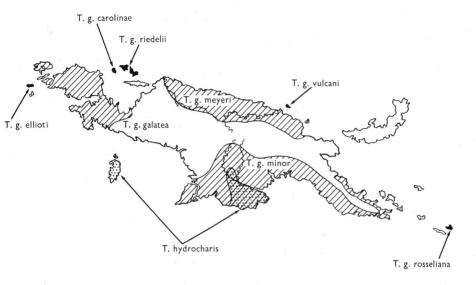

Fig. 78. Distribution of the races of the kingfisher, Tanysiptera galatea, *and of the related species* T. hydrocharis, *on New Guinea*

Redrawn from E. Mayr, *Systematics and the Origin of Species* (New York, Columbia University Press, 1942).

Nevertheless the island kingfishers are well differentiated from the mainland forms. It is clear, therefore, that environmental selection alone cannot account for the variation pattern of the New Guinea kingfishers. This variation pattern is what we would expect, however, if drift were also involved during the establishment of the outlying colonies.[32]

THE ORIGIN OF ISOLATING MECHANISMS AS A BY-PRODUCT OF RACE FORMATION

It is an important fact that all of the types of isolating mechanisms that separate species can be found also between population systems at the racial stage of divergence. Ecological isolation is present by definition between ecological races. Mechanical isolation prevents the crossing of very large with very small breeds of dogs and horses and could operate in the same way between large-bodied and small-bodied

[32] Mayr, 1942, 152–53, 236; 1954.

geographical races of shrews and reed warblers.[33] The shape of the genitalia is subject to geographical variation in butterflies and beetles and could bring about some mechanical hindrances to interracial crossing in these insects.[34] We have seen instances of seasonal isolation between races of herrings, tarweeds (in Chapter 12), and pines (in Chapter 13).

Ethological isolation between races of flowering plants pollinated by flower-constant bees has been described in *Gilia capitata*, as noted in Chapter 13. What we now call ethological isolation was discussed by Darwin in relation to races of dogs, horses, sheep, pigeons, and other domesticated animals. He concluded[35]:

Certain domestic races seem to prefer breeding with their own kind; and this is a fact of some importance, for it is a step towards that instinctive feeling which helps to keep closely allied species in a state of nature distinct. We have now abundant evidence that, if it were not for this feeling, many more hybrids would be naturally produced than is the case.

Internal reproductive isolating mechanisms are often found within a species. Thus certain southern and northern populations of *Gilia achilleaefolia* cannot be crossed successfully in hybridization experiments; but since two incompatible populations both cross freely with a third population, the latter can serve as an indirect pathway of gene exchange between the former.[36] Among a large number of interracial hybrids that have been produced in various species of Gilia, most have been semisterile, and quite a few have been weak. The F_2 generations derived from these hybrids generally include a proportion of inviable individuals, thus exhibiting partial hybrid breakdown.[37]

Where isolating mechanisms arise as a by-product of evolutionary divergence, we would expect the most extreme races of a species to exhibit the strongest degree of reproductive isolation. This expectation is realized in a number of cases.

The annual plant *Streptanthus glandulosus* (Cruciferae) occurs in a series of disjunct colonies forming several geographical races in the California Coast Ranges. Kruckeberg studied the hybrid fertility of 334 hybrid combinations derived from the intercrossing of 32 geographically different populations of this species. He found a statistically

[33] Rensch, 1959, 52.   [34] Rensch, 1959, 52.   [35] Darwin, 1868, ch. 16.
[36] Grant, 1954*a*.   [37] Grant, 1950*b*, 1954*b*; Grant and Grant, 1954, 1960.

significant correlation between hybrid fertility and the geographical distance separating the parental strains. The farther apart the populations in miles, the more sterile their hybrids.[38]

A similar relationship between the geographical remoteness of races and the fertility of their hybrids has been reported in *Clarkia biloba* in the Sierra Nevada of California.[39] The species of spruce (Picea), which are widely distributed in the Northern Hemisphere, differ in the ease with which they can be intercrossed. On the basis of a large number of crossings, it has been found that species of spruce with neighboring ranges are mostly compatible with one another, whereas the incompatibility barriers are strongest and most common between species that are widely separated geographically.[40] Likewise in several species of frogs, toads, and salamanders, hybrid inviability is most pronounced in hybrids between geographically remote races, and little or not at all evident between adjacent races.[41]

That the divergence of races with respect to various phenotypic traits should sometimes lead to ecological, mechanical, ethological, seasonal, and incompatibility barriers to crossing is only to be expected. Nor is it surprising that partial sterility and inviability is a common occurrence in F_1 and F_2 combinations of genotypically differentiated races. To the extent that environmental and reproductive isolation arises as a by-product of racial divergence, the races can behave as non-interbreeding species if and when they extend their ranges and become sympatric.

According to the geographical theory of speciation, some races could become species merely by range extensions or by the extinction of intermediate forms. The overlapping rings of races as found in *Parus major*, Diplacus, and many other plants and animals show that the postulated situation indeed occurs in nature. Darwin's thesis that races can be incipient species is justified by these considerations.

REVERSALS IN THE PROCESS OF DIVERGENCE

The course of evolutionary divergence is not inexorable. If the evolutionary forces which bring about the divergence change their direction of action at some point, the course of divergence can be

[38] Kruckeberg, 1957.   [39] Roberts and Lewis, 1955.   [40] Wright, 1955.
[41] Volpe, 1954, 1955, and other references cited therein.

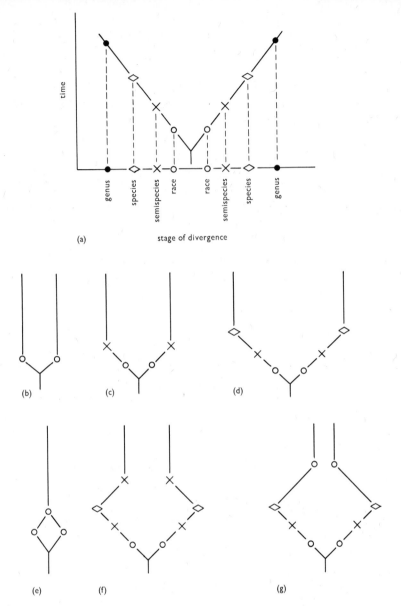

Fig. 79. Modes of arrested divergence

(a) Continuous divergence through the genus level. This figure is a standard of reference for the other diagrams.

(b)–(d) The phyletic lines cease to diverge at various stages, and then remain as different races (b), semispecies (c), or species (d).

(e)–(g) The phyletic lines converge from a higher to a lower level of differentiation as a result of hybridization. Reversal from separate races to a single race (e), from species to semispecies (f), or from species to races (g).

442

altered or even reversed. Some of the possible changes of course are shown diagrammatically in Fig. 79.

The simplest case is that of two phyletic lines which diverge up to some stage, then cease to diverge further, and run parallel for a long time thereafter. The parallel evolution may take place in related phyletic lines at any stage of separation: races (Fig. 79b), species (79d), genera, etc. Parallel evolution between genera and other higher categories constitutes a problem of macroevolution and will not be considered further here. Parallel evolution between the minor systematic categories, on the other hand, is of interest to us here in showing that the process of divergence does not necessarily always go on to the stage of species formation.

The skunk cabbage, *Symplocarpus foetidus*, occurs in swamps and wet meadows in temperate deciduous forests of two widely separated geographical areas, eastern North America and eastern Asia. On paleobotanical and paleoclimatological grounds, the American and Asiatic population systems of the skunk cabbage have probably been separated for 15 to 20 million years.[42] Yet in all this time they have scarcely reached the racial level of divergence, being very similar morphologically but different enough to be given different varietal names.[43] Figure 79b describes the incomplete divergence of *Symplocarpus foetidus*.

Diverging phyletic lines may at some point begin to hybridize, and if the hybridization is extensive enough, the course of divergence may be reversed. Several patterns of convergent evolution are shown diagrammatically in Fig. 79e–g. For example, two disjunct and morphologically discontinuous geographical races may reestablish contact and interbreed. This interbreeding will give rise to extraordinarily variable populations in the zone of overlap, a situation often referred to as secondary intergradation. At this stage there are two continuously intergrading contiguous races. If the hybridization is extensive and long continued the original races may become swamped out and amalgamated into a single new race of hybrid origin (Fig. 79e).

Or phyletic lines may return from the stage of species to that of semispecies or races as a result of extensive hybridization (Fig. 79f–g). *Gilia latiflora*, *G. leptantha*, and *G. tenuiflora*, for example, have reached

[42] Stebbins, 1950, 540–49.   [43] Stebbins, 1950, 540–49; Fernald, 1950, 384.

the stage of distinct sympatric species in some parts of their distribution area, while in other places they are linked together by hybridization and secondary intergradation, and are consequently regarded as semi-species.

Gilia capitata is today a single species composed of eight contiguous geographical races on the Pacific slope of North America. The most extreme forms in the species on morphological, ecological, and genetical characteristics are, on the one hand, the races *staminea* and *chamissonis* which are plants of sandy valleys and coastal sand dunes, and, on the other hand, race *capitata*, a plant inhabiting rocky hillsides in the Coast Ranges. If only these population systems were in existence today, they would be seen to have reached a stage of divergence more advanced than races, for *staminea-chamissonis* and *capitata* are separated by a well-marked morphological discontinuity which holds up even in various areas of sympatric overlap. The extreme entities, *staminea-chamissonis* and *capitata*, are, however, actually linked together at the present time into a single species by intermediate races which are of hybrid origin between them and which occupy intermediate habitats of recent origin. On the basis of various indirect correlations it is believed that *staminea-chamissonis* and *capitata* reached their maximum stage of divergence in the Pliocene and rejoined into a single continuously intergrading species during the Pleistocene epoch.[44]

Although divergence may turn in the direction of parallel evolution at any stage whatever in the hierarchy of taxonomic categories, the reversal from a more isolated to a more closely related condition can take place only below a certain stage of divergence. The stage of inter-sterile species marks a critical point in divergence. Once the internal reproductive isolation between two species is complete, so that hybrids cannot be produced, or if produced are completely sterile, the possibilities of gene exchange between these species and hence of their return to more interfertile species or hybridizing semispecies or inter-breeding races are cut off. When two or more species become separated by barriers of complete incompatibility or complete hybrid sterility, they must thenceforth evolve independently of one another's supply of genetic variations. The importance of this point of no return in the

---

[44] Grant, 1950*b*.

evolution of species has been emphasized by Clausen, Keck, and Hiesey among others.[45]

THE SPECIES AS A STAGE IN CONTINUOUS DIVERGENCE

Insofar as divergence is a continuous process, one stage of differentiation will pass gradually into the next higher stage. The differentiation between related colonies merges into racial differentiation. Two evolutionary lines may be found, not only at the racial stage or the specific stage of differentiation, but also at some transitional stage as exemplified by ecological races, semispecies, and allopatric species (Fig. 77). In a continuous and gradual sequence of evolutionary divergence, the dividing line between colonies and races, or between races and species, is necessarily arbitrary.

One may wonder whether the foregoing statement is in contradiction to the conclusion stated in Chapter 12 that objectively real species exist in nature. Was Darwin right in holding that species do not differ essentially from races, so that "the term species . . . is one arbitrarily given for the sake of convenience . . . ?" Or was Karl Jordan right in stating that "the living inhabitants of a region . . . are composed of a finite number of distinct units [viz., species] which are sharply delimited" by bridgeless gaps?[46] The proper question is not, was Darwin right and Jordan therefore wrong, or vice versa, but what measure of truth exists in the conclusions expressed by each of these keen observers of natural population systems?

In order to compare the apparently contradictory viewpoints of Darwin and Jordan properly we must place them in a common context. There are two such contexts in which these ideas can be compared. We can consider Darwin's and Jordan's statements as relating to the situation in nature at any given moment of time; or as relating to the course of evolutionary divergence in a four-dimensional field of space and time.

In Chapter 12 we attempted to reconcile the two views with regard to the situation prevailing in nature at the present time level. We saw that Darwin's statement is a generalization encompassing the whole array

[45] Clausen, Keck, and Hiesey, 1939, 1943, 1945, ch. 5; Clausen, 1951, 170–78 *passim.*

[46] Full quotations in ch. 12.

of population systems in the world, while Jordan's statement refers particularly to non-interbreeding population systems living together in the same biotic community. Sympatric species with well-defined limits undoubtedly exist, as Jordan stated, but so do various poorly defined intermediate population systems, as Darwin emphasized.

Let us now consider the two ideas as they apply to evolutionary divergence in a field of space and time. If the evolutionary changes are gradual and continuous, the divergence also passes gradually from one taxonomic level to another. Even though evolution is continuous, however, evolution crosses boundary lines. The boundary line separating the racial from the specific stage of divergence is defined by the development of reproductive isolation. The latter development requires time in itself, as we shall see, and the boundary line between races and species is therefore more accurately visualized as a border *zone* of varying width on the time line. This border zone corresponds to the semispecies level of divergence in Fig. 77.

We are inclined to believe that portraying a series of events in the time dimension gives us a more complete picture of the process than does a description of the phenomena observed at some particular moment of time. One conclusion that can be drawn from this premise is that the Darwinian concept of species as arbitrarily defined entities has, in the larger perspective of time, a higher order of validity than the concept of species as non-arbitrarily delimited units. However, this is not the only conclusion that can be deduced from the premise. There is a sense in which the relations between population systems at any given moment of time have a special claim to reality; in this sense the consideration of the divergence between the population systems in a space-time continuum is academic, for population systems interact with other contemporary population systems and not with remotely ancestral ones.

These interactions take various forms—genetical and ecological. Referring to the levels marked as semispecies or sympatric species in Fig. 77, the population systems A and B either interbreed or prevent interbreeding by isolating mechanisms, and they either compete directly for raw materials and energy sources or avoid direct competition by diverging in their ecological requirements. But Species or Semispecies A does not interact in these genetical and ecological ways with Colony A that existed in past time. The genetical and ecological relations between contemporary population systems where they come into contact with one another are very real and very important. These genetical and ecological relations, in the form they take between sympatric species, set such species apart in nature as objectively real and non-arbitrarily delimited units of organization.

SIXTEEN

Modes of Speciation

THE GOAL AND THE STEPS

EACH SPECIES POSSESSES its particular adaptive properties which enable it to live in some habitat or niche. The adaptations of a species consist of combinations of phenotypic characters which are determined by combinations of genetic factors. The gene combinations which underlie the essential adaptations of the species are protected from disintegration by reproductive isolating mechanisms. The formation of a new species is basically the fixation of a new isolated adaptive gene combination. This process can be visualized as taking place in a series of stages.

Let us assume that an ancestral species which is successful in some habitat has the genetic constitution $a_1a_1b_1b_1$ for two independent genes **A** and **B**. Assume further that the allele combination $a_2a_2b_2b_2$ would be adaptively valuable in some available unoccupied habitat. The adaptive goal is the formation by the ancestral species $a_1a_1b_1b_1$ of a new daughter species with the constitution $a_2a_2b_2b_2$, the ancestral species remaining extant, that is, the parental species ($a_1a_1b_1b_1$) branches into two daughter species ($a_1a_1b_1b_1$ and $a_2a_2b_2b_2$). The steps by which the adaptive goal $a_2a_2b_2b_2$ can be reached are (1) production of multiple-gene variation; (2) formation of the adaptive allele combination; (3) fixation of the allele combination in a population; and (4) preservation of the allele combination.

1. In an ancestral population composed of $a_1a_1b_1b_1$ individuals the production of some a_2 and b_2 alleles is clearly the first step toward the eventual formation of a new isolated $a_2a_2b_2b_2$ population. The a_2 and b_2 alleles can be introduced into the ancestral population by mutation or gene flow. By these processes individuals of the constitution $a_1a_2b_1b_1$

and $a_1a_1b_1b_2$ arise in the ancestral population. The alleles a_2 and b_2 and the genotypes containing them will usually be present in low frequencies in the original population.

2. The next step is the formation of the new allele combination $a_2a_2b_2b_2$. The process involved in this step is recombination. The crossing of individuals $a_1a_2b_1b_1$ and $a_1a_1b_1b_2$ will lead to the production of some $a_2a_2b_2b_2$ types in the second generation. The $a_2a_2b_2b_2$ genotype will be rare in the population at first.

3. The sexual process which forms the allele combination $a_2a_2b_2b_2$ can easily break it up again. Crossing of the $a_2a_2b_2b_2$ individuals, which are rare initially, with the other more common genotypes such as $a_1a_1b_1b_1$ will quickly dissolve the new adaptive allele combination. In some respects the most critical step in the speciation process, therefore, is the fixation of the new allele combination in a derivative population. There are in general two ways in which an allele combination can be established: selection, and inbreeding.

Selection in some geographical race of the original species in favor of the $a_2a_2b_2b_2$ types could gradually increase the frequency of the a_2 and b_2 alleles in the population. In time the race undergoing such selection would change in composition from the initial high frequency of $a_1a_1b_1b_1$ individuals to a high frequency of the new genotype $a_2a_2b_2b_2$. But selection for combinations of independent genes under conditions of random mating is very costly to a population in terms of genetic deaths, as we learned in Chapter 9. In the hypothetical case before us, selection would be discriminating not only against the carriers of the a_1 and b_1 alleles, but also against such carriers of the a_2 allele as did not possess the b_2 allele (thus $a_2a_2b_1b_1$), and against certain carriers of b_2 as well (viz., $a_1a_1b_2b_2$), all of which are produced by recombination in the population. And a real adaptation of specific magnitude would involve not two but many separate genes.

Inbreeding provides an economical way out of this difficulty. If the $a_2a_2b_2b_2$ types cross *inter se*, they can multiply quickly in numbers under positive selection, without an excessive loss of individuals carrying the favored alleles in unfavorable recombinations (as $a_2a_2b_1b_1$ or $a_1a_1b_2b_2$). *The new adaptive allele combination can be fixed in a population most quickly and with the minimal number of genetic deaths by inbreeding.*

Inbreeding can be brought about in various ways. In a small population mating tends to be predominantly between cousins or siblings, hence between genotypically similar individuals, so that in a relatively few generations a particular homozygous gene combination can become established with or even without selection. The idea of genetic drift is derived from a consideration of the effects of inbreeding in small populations. The same result can also be achieved in a large population by assortative mating, that is, by a tendency of particular genotypes to mate preferentially with one another rather than randomly. The closest possible form of inbreeding is self-fertilization, which can lead to the rapid multiplication of a new gene combination.

Selection and inbreeding are not to be thought of as mutually exclusive processes. The adaptive gene combination $a_2a_2b_2b_2$ is favored by selection whether it arises infrequently in a large randomly mating population or is multiplied rapidly by inbreeding. The processes to be contrasted, then, are selection operating without the help of inbreeding in a widely outcrossing population and inbreeding combined with selection of the inbred products.

4. The new adaptive allele combination, once formed and established in a derivative population, must be protected from disintegration due to intercrossing if it is to persist for long in the vicinity of related populations. This protection is secured by reproductive isolating mechanisms. Now it is possible that the divergence in genetic constitution of two populations may in itself bring about reproductive isolation between them, as where $a_1a_1b_1b_1$ and $a_2a_2b_2b_2$ bloom or mate at different seasons or produce sterile hybrids, etc. The formation of isolating mechanisms as a by-product of racial divergence has been considered in the preceding chapter. It remains to add here that the isolating mechanisms arising as by-products of divergence, if not sufficiently effective for their task of preventing hybridization, can be reinforced by the addition of new or stronger barriers built up by selection for isolation per se. This case will be examined in the next chapter.

To summarize, the steps leading to an adaptive goal defined by a favorable new allele combination are (1) production of multiple-gene variation in the ancestral population by gene flow or mutation; (2) formation of the allele combination by the sexual mechanism; (3)

fixation of the allele combination in a population by selection under wide random outcrossing, or by inbreeding combined with selection; and (4) protection of the new genotype by reproductive isolating mechanisms arising as (4a) by-products of divergence or as (4b) special contrivances.

Some of these steps may be combined in certain cases. Thus hybridization might introduce new genetic factors into a population and recombine them in practically a single step. Step (4a) may take place simultaneously with step (3). In some methods of hybrid origin of species in plants, as we will see later, (1), (2), (3), and (4a) are all compressed into a few generations.

THE PATHWAYS

The genetic variations necessary for differentiation of species may exist in a latent state in a polymorphic ancestral population system, or may arise in sporadic bursts as a result of unusual events of mutation or hybridization. Hybridization may occur between existing races, or between interfertile species, or between intersterile species, with pronounced effects on variation. Bursts of new mutations may be brought about by internal mutator genes or by external environmental factors, and may involve changes in the genetic material at various levels of organization from genes, subgenes, and small chromosome segments to whole chromosomes and chromosome sets.

The sources of variation that are believed to be significant for speciation are indicated in Fig. 80. These sources, corresponding to steps (1) and (2) of the process, are (v) the store of variability already in existence in the population system as a result of previous mutation, gene flow, and recombination; (h) bursts of new variation arising from hybridization; and (u) sporadic avalanches of mutations, some of which could have drastic phenotypic effects.

The completion of the process of speciation, the carrying out of steps (3) and (4a), depends on the fixation of the different variations in different isolated populations. The establishment of variations in the derivative populations may be brought about by: (s) the differential action of natural selection in different populations or subpopulations under conditions of wide and random outcrossing; or (i) by inbreeding.

The factors determining whether inbreeding will or will not occur on a scale sufficient to fix some variations in some populations depend upon the reproductive behavior and population structure of the organisms concerned. Inbreeding may be brought about in various ways: by small population size, assortative mating, or self-fertilization. In some groups of organisms one method of inbreeding may prevail, in another group another method may be possible, and of course in

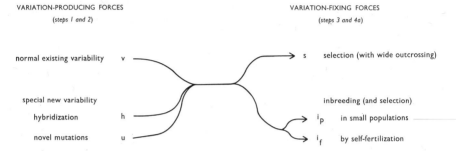

VARIATION-PRODUCING FORCES
(steps 1 and 2)

VARIATION-FIXING FORCES
(steps 3 and 4a)

normal existing variability   v

s   selection (with wide outcrossing)

special new variability

inbreeding (and selection)

hybridization   h

i_p in small populations

novel mutations   u

i_f by self-fertilization

Fig. 80. Combinations of evolutionary forces that can bring about speciation.

still other groups the conditions for inbreeding may be lacking altogether. In Fig. 80 we have depicted two of the most common and important modes of inbreeding, namely, mating between close relatives in small populations (symbolized by i_p) and self-fertilization (i_f).

The possible combinations of variation-producing and variation-fixing forces are shown diagrammatically in Fig. 80. We can see from this diagram that the process of speciation may follow various pathways. One such pathway is v—s, an alternative path is v—i_p, still another is h—i_f, and so on. Altogether nine pathways are indicated in the generalized scheme of Fig. 80.

THE PATTERNS AND THE FIELD

The change from one reproductively isolated population system to two or more genetically differentiated, reproductively isolated population systems involves a branching of phyletic lines. There are in general two patterns of branching: a dichotomous or Y-shaped pattern, and an excurrent or *V*-shaped pattern. In other words, all the main races in the ancestral species may diverge and become new species, the original species splitting up as a whole into two or more daughter species; or the ancestral species may persist as such but give rise to offshoot populations which develop into daughter species. The dichotomous pattern of speciation can be likened to the simultaneous branching of a stem into two or more approximately equal twigs; the excurrent pattern to the budding off of side shoots from a persistent main stem.

The spatial field in which speciation takes place differs from case to case. This field may be a large area inhabited by two or more geographical races. It may be a small area inhabited by neighboring local populations, or by ephemeral small populations. Or the new species may develop within the spatial limits of a single ancestral population. We have then the possibilities of allopatric, neighboringly sympatric, and biotically sympatric speciation. The process of speciation may follow a course through an intermediate stage of large geographical races, through small local and often ephemeral races, or may bypass the stage of races altogether.

THE MODELS

The 2 patterns of branching (dichotomous and excurrent), the 3 types of spatial setting (allopatric, neighboringly sympatric, and biotically sympatric), and the 9 pathways (v—s, v—i_p, etc.) can be combined in 54 different ways. About 16 of these ways are theoretically unlikely or even impossible. Thus the 2 combinations, v—s, dichotomous or excurrent, and biotically sympatric, are very unlikely to be realized in practice, and the combinations of v—s with neighboringly sympatric field are also improbable. Deducting the improbable combinations leaves us with about 38 possible models of speciation to consider.

Certain ones of these 38 possible models of speciation are undoubtedly more common and important in evolution than others. Particular pathways tend to be correlated with particular patterns and fields. The pathway v—s, for example, usually takes place in an allopatric field, via large geographical races, and with a dichotomous pattern of branching. This is the mode of speciation described in Chapter 15. The pathway v—i_p is usually associated with an allopatric field, local and perhaps ephemeral races, and an excurrent branching pattern. Speciation by path h—i_f is likely to be excurrent and biotically sympatric without any intermediate stage of race.

In view of the rich diversity of possibilities before us, the single-minded devotion of some evolutionary biologists to one-sided theories of speciation is remarkable. Thus Goldschmidt devotes a book to the exposition of one-half of one model of speciation, dealing at length with the origin and nature of a postulated class of species-creating mutations, without however even considering the problem of their establishment in populations.[1] In his book Goldschmidt not only presents his half model (u—), for which the empirical evidence is slight, but also attempts to explain away the existence of another pathway (v—s), which is supported by good empirical evidence. Some vertebrate taxonomists have regarded pathway v—s as typical and certain special cases of v—i as theoretically impossible, while some botanists like Lotsy, on the other hand, claim a virtual monopoly for path h—i.

An alternative manner of treating speciation, and one which would serve no useful purpose, would be to analyze every hypothetically possible method of species formation, irrespective of its relevance to the situation in nature. Eschewing the task of laying a large number of purely hypothetical models before the reader, but compelled by the facts to recognize a variety of speciation phenomena, we shall attempt in this chapter to describe several ways in which species can and probably do originate, and then to make the story more complete we shall mention one or two classical and modern ways in which species probably do not originate.

ALLOPATRIC SPECIATION IN ORGANISMS WITH WIDE OUTCROSSING

In organisms with wide outcrossing, species may develop out of large geographical races. Speciation in such cases is allopatric and follows the course v—s. This is the mode of speciation described in Chapter 15 and exemplified by *Parus major*, Diplacus, and the cross-pollinating Cobwebby Gilias (see also Chapter 12). This form of species formation is in some ways the easiest to understand; it was at any rate the first to be widely generalized or even overgeneralized.

The goal of speciation is the formation of new complex adaptations. Complex adaptations are determined by gene combinations. Earlier in this chapter we assumed that the gene combinations constituting the

[1] Goldschmidt, 1940.

goal of speciation are homozygous, like $a_1a_1b_1b_1$ and $a_2a_2b_2b_2$. But adaptations are often based on heterozygous gene combinations, as we have seen in previous discussions (Chapter 10). The usual way in which permanent heterozygosity and hence high fitness is maintained in organisms making use of the physiological advantages of heterozygous gene combinations is by random mating in large polymorphic populations.

The fitness of the ancestral species may depend on various heterozygous combinations of a series of coadapted alleles of the genes **A** and **B**. Suppose that the alleles a_1, a_2, a_3, and b_1, b_2, b_3 in their various heterozygous combinations ($a_1a_2b_2b_3$, etc.) bring about phenotypic reactions with high adaptive value in the environment occupied by the ancestral species. These gene combinations are assembled generation after generation by wide outcrossing between individuals carrying these alleles and contributing them to the gamete pool of the population. Owing to mutation or gene flow, other alleles (a_4 and b_4) of no immediate adaptive value may exist in low frequencies in the polymorphic population.

Now if a new heterozygous gene combination, say $a_3a_4b_3b_4$, is adaptive in some available habitat not yet occupied by the species, the formation and establishment of this gene combination will enable the ancestral population to extend its range. The new allele combination is formed by recombination of existing genetic variations and is established by selection for the favorable gene combination in one racial branch of the species. The course of divergence follows the pathway v—s in a large allopatric field.

A system of random mating which produces well-adapted heterozygotes in a polymorphic population also produces a proportion of less fit homozygotes. The loss of the latter is the price which the population must pay in order to maintain a high average fitness based on a heterozygous genetic condition. Stabilizing selection takes a definite toll in a population composed of permanent heterozygotes. The progressive selection involved in the change from one heterozygous gene combination to another is even more costly.

The cost of selection and hence the luxury of permanent heterozygosity can be borne most easily by organisms with a net fecundity well in excess of their opportunities for expansion in numbers. This condition of an excessive net fecundity is realized in species of animals and plants which form regular elements in stable or closed communities. It is best realized in the dominant species of a region.

A tree in a stable climax forest may produce many thousands of seeds year after year for a century or more, most of which are doomed to extinction owing to lack of *Lebensraum* in the forest. When a windfall creates a temporary clearing in the forest, its seeds and those of other surrounding parent trees sprout up by the hundreds, but only one or a few of the seedlings will ever survive to reach maturity. Since a great loss of zygotes is inevitable under such ecological circumstances, that loss may as well have a selective component.[2] Selection, either stabilizing or progressive, for heterozygous gene combinations can be tolerated. The survival of the fittest in such cases is likely to be synonymous with the perpetuation of heterozygotes.

Random mating in large polymorphic populations is characteristic of species which form constituents of stable or closed biotic communities. This type of breeding system and population structure is correlated with this type of ecological setup in the wild forest-inhabiting species of Drosophila and in the chaparral-inhabiting Diplacus, which we considered earlier. The same combination of features is found among many birds and mammals and many perennial plants, especially where the species are dominant or subdominant in their communities. In the same organisms speciation is frequently allopatric and dichotomous, following the pathway v—s. It is perhaps because this mode of speciation is found in the dominant forms of life that it has sometimes been regarded by evolutionists as *the* typical mode.

SPECIATION BY THE BUDDING OFF OF PERIPHERAL POPULATIONS IN CROSS-FERTILIZING ORGANISMS

A widespread polymorphic outcrossing species may give rise to new species, not only by fractionation and selection as described in the preceding section, but also in another way, namely, by the isolation of a few individuals forming a new peripheral colonizing population. This mode of speciation has been described by Mayr.[3]

Gene flow in the ancestral population system—large, continuous, polymorphic, and heterozygous—tends to have a conservative effect,

[2] Grant, 1958, 349, based on a suggestion of Stebbins, 1950, 177.
[3] Mayr, 1954.

buffering the organisms against the effects of new types of alleles. But the isolation of a few individuals from the continual stream of gene flow in such a population, and the enforced inbreeding in the ensuing generations in the new isolated small population, has drastic genetic effects. This budding off of a small peripheral colony has the effect of converting the organisms from a state of heterozygosity to one of homozygosity and from an ancestral selected heterozygous allele combination to some new homozygous allele combination.[4]

If, as in the setup hypothesized in the preceding section, the ancestral population is polymorphic for the alleles a_1, a_2, a_3, and a_4 of gene **A** and for b_1, b_2, b_3, and b_4 of **B**, and produces the various heterozygous and homozygous gene combinations of these alleles in proportion to their relative frequencies in the gamete pool, the migration of a few individuals out of this population to some outlying area and their establishment there as a new colony will lead, by genetic drift, to the fixation of a small sample of the original polymorphic variation in a homozygous condition. The alleles a_1, a_2, b_2, and b_3 might be carried into the founder population, and after a few generations the homozygous allele combination $a_2a_2b_3b_3$ might reach a high frequency or complete fixation in that population.

There is no assurance, of course, that the allele combination $a_2a_2b_3b_3$ produced by inbreeding in this founder population will be adaptively valuable. Most of the homozygous allele combinations that could be assembled by inbreeding in an isolated segment of the original population will in fact probably possess a low adaptive value. But some one gene combination may prove successful. If, therefore, many founder populations bud off the ancestral population, most of them are probably doomed to extinction, but one such derivative population might acquire a valuable allele combination, which could never persist in the midst of wide outcrossing in the ancestral population. And when a founder population does acquire a new adaptive allele combination, it could shift rapidly in its adaptive characteristics.[5]

These relationships are exemplified by the geographical variation in the kingfishers of the *Tanysiptera galatea* group on New Guinea (see Chapter 15). It will be recalled from our earlier discussion that the mainland races of *Tanysiptera galatea* are very much alike from one end

[4] Mayr, 1954.   [5] Mayr, 1954.

of this 1,500-mile-long island to the other. But the surrounding islands, which were probably colonized by a few founders migrating from the large continuous mainland population, are inhabited by five races and one species (*T. hydrocharis*), all of which differ markedly from the mainland birds and from one another.[6]

Mayr suggests that speciation by the fractionation of a large continuous polymorphic population system and speciation by the isolation of small colonies derived from such a population system will have markedly different effects as regards the degree of divergence of the daughter species. Radical departures from the ancestral type cannot become established in the former case owing to the swamping effect of continual gene flow and to the effect of selection for stable buffered heterozygous genotypes. Radical deviations from the ancestral form can, however, arise by recombination and become established by inbreeding in small founder populations, and if the new type is well adapted it will then multiply in numbers and spread in area.[7]

In its later history the new species derived from an inbred and homozygous founder population may again build up its store of variability and become permanently heterozygous. However, the new series of alleles is now selected to provide a well-buffered development of the new adaptive characteristics of the daughter species in its stage of expansion.[8]

It is noteworthy that a number of evolutionists, independently and by different routes, have arrived at the conclusion that small isolated populations may be the site of rapid evolutionary changes and of the formation of novel types. Gulick and other early modern naturalists appreciated the effect of small population size on racial variation. The early breeders were well aware of the drastic effects of inbreeding in normally outcrossing plants and animals. On these foundations and particularly from the consideration of inbreeding Wright developed the theory of genetic drift.[9]

Simpson recombined the concept of drift with the data of paleontology to produce the idea of quantum evolution. Simpson argued that the absence or rarity of fairly complete fossil series connecting new major groups of organisms with their ancestral stocks would be difficult

[6] Mayr, 1942, 1954; see Chapter 15 of this book.   [7] Mayr, 1954.
[8] Mayr, 1954; Carson, 1959.   [9] Wright, 1931.

to explain if their population size were as large in the period of their origin as it was in their later history, when the fossil representation becomes more adequate. Furthermore, the geological time available for the divergence of the new major group from its parent stock requires a much more rapid evolution during the period of origination than during the subsequent period of expansion. These facts can be accounted for on the genetically plausible hypothesis that the new major groups—genera, families, orders, etc.—originate from small isolated populations undergoing rapid shifts from the ancestral to a new adaptive state, that is, by quantum evolution.[10] Now divergence at the species level is not necessarily involved in quantum evolution, but may be and probably is frequently involved in this process, in which case we can speak of quantum speciation.

Mayr arrived at the same conclusion and suggested a genetical basis for it from his studies of modern populations of kingfishers, Drosophila, and other animals.[11] Carson and White have recently elaborated on Mayr's hypothesis and have summarized much interesting evidence bearing on speciation in Drosophila.[12] And Lewis and Raven have presented a parallel model that is applicable to the situation in annual plants, which we will discuss later in this chapter.[13]

Quantum speciation, by the budding off of peripheral populations in cross-fertilizing organisms, is allopatric, but otherwise differs from the allopatric speciation discussed in the preceding section in almost every respect. The new species develop from ephemeral local races rather than from large geographical races; the pattern of branching is excurrent rather than dichotomous; and the combination of forces involved is $v—i_p$ instead of $v—s$. Finally, the new species developing from founder populations may diverge rapidly and markedly from the ancestral form, whereas species developing as outgrowths of large geographical races seem to be more conservative in their evolutionary rates and trends.

ALLOPATRIC SPECIATION IN SEDENTARY ORGANISMS

In considering the relevance and validity of the geographical theory of species formation in different groups of organisms, it is necessary to

[10] Simpson, 1944. [11] Mayr, 1954. [12] Carson, 1959; White, 1959.
[13] Lewis and Raven, 1958.

bear in mind that organisms differ greatly in their powers of dispersal. In sedentary organisms with very slight amounts of gene flow, individuals living in closely adjacent areas might be able to diverge to the racial and eventually the specific level.

Lewis has reported a case of two adjacent populations of the annual plant *Clarkia xantiana* (Onagraceae), each consisting of several hundred individuals, and separated by a distance of 100 feet. During three years of observation one of the populations was composed uniformly of individuals with bright lavender flowers and the other entirely of white-flowered individuals.[14] The divergence between closely adjacent populations in a sedentary plant like Clarkia could easily involve characters that bring about reproductive isolation. In such a case new species could originate rapidly in a microgeographical field.[15]

Although birds, with their relatively wide radius of dispersal, can diverge into numerous endemic species only on island archipelagos and not on single islands, the situation is otherwise among insular land snails and insects, which have a shorter dispersal range and can consequently form geographically isolated populations in a small area. In these more sedentary organisms an extensive formation of new species is known to have taken place on solitary islands like St. Helena.[16]

Among birds belonging to the same group, as Mayr has pointed out, the wide-ranging and randomly mating ducks do not form many well-marked races, whereas the colonial and partially inbreeding species of geese are broken up into numerous distinct races, some of which approach or actually attain the status of species. Thus in the Canada goose (Branta) there are three or four geographical races that are morphologically different enough to be considered species by some authors, as well as some sympatric and reproductively isolated population systems.[17]

It is instructive to compare the pattern of geographical variations in the two plant genera, Cupressus and Juniperus, which are phylogenetically related, have similar vegetative characteristics, and the same method of pollination by wind, but differ markedly in their dispersal potential. The seeds of junipers are contained within succulent blue berries which are extensively eaten and disseminated by birds. The seeds of the

[14] Lewis, 1953, 8.   [15] Lewis, 1953; Lewis and Raven, 1958.
[16] Lack, 1947, 150–58.   [17] Mayr, 1942, 242.

cypresses, by contrast, are borne in a heavy woody globose cone which normally travels only as far from the parent tree as it can roll downslope or be carried by squirrels upslope. This difference in range of seed dispersal is associated with and probably responsible for a striking difference in the pattern of species distribution in the cypresses and junipers.[18]

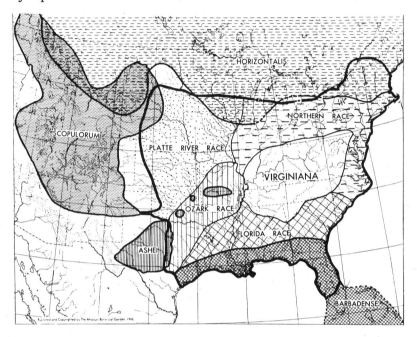

Fig. 81. Geographical distribution of the species of Juniperus in the central and eastern United States

From Anderson, *Biol. Revs.*, 1953, with permission of the Cambridge Philosophical Society.

The junipers form extensive populations scattered widely across North America. Their variation pattern in the area from the Rocky Mountains to the Atlantic coast consists of a sequence of intergradations on several broad fronts (see Fig. 81). One such sequence involves a tall arboreal form of *Juniperus airginiana* in Kentucky and Tennessee,

[18] Grant, 1958, 345–46.

Fig. 82. Geographical distribution of species in the genus Cupressus in California

Grant, 1958, drawn from data of Wolf, 1948.

the low bushy *Juniperus ashei* in Texas and Oklahoma, and an intermediate race of *J. virginiana* in the intervening area. To the north *J. virginiana* intergrades with *J. horizontalis* of Canada, to the west through the Great Plains with *J. scopulorum*, and to the south with *J. barbadense* (Fig. 81).[19]

The variation in the more sedentary genus Cupressus is channelized within a large number of more discrete and endemic species. The North American cypresses form a series of disjunct and more or less endemic populations ranging from Oregon through California to Texas and thence through Mexico to Central America. Fifteen species are recognized in this region by the latest monographer, Wolf. Although some species like *Cupressus arizonica* form extensive stands, most of the species exist in a small number of isolated groves. In 11 of the 15 species the number of groves per species ranges from 1 to 7 and averages 3.[20] The localized distribution of 12 of the cypress species is shown in Fig. 82.

NEIGHBORINGLY SYMPATRIC SPECIATION IN CROSS-FERTILIZING ORGANISMS

The amount of spatial isolation that will allow an initial divergence between populations to take place is, as we have seen in the preceding section, relative to the powers of dispersal of the organisms concerned. A distance of 100 feet might provide as much isolation for a sedentary Clarkia as a much greater distance would furnish in a more motile organism. And there are organisms even more circumscribed in their movements than this. Individual insects restricted to different food niches and individual parasites living on different hosts or different body parts of the same host, for instance, may be effectively isolated spatially by distances of only a few feet.

Many of the solitary bees in arid regions are species specific in their pollen sources, each species of bee collecting pollen from the flowers of a single species or species group of plants. The flowers visited for food are also a site of mating. Most of the species of Diadasia (Anthophoridae) collect pollen from various species of Sphaeralcea and other members of the family Malvaceae. Some species of Diadasia, however,

---

[19] Fassett, Hall, E. Anderson; see Grant, 1958, 345–46.
[20] Wolf, 1948, 11.

are restricted in foraging to such unrelated plants as sunflowers, Clarkia, Convolvulus, and cacti, which suggests the possibility of abrupt changes in food niches associated with species formation. In times of food scarcity the individuals of a bee species normally restricted to one kind of plant will explore new sources of food. Thus Linsley and MacSwain found *Diadasia australis*, a cactus feeder, on the flowers of Phacelia (Hydrophyllaceae) when the local cacti were past blooming in southern California. These authors point out that if some individuals of a bee species with a narrow range of food sources change their feeding habits in a time of scarcity, and if the newly adopted flower is a mating site, then divergence might proceed to the stage of species within a small local area.[21]

SYMPATRIC SPECIATION DUE TO DISRUPTIVE SELECTION

A theoretical problem on which there is considerable difference of opinion among evolutionists is whether a new, reproductively isolated, cross-fertilizing population can arise within the limits of a parental cross-fertilizing population. Do new species always develop from geographical or microgeographical races in outbreeding organisms? Or can they sometimes originate directly and sympatrically under conditions of outcrossing?

Very strong disruptive selection could in theory separate two or more classes of genotypes in a population in the face of random interbreeding, and has done so in laboratory experiments with Drosophila (see Chapter 10).[22]

In the western white pine (*Pinus monticola*), which is widely cross-pollinated by wind, well-differentiated races are known to exist in adjacent sites. In mountainous country in northern Idaho two populations growing one-half mile apart, one on a moist and the other on a dry slope, were found to differ in growth rate, dry-matter content of the leaves, and osmotic pressure. These characters are among the adaptations of the plants to the moisture factor in their habitat. Selection against the unfit seedlings, taking one course in moist sites and another course in adjacent dry sites, could and probably does keep up the racial

[21] Linsley and MacSwain, 1958.
[22] Thoday and Boam, 1959; Streams and Pimentel, 1961.

adaptations of neighboring populations of white pines in spite of gene flow.[23]

Whether disruptive selection is sufficiently intense and long sustained to bring about the formation, not only of races, but of species under conditions of cross-fertilization and gene flow—in pines, Drosophilas, or other organisms—remains to be seen.

BIOTICALLY SYMPATRIC SPECIATION DUE TO ASSORTATIVE MATING

Assortative mating, or a tendency for genetically like individuals in a population to mate with one another, forms the basis of another possible method of sympatric species formation. A model applicable to flowering plants pollinated by bees and other insects with strong instincts of flower constancy has been developed and discussed by several authors.[24]

Let mutations or recombinations affecting the shape, marking, color, or odor of the flowers arise in a population of plants pollinated by flower-constant insects, for instance, bees. Some individual bees would then tend to cross-pollinate the ancestral type of flowers preferentially, while other bees, becoming fixed on the mutant flower forms, would tend to cross-pollinate these preferentially with one another. The resulting system of non-random mating would be, in effect, an incipient etholog-ical isolating mechanism developing within a once randomly mating population. In the shelter of the new ethological isolating mechanism, other genetic differences could conceivably accumulate in the separate populations, which might in time reach the stage of well-isolated species.

In the original model the genetic changes initiating such a divergence were assumed to be mutations affecting the shape or markings of the floral organs.[25] Manning has shown that flower-constant behavior in bees is elicited by scent more than by visual stimuli, from which it follows that mutations affecting the floral odor would be most effi-cacious in initiating a sympatric divergence.[26] And Straw has suggested

[23] Squillace and Bingham, 1958.
[24] Grant, 1949; Straw, 1955; Manning, 1957; Heslop-Harrison, 1958; Dronam-raju, 1960.
[25] Grant, 1949, 91.
[26] Manning, 1957.

that gene recombination is a more likely source than mutation of variant flower types.[27]

The great difficulty of this scheme lies in the assumption that the ethological isolation will remain complete enough for a long enough time to carry the process of species formation through to completion. The constancy of flower-feeding insects—bees, Lepidoptera, etc.—to a particular type of flower is, for the insects, quite secondary to the goal of getting a supply of food, and whenever the source of pollen or nectar becomes inadequate in one type of flower, lapses in flower-constant behavior by the insects and hence in the ethological isolation of the plants become inevitable. Just one such lapse in the course of several seasons of otherwise rigidly selective pollination could set the divergence back to the initial stage of genetic polymorphism within a single population. The theoretical model may apply only rarely to real cases in nature therefore.[28]

The foregoing model does, however, gain some semblance of reality by a case described by Straw in the genus Penstemon (Scrophulariaceae). *Penstemon centranthifolius*, with pendant red trumpet-shaped flowers, is pollinated by hummingbirds, while a related and marginally sympatric species, *P. grinnellii*, with flesh-colored, two-lipped, wide-throated flowers, is pollinated by carpenter bees (Xylocopa). A third species in the same group, *P. spectabilis*, with smaller bluish flowers, is pollinated by pseudomasarid wasps. The three species of Penstemon are mechanically and ethologically isolated by their pollinating agents in some areas in southern California, though not in others, where they occur sympatrically.[29]

Penstemon spectabilis is believed to be of hybrid origin between *P. centranthifolius* and *P. grinnellii* on the basis of its morphological and ecological intermediacy and its resemblance to natural hybrids of the latter two species. Straw suggests that in the hybrid swarm out of which the new species emerged, the *centranthifolius* types were still pollinated by hummingbirds and the *grinnellii* types by carpenter bees, but uncommitted wasps took over the pollination of the progenitors of *P. spectabilis*. In almost a single step, then, a new flower type produced by hybridization and preadapted to a different type of pollinator could gain the reproductive isolation needed for development into a distinct

[27] Straw, 1955.   [28] Grant, 1949, 91–92.   [29] Straw, 1955, 1956.

species. But, as Straw comments, "That sympatric speciation by this method has not been more effective in the production of species is probably due partly to the fact that relatively few such distinct [pollinator] niches are available in nature."[30]

BIOTICALLY SYMPATRIC SPECIATION DUE TO SELF-FERTILIZATION

In organisms which do not form cross-fertilizing populations, but reproduce by self-fertilization, as in many groups of plants and some lower animals and protista, the condition which makes sympatric species formation difficult in outcrossing organisms of course ceases to exist, and new reproductively isolated lines can and sometimes do arise within the limits of a parental population.

The species groups known as *Paramecium aurelia, bursaria,* and *caudatum* comprise a large number of species isolated by incompatibility barriers. Reproduction in these Paramecia takes place in various ways: asexually by fission of the cells, sexually by conjugation between different individuals and cross-fertilization, and autogamously by self-fertilization. The relative amount of autogamy and inbreeding varies from one species to another. Thus the species belonging to the *P. bursaria* group are strongly outcrossing, whereas most of the members of the *P. aurelia* and *P. caudatum* groups are more strongly inbreeding. The degree of racial and specific differentiation is correlated with the amount of inbreeding.[31]

In Species 1 of the *P. bursaria* group, an outcrosser, the widely scattered populations are much alike genetically and chromosomally, so that the interstrain hybrids are viable. In the self-fertilizing Species 4 of the *P. aurelia* group, on the other hand, geographically separate populations differ in the size, form, and number of the chromosomes, and the F_2 progeny of the interracial crosses exhibit much inviability. Between the extreme conditions found in *P. bursaria*-1 and *P. aurelia*-4 there are various intermediate conditions, as exemplified in many species of the *P. aurelia* group, where an intermediate amount of inbreeding is

[30] Straw, 1955.

[31] Sonneborn, 1957, except as to taxonomic terminology and interpretation; the reproductively isolated population systems which are referred to here as species are called varieties by Sonneborn.

correlated with an intermediate degree of genetical differentiation between populations.[32]

The effect of inbreeding on divergence is seen not only at the racial but also at the species level. Speciation is retarded in the outcrossing *P. bursaria* group by comparison with the more autogamous *P. aurelia* group. In the United States there are probably only three species in the *P. bursaria* group as compared with at least fourteen and perhaps more species in the *P. aurelia* group.[33]

In various groups of annual plants also an accelerated formation of species is associated with an autogamous breeding system. Thus the Cobwebby Gilias comprise about twenty-one autogamous species as compared with about eight cross-fertilizing species (some of which are only semispecies). A similar high rate of speciation is found in the autogamous members of Erophila, Elymus, Galeopsis, Clarkia, and other plant genera.[34]

In a number of self-fertilizing plant groups the variation which is preliminary to the formation of a new species is known to arise from hybridization between preexisting species. The speciation process thus follows the course h—i_f in a biotically sympatric field. This mode of speciation will be described more fully in the next section.

We have so far considered the sympatric origin of new species as a genetical problem, involving the development of barriers to interbreeding. It is also an ecological problem, for the new sympatric types are subject to all the forms of competition and selective replacement that we discussed in Chapter 14. Even assuming that a population succeeds in overcoming the obstacles to the formation of reproductive isolating mechanisms under conditions of sympatry, it (or the parental population) may well be crowded out of its sympatric area and displaced to a new and different area at an early stage by interspecific competition and interspecific selection. It may be for this reason that plant species of sympatric and hybrid origin often inhabit areas that lie at least partly outside the territories of the parental species. Thus *Penstemon spectabilis* inhabits an ecological zone which is largely separate from the zones occupied by *P. centranthifolius* and *P. grinnellii*.

In summary, species formation in many animals and plants is inferred

[32] Sonneborn, 1957, 297.   [33] Sonneborn, 1957, 307.
[34] See Grant, 1958, 355, for references.

to follow a pathway from allopatric races to sympatric species. In some plants an alternative pathway is followed from biotically sympatric types to allopatric or neighboringly sympatric species. No doubt there are cases in small autogamous plants and protistans of populations which diverge sympatrically and remain sympatric thereafter.

THE HYBRID ORIGIN OF PLANT SPECIES
ON THE DIPLOID LEVEL

The idea that new species of plants can originate from the products of hybridization between preexisting species has been stated by many botanists since Linnaeus. Linnaeus suggested a hybrid origin of plant species in several notes from 1744 to 1767.[35] In a thesis on Peloria (Linaria, Scrophulariaceae) written in 1744 by a student, D. Rudberg, who clearly expresses the ideas of the master, we read[36]:

It seems on first sight a paradox that new species, nay, even new genera, should arise from the copulation of different species in the vegetable kingdom; meanwhile observations persuade us that this very thing takes place.... If it can be decided with certainty that the *Peloria* must have arisen as a hybrid species from the *Linaria* and some other plant, a new truth would come to light in the vegetable kingdom, and that process [hybridization] in the case of plants would be further advanced than in the case of animals in that animal hybrids, like Mules and others, lack the ability to propagate themselves. But we see that the *Peloria* propagates, for it has perfect seeds and in its natal place multiplies copiously and of its own accord.

Linnaeus himself in the *Disquisitio de Sexu Plantarum* (1760) described various plant species—*Veronica spuria, Delphinium hybridum, Hieracium taraxaci, Tragopogon hybridum,* etc.—as hybrid derivatives of other known species: *Veronica maritima* and *V. officinalis, Tragopogon pratense* and *T. porrifolium,* etc. And about these plants he says[37]:

There can be no doubt that these are all new species produced by hybrid generation.... For it seems probable that many plants, which now appear different species of the same genus, may in the beginning have been but one plant, having arisen merely from hybrid generation.

Similar views, which were of course inconsistent with Linnaeus' notions of the fixity of species (as quoted in Chapter 12), were stated in the *Species Plantarum* (1762-63) and the *Systema Naturae* (1767).[38]

---

[35] Ramsbottom, 1938.   [36] Quoted in translation by Ramsbottom, 1938.
[37] Ramsbottom, 1938.   [38] Ramsbottom, 1938; Sirks, 1952.

The hypothesis that hybrids can be the starting point of new species of plants was restated more precisely in the nineteenth century by Naudin (1863), Kerner (1891), and others.[39] These and other early twentieth century authors like Lotsy (1916) considered the process of hybrid origin of species from the standpoint of morphology and ecology but not from that of sterility barriers. Hybridization was for the older authors in effect a mechanism for producing new combinations of the morphological characters and ecological tolerances of the parental species.

It is well known that hybridization produces new recombination types. Some of these may represent radical departures from the normal condition in either parental population. Such macrorecombinations, as they may be called, have been observed in experimental hybrid progenies in Antirrhinum, Digitalis, Geum, Gilia, Gossypium, Lycopersicon, and Madia × Layia.[40]

The two species *Antirrhinum majus* and *A. glutinosum* both have typical snapdragon flowers. Some individuals in the F_2 generation derived from the cross between these two species have flowers which resemble in form those of another genus, Rhinanthus, belonging to a different tribe in the same family. Plants with Rhinanthus-shaped flowers became frequent or predominant in several selected F_4 lines.[41]

Both *Gilia achilleaefolia* and *G. millefoliata* possess flowers with the five calyx lobes, five corolla lobes, five stamens, and three carpels that are characteristic of the family Polemoniaceae. Some of the hybrid progeny of these two species in the F_3 and F_4 generations produced flowers with three parts in each whorl instead of the normal five. (In this particular hybridization experiment, the hybrid derivatives with aberrant flowers happened to be tetraploid.) Now the three-merous flowers of the Gilia hybrids deviate markedly from the floral plan of the family Polemoniaceae, and resemble in gross appearance as well as in basic design the flowers of the family Commelinaceae. That similar floral aberrations can arise and be fixed on the diploid level in nature is shown by the occurrence of natural populations of the related genus Leptodactylon with four-merous flowers.[42]

---

[39] Naudin, 1863; Kerner, 1894–95, II, 576 ff.
[40] Lotsy, 1916a; Gajewski, 1953; Rick and Smith, 1953; Grant, 1956b, 72–76; Schwanitz, 1957; Clausen and Hiesey, 1958, 266.
[41] Lotsy, 1916a.   [42] Grant, 1956b, 72–76.

New morphological structures can thus be produced by hybridization on the diploid level and, if the new structures have adaptive value, the gene combinations determining them will be favored by selection. A rather large change in phenotypic characteristics might take place in a few generations as a result of selection acting on the novel variations produced by hybridization.[43]

The problem of the hybrid origin of species has to be considered, not only in relation to gene-controlled phenotypic traits, but also in relation to sterility barriers in those cases where the parental species are inter-sterile. How can a new population, which is fertile *inter se*, arise from a sterile interspecific hybrid? And how can a new fertile population which is intersterile with its parents arise by hybridization? These problems have been considered and solved by Müntzing, Gerassimova, Stebbins, and others.

Related species of plants in many groups, particularly in the herbaceous genera, are mostly intersterile (see Table 18 in Chapter 17). The type of sterility barrier which predominates in herbaceous genera containing many intersterile species is chromosomal gametic sterility. Cytogenetic studies of hybrids have established the existence of gametic sterility due to heterozygosity for structural rearrangements of the chromosomes in such genera as Brassica, Bromus, Clarkia, Crepis, Elymus, Galeopsis, Gilia, Gossypium, Helianthus, Layia, Madia, Nicotiana, and Viola. Genic sterility may well be present too, but generally plays a subordinate role in such plant groups.[44] Chromosomal gametic sterility is the principal type of sterility which must be first surmounted, then reformed in new ways, in the hybrid origin of intersterile plant species.

The basis for formulating a hypothesis of the origin of chromosomal sterility barriers by recombination was provided by Müntzing's work on Galeopsis (Labiatae) from 1929 to 1938.[45] Müntzing found that *Galeopsis tetrahit* and *G. bifida* are separated by a barrier of hybrid semisterility which is caused most probably by small structural differences in the chromosomes of the parental species. The F_2 generation derived from the cross *G. tetrahit* $\times$ *G. bifida* segregates for pollen fertility, the different F_2 individuals producing varying proportions of

[43] Rick and Smith, 1953, Anderson and Stebbins, 1954; Grant, 1956*b*.
[44] Grant, 1958, 354. [45] Müntzing, 1929, 1930, 1938.

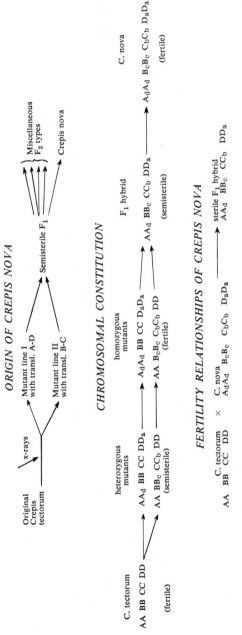

Fig. 83. Chromosomal constitution of various strains of Crepis tectorum and their hybrids in an experiment on the origin of chromosomal sterility barriers by recombination

From data of Gerassimova, 1939.

viable pollen grains; and therefore the causal factors, the postulated differential chromosome segments, segregate according to the laws of Mendelian inheritance.[46]

Chromosomal sterility barriers of a similar nature also exist within *Galeopsis tetrahit*, the strains of which are interfertile in some combinations but intersterile in others.[47] How have the partial sterility barriers originated within *G. tetrahit?* *Galeopsis tetrahit* hybridizes naturally with *G. bifida*. The progeny of the natural hybrids when grown in the garden segregate for both the morphological differences and the sterility barrier between the parental species.[48] These facts suggest that the partial sterility between strains of *G. tetrahit* is due to previous natural hybridization with *G. bifida*, followed by the segregation in later generations of different structurally homozygous lines possessing the morphological characters of *G. tetrahit*. Insofar as the different segregate lines of *G. tetrahit* originating from the hybrid differ in chromosome structural arrangement, they will produce semisterile hybrids, as is the case.[49]

The first experimental production of new fertile lines intersterile with the parental forms as a result of the recombination of chromosomal sterility factors was reported by Gerassimova in Russia in 1939. This worker crossed two mutant lines of *Crepis tectorum* which were homozygous for different translocations and were fertile. One line had a translocation involving the A and D chromosomes, and the other a translocation in the B and C chromosomes. The derivation and constitutions of the two homozygous mutant lines are shown in Fig. 83. The interline hybrid was, as expected, semisterile with 36 percent good pollen and 31 percent good seeds on selfing. It produced an F_2 generation comprising the expected range of translocation heterozygotes and homozygotes, among which was a plant homozygous for both translocations. This plant, which was designated *Crepis nova*, had normal meiosis and was fully fertile (see Fig. 83). *Crepis nova* was backcrossed to the original strain of *Crepis tectorum*. The backcross hybrids had irregular meiosis and were about 30 percent fertile as to seeds.[50]

In a series of papers from 1942 to 1959 Stebbins has developed a definite genetical model to account for the formation by recombination

[46] Müntzing, 1930. [47] Müntzing, 1929, 1938. [48] Müntzing, 1930.
[49] Müntzing, 1929, 1938. [50] Gerassimova, 1939.

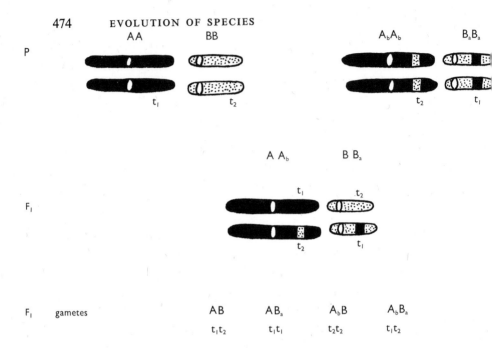

Fig. 84. *The chromosomal constitution of two parental types differing for one small translocation, and the constitution of their F₁ and their fertile F₂ progeny*

The centromere in each chromosome is indicated by an oval ring, the translocated segments (t) by black bands or stippled bands in the right arms.

of new fertile lines which are chromosomally intersterile with both parents.[51] Let us examine this model.[52]

Assume that two plant species differ with respect to a reciprocal translocation involving short segments of chromosomes A and B, so that one species has the chromosomal constitution AA BB, and the other the constitution A_bA_b B_aB_a (see Fig. 84). Their F₁ hybrid is a translocation heterozygote (AA_b BB_a). Its chromosomes may pair normally in bivalents at metaphase of meiosis, since the differential

[51] The key references are Stebbins, 1950, 287–89, and 1957a, 349–50.
[52] Following the later and amplified exposition of Grant, 1958, 355–56.

segments are short, and will separate to the poles at anaphase to form the four classes of daughter nuclei indicated in Fig. 84.

The two parental-type nuclei (AB and $A_b B_a$) each contain a representation of every chromosome segment in a normal haploid complement, and therefore can develop into viable pollen grains and female gametophytes. But the recombination-type daughter nuclei containing chromosome A in combination with chromosome B_a carry a deficiency for one segment (t_2) and a duplication for another (t_1). And the recombination-type nuclei $A_b B$ also carry a deficiency and duplication, having two doses of the segment t_2 and none of t_1 (see Fig. 84).

The pollen grains and female gametophytes destined to develop from the deficiency-duplication nuclei are unlikely to possess the genic balance needed for proper growth and functioning and so will be abortive or inviable. Their abortion or inviability gives rise to the gametic sterility of the hybrid. Since the four classes of gametes are produced in equal numbers by independent assortment in the heterozygote, and only two of these are viable, the gametic fertility of the hybrid is about 50 percent.

It will be noted that the translocated segments on the A and B chromosomes, that is, the segments t_1 and t_2, act as a pair of complementary factors which must both be present for normal functioning of a pollen grain or egg, whereas the absence of one of them has a lethal effect on the gametophyte. We may regard t_1 and t_2 as a pair of complementary fertility factors or, conversely, we can regard the segments $t_1 t_1$ and $t_2 t_2$ as forming pairs of complementary gamete lethals.

The random union of the two classes of viable gametes produced by the F_1 hybrid in this case gives rise to three types of zygotes in the F_2 generation (Fig. 84). The first parental type, the translocation heterozygote, and the second parental type are expected to appear in the F_2 generation in a $1:2:1$ ratio. The translocation heterozygote in F_2 will, of course, be semisterile like the F_1 hybrid, whereas the parental types, which are structurally homozygous, will be fertile.

Our second case involves a pair of species differentiated with respect to two independent translocations on four chromosomes, A, B, C, and D. The genetic constitutions of the parental species and their F_1 hybrid are as indicated in Fig. 85. The F_1 hybrid will form sixteen classes of gametes, only four of which are viable, and the gametic fertility is therefore 25 percent (Fig. 85). The F_2 generation will contain various translocation heterozygotes and homozygotes. The former are semi-sterile, but the latter are fertile. The fertile homozygotes consist of the two parental types, which do not interest us here, and two new recombination types, which are fully fertile within the line, but partially intersterile with one another and with the original parents.

P	AA BB CC DD		A_bA_b B_aB_a C_dC_d D_cD_c	
F_1		AA_b BB_a CC_d DD_c		
gametes	# ABCD	AB_aCD	A_bBCD	# A_bB_aCD
	$ABCD_c$	AB_aCD_c	A_bBCD_c	$A_bB_aCD_c$
	ABC_dD	AB_aC_dD	A_bBC_dD	$A_bB_aC_dD$
	# ABC_dD_c	$AB_aC_dD_c$	$A_bBC_dD_c$	# $A_bB_aC_dD_c$
F_2	P types	AA BB CC DD		A_bA_b B_aB_a C_dC_d D_cD_c
	R types	AA BB C_dC_d D_cD_c		A_bA_b B_aB_a CC DD

Fig. 85. The chromosomal constitution of two parental types differing in two independent translocations, the constitution of their F_1 hybrid, and of their fertile F_2 progeny

The viable gametes produced by the F_1 are marked with a crosshatch. The fully fertile F_2 genotypes, consist of two parental and two recombination types as shown.

Now suppose that the two species differ for four independent translocations. Their hybrid then produces 256 classes of gametes, only sixteen of which are viable, and the gametic fertility is thus 6.25 percent. The F_2 generation now contains a smaller proportion of fertile homozygotes, but a higher proportion of these homozygotes are new recombinations.

The general relationships are summarized in Table 16. (1) As the chromosomal sterility of a hybrid increases, the chances of extracting any fertile progeny from it decrease. (2) As the frequency of fertile F_2 types decreases, the proportion of the fertile individuals that are new increases. Therefore (3) the number of fertile but intersterile lines that can be extracted from a hybrid rises in direct proportion to the degree of gametic sterility of the hybrid.[53]

New fertile groups have been shown to arise from sterile hybrids in experiments with Crepis, Nicotiana, and Elymus.[54] The grass hybrid

[53] Stebbins, 1950, 288; 1957a, 350.
[54] Gerassimova, 1939; H. H. Smith, 1953, 1954; Stebbins, 1957b.

Elymus glaucus × *Sitanion jubatum* has a high degree of chromosomal sterility. The sterile F_1 when backcrossed to the *Elymus glaucus* parent in numerous trials yielded a single semifertile backcross individual. This individual when selfed gave rise to fully fertile and presumably structurally homozygous progeny. But the new fertile line was highly intersterile with *Elymus glaucus*. It apparently inherited a new homozygous combination of the chromosomal sterility factors which separate the parental species.[55]

Table 16. *The proportion of fertile recombination types possible for different numbers of translocations*[a]

Number of independent translocations	Fertility of F_1 hybrid (in percent)	Proportion of fertile individuals in F_2 (in percent)	Proportion of fertile F_2 types that are new (in percent)
1	50	50	0
2	25	25	50
3	12.5	12.5	75
4	6.25	6.25	87.25
6	1.56	1.56	96.88
n	$1/2^n$	$1/2^n$	$(2^n - 2)/2^n$

[a] Grant (1958, 356).

A large number of intersterile species exist in the *Elymus glaucus* group in nature. These species occasionally hybridize naturally. It is probable, therefore, that many of the natural occurring fertile but intersterile species have arisen by hybridization in the same way as the new line produced in a breeding experiment.[56]

A species is of course more than a group of individuals surrounded by a sterility barrier. Each species faces the pragmatic test of survival in the midst of rigorous competition. We have accounted for the origin of new chromosomal sterility barriers following hybridization. At the beginning of this section we considered the formation of new adaptive properties by recombination following hybridization. Our next problem is to explain the association of these adaptive properties with the new sterility barriers.

[55] Stebbins, 1957*b*.   [56] Stebbins, 1957*b*.

In the ancestral hybrid swarm the genes determining the morphological and physiological differences between the parental species segregate more or less at random. The chromosome structural arrangements differentiating the parental species and causing the sterility of their hybrids also segregate. By chance a combination of genes having a superior fitness in some available environment will sometimes arise, and when it does it will be associated with a particular chromosomal arrangement. It is well known that translocations and inversions in a heterozygous condition enforce gene linkage. The particular structural arrangement possessed by the new hybrid segregate will therefore shelter its favorable gene combination from disintegration. This gives the new species a chance to increase and spread, first within the hybrid swarm, and later throughout an area occupied by related species, while preserving its favorable gene combination intact.[57]

The new structural arrangement of the chromosomes does not inhibit gene recombination within the structurally homozygous species. In a sense its persistence during the later life of the species can be regarded as a vestige from the period of origin of this species when, in the presence of numerous other structural arrangements in the ancestral hybrid swarm, it did serve a function of linkage. The chromosome structural arrangement is the umbilical cord of the species.[58]

The hybrid origin of new fertile lines which are chromosomally sterile with their parents and siblings depends upon the hybrid progeny becoming structurally homozygous. The formation of homozygotes depends in turn on the union of like gametes. And the chances of like gametes uniting are strongly affected by the breeding system of the plants in question. Structural homozygosity will be attained most rapidly in the progeny of a hybrid under conditions of self-fertilization and will develop more slowly with obligate outcrossing.[59]

Consider a mixed population consisting of many fertile plants belonging to two species and one or a few sterile F_1 individuals. With outcrossing, the few viable gametes produced by the hybrids will unite predominantly with the abundant gametes of the parental species to form backcross zygotes, which are heterozygous. Sib crossing of sister

[57] Grant, 1956a, 93, and 1958, 357; as based on previous suggestions by Stebbins, 1950, 244–47 passim.
[58] Grant, 1956a, 93, 103; 1958, 357.   [59] Stebbins, 1957a; Grant, 1958.

F_1 individuals to yield F_2 zygotes takes place less frequently than back-crossing, perhaps much less frequently or not at all if the number of F_1 individuals is small and their sterility high, and leads to the formation mainly of more heterozygotes anyway. Structural homozygotes can arise by chance, of course, in the second or any subsequent hybrid generation, whenever like gametes get together; but a breeding system which enforces the union of gametes produced by different plants, generation after generation, will greatly delay this event in a hybrid swarm composed of genotypically dissimilar individuals.

The case is otherwise if the structural heterozygotes in the hybrid swarm are capable of self-fertilization. Most of the zygotes produced by selfing in the F_1 or B_1 generation will be heterozygous, to be sure, but proportionally more of the F_2 or B_2 types will be structurally homozygous under conditions of inbreeding than under outcrossing. And a breeding system which promotes inbreeding in the first (F_1) or second (F_2 or B_1) hybrid generation will have the same effect in later generations, thus increasing the chances of homozygote formation as time goes on. Among the structural homozygotes formed, some may be new recombination types.

It is significant that the examples of new fertile lines of hybrid origin mentioned earlier all involve plants which are regularly or facultatively self-pollinating. *Galeopsis tetrahit* and *G. bifida* are autogamous.[60] The new fertile lines originating by recombination in Crepis and the *Elymus glaucus* group were produced experimentally by self-pollination.[61]

If more than one homozygous recombination type is produced in the hybrid swarm, the various new structural homozygotes will be likely to differ from one another. Inbreeding in a hybrid swarm thus causes the progeny of a chromosomally sterile hybrid to become sorted out into different lines, each of which becomes structurally homozygous in a few generations, so that a swarm of new daughter species, fertile *inter se* but intersterile with one another and with the parental species, is produced (Fig. 86). The recurrence of hybridization between the daughter species, or between them and the parental species, can then initiate a new cycle of species formation. Chromosomal sterility

[60] Müntzing, 1930. [61] Gerassimova, 1939; Stebbins, 1957*b*.

barriers can multiply most rapidly under conditions of self-fertilization.[62]

As noted earlier, chromosomal sterility barriers are of widespread occurrence in many groups of herbaceous plants. Such barriers are more frequent in partially self-pollinating annual plants than in the more exclusively outcrossing perennial groups (Table 18 in Chapter 17). They reach their most extreme development in autogamous groups like *Galeopsis tetrahit, Erophila verna,* the *Elymus glaucus* group, the *Gilia*

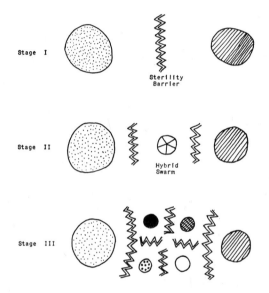

Fig. 86. Multiplication of intersterile species by hybridization under conditions of self-fertilization

From Grant, *Cold Spring Harbor Symposia Quant. Biol.,* 1958, with permission of the Long Island Biological Association.

inconspicua group, and others, in which morphologically similar populations are often intersterile.[63] Natural hybridization also occurs fairly frequently in the same plant groups. The mechanism of sterility barrier formation described above provides a partial explanation for

[62] Stebbins and Vaarama, 1954; Stebbins, 1957a, 1957b; Ray and Chisaki, 1957; Grant, 1958.
[63] Stebbins, 1957a, 347; Grant, 1958, 355, 358.

the occurrence of chromosomal rearrangements between species in many herbaceous genera, for the species originating by hybridization between preexisting species which differ in chromosome structure will themselves come to inherit new structural arrangements.[64]

ALLOPOLYPLOIDY

A chromosomally sterile hybrid may give rise to a new structurally homozygous and hence fertile but intersterile species in another way. The doubling of the chromosome number of a chromosomally sterile hybrid, a process and condition known as allopolyploidy, provides a way out of the impasse of hybrid sterility.

Let two species with structurally differentiated chromosome sets, represented in the diploid condition as AA and BB, respectively, produce a chromosomally sterile F_1 hybrid AB. If now the chromosome number in the hybrid is doubled to give the tetraploid constitution AABB, every chromosme will have a homologous partner for normal pairing and separation at meiosis. The products of meiosis consequently receive a full complement of chromosome segments and genes and are viable. The new allotetraploid species, in short, will be fertile, but will form chromosomally sterile (triploid) hybrids with the parental diploid species.

The hypothesis of the origin of new fertile plant species by a combination of hybridization and chromosome doubling was proposed by Winge in 1917 and verified experimentally in Nicotiana by Clausen and Goodspeed (1925). Other early studies were those of Digby (1912) on Primula, Karpechenko (1927, 1928) on Raphanobrassica, Müntzing (1930) on Galeopsis, and Huskins (1931) on Spartina.[65] The phenomenon of allopolyploidy has been extensively studied since that early period and is now rather well understood.[66] Only a brief sketch of the subject can be presented here.

There are in general two ways in which a sterile plant hybrid can undergo chromosome-number doubling. In a diploid hybrid plant with mainly sterile flowering branches, aberrant cell divisions in a bud may lead to a tetraploid flowering branch, which is fertile. Or in a diploid

[64] Grant, 1956a, 1958.
[65] Stebbins, 1950, 298–99, for brief early history with references.
[66] Excellent review in Stebbins, 1950, chs. 8, 9.

and sterile flower the meiotic process may end, not in chromosome reduction, but in the formation of unreduced diploid pollen grains and eggs, the union of which gives a tetraploid. The two main sources of allopolyploid derivatives are thus somatic doubling and the formation and union of unreduced gametes.

In long-lived perennial plants, either herbaceous or woody, somatic doubling has a long time and numerous chances to occur in a hybrid. If the plants are outcrossing, as is often the case in perennial groups, somatic doubling may well surpass the union of unreduced gametes as a source of allopolyploids. In any case, somatic doubling has a greater chance of occurring in perennial plants than in annual plants. The sterile diploid primrose hybrid, *Primula verticillata* × *P. floribunda*, a perennial herb, gave rise by somatic doubling to a tetraploid branch bearing fertile flowers from which the new allotetraploid, *Primula kewensis*, arose; but the tetraploid branch did not develop during the first year of the life of the hybrid plant, and it is fair to assume that *Primula kewensis* might never have originated if the parental species had been annuals.[67]

Polyploid species are known to occur in a higher percentage frequency in herbaceous perennial plants than among annuals.[68] The long time available for the establishment of the polyploid condition in the perennial herbs, especially by somatic doubling but also by union of unreduced gametes in special cases, may well be the reason for the high frequency of polyploidy in perennial herbaceous groups.[69]

In hybrid plants with an annual life cycle the time available for vegetative growth is limited, and somatic doubling is consequently unlikely to occur frequently, but unreduced gametes can still be formed. The chances of the union of two such unreduced gametes during the single flowering season of the annual hybrid will then be influenced by the breeding system of the plant in question. Outcrossing reduces the chances of formation of a polyploid zygote, whereas self-fertilization increases those chances. It is probably for this reason that polyploid species are far more frequent among autogamous annuals than among cross-fertilizing annual plants.[70] In the annual sections of the genus

[67] Grant, 1956c, based on Stebbins, 1950, 354–56.
[68] Müntzing, 1935; Stebbins, 1938; Gustafsson, 1948.
[69] Stebbins, 1950, 354–56.   [70] Grant, 1956c.

Gilia, for example, all but one of the polyploid species are themselves autogamous and are related to autogamous diploid species, and a similar association between inbreeding and polyploidy is found in other annual groups such as Clarkia, Mentzelia, Escholtzia, Microseris, Layia, and Madia.[71]

The same correlation between inbreeding and polyploidy is found in some perennial herbaceous genera like Primula. In the section Vernales of Primula the species are all obligately outcrossing (by heterostyly) and are all diploid. The section Farinosae of this genus contains both diploid and polyploid species. The diploid species are heterostylous and outcrossing, whereas the polyploids are homostylous, self-compatible, and hence capable of self-fertilization.[72]

The proportion of angiosperm species that are polyploid is very high. In 1938 Stebbins estimated roughly that the proportion of polyploid species in the flowering plants was about 30 to 35 percent.[73]

Some years ago Dr. H. Latimer and I tabulated the number of angiosperm species possessing various chromosome numbers on the basis of the data given by Darlington and Wylie in the *Chromosome Atlas of Flowering Plants* and by Darlington and Janaki-Ammal in the earlier edition of this work (in the cases where information presented earlier was not repeated in the second edition).[74] The statistical variate used in our tabulation was as a rule the taxonomic species, cultivated forms and other infraspecific groups not being listed separately. A taxonomic species containing two or more different chromosome numbers was, on the other hand, usually treated as more than one species for statistical purposes. The presence of supernumerary or B chromosomes was disregarded. Odd somatic numbers were expressed in the table of gametic numbers as the arithmetic half-value, $2N = 15$ becoming $N = 7.5$.

The number of species in the sample is 17,138; there are 11,987 dicotyledons and 5,151 monocotyledons. The frequency distribution of haploid chromosome numbers in this sample is given in Table 17. Most

[71] For literature references pertaining to the different plant groups, see Grant, 1956c.

[72] Valentine, 1961, 84.   [73] Stebbins, 1938.

[74] Darlington and Wylie, 1955; Darlington and Janaki-Ammal, 1945; Grant and Latimer, unpublished.

Table 17. The number of species of flowering plants with a given haploid chromosome number[a]

Haploid number	Dicots	Mono-cots	Both	Haploid number	Dicots	Mono-cots	Both
2	1		1	25	31	33	64
3	7	8	15	25.5	14	1	15
4	99	35	134	26	123	51	174
5	159	66	225	26.5	1	1	2
5.5	1		1	27	82	96	178
6	291	117	408	27.5	5	1	6
6.5	1	1	2	28	156	110	266
7	775	598	1,373	28.5	9	1	10
7.5	11	5	16	29	14	24	38
8	1,117	181	1,298	29.5	1		1
8.5		5	5	30	87	147	234
9	1,151	235	1,386	30.5		3	3
9.5	3	2	5	31	7	24	31
10	533	283	816	31.5	8	7	15
10.5	58	14	72	32	78	40	118
11	1,081	174	1,255	32.5		1	1
12	810	328	1,138	33	45	25	70
12.5	2	4	6	33.5	1		1
13	708	112	820	34	93	24	117
13.5	20	7	27	34.5		4	4
14	756	408	1,164	35	21	45	66
14.5		6	6	36	97	57	154
15	215	154	369	37		11	11
15.5		1	1	37.5		5	5
16	491	162	653	38	35	15	50
16.5	6	19	25	38.5	2	1	3
17	353	43	396	39	27	6	33
17.5	65	11	76	40	36	46	82
18	502	276	778	40.5	4	2	6
18.5	1	2	3	41	7	9	16
19	194	68	262	42	37	22	59
19.5	5	11	16	42.5	1	1	2
20	326	356	682	43	5	7	12
20.5		2	2	43.5	1	2	3
21	238	222	460	44	16	6	22
21.5	1	6	7	45	23	15	38
22	271	69	340	45.5	1	3	4
22.5	12	12	24	46	10	2	12
23	89	28	117	47	1		1
23.5	1	3	4	47.5		2	2
24	307	147	454	48	28	13	41
24.5	2	6	8	48.5		1	1

Table 17 (continued)

Haploid number	Dicots	Mono-cots	Both	Haploid number	Dicots	Mono-cots	Both
49	2	1	3	74.5		1	1
50	10	9	19	75	1		1
50.5	1	1	2	76	5		5
51	26	5	31	77	2		2
51.5		1	1	78	2		2
52	9	2	11	79			
52.5	1	2	3	80	5	2	7
53				81			
54	10	10	20	82	4		4
54.5	1		1	83			
55	2	3	5	84	1		1
56	26	5	31	85	4		4
57	16	3	19	85.5	1		1
58	3	2	5	86			
58.5		1	1	87			
59				88	2		2
59.5		1	1	89			
60	16	23	39	90	3	4	7
60.5		1	1	91			
61		2	2	92	1		1
61.5		1	1	93			
62	1	2	3	94			
63	4	4	8	95	1		1
63.5	1		1	95.5	1		1
64	6	2	8	96	1		1
65	5	1	6	97			
66	7		7	98			
66.5	1		1	99	1		1
67				100	6		6
67.5	1		1	112	1		1
68	7		7	113		1	1
69	3	2	5	119	1		1
70	2	4	6	132	1		1
71				150	1		1
72	7	3	10	154	1		1
73				250	1		1
73.5		1	1		———	———	———
74	2		2		11,987	5,151	17,138

a From compilations of Darlington, Janaki, and Wylie, as analyzed by Grant and Latimer.

of the data in the table, except those for the species with the very highest numbers and the odd somatic numbers, are shown graphically in Fig. 87.

The average haploid number of chromosomes per species for different groups of angiosperms calculated from this sample turns out to be as follows: temperate and arctic dicots, $N = 15.26$; tropical dicots, $N = 16.07$; all dicots, $N = 15.36$; monocots, $N = 17.46$; all angiosperms, $N = 15.99$.

The modal basic chromosome numbers are $N = 7$, 8, and 9 in the subsample of herbaceous dicots, and $N = 7$ in the monocots. The frequency distribution of basic chromosome numbers in the subsample of woody dicots is bimodal with one peak at $N = 8$ and 9, and a second and higher peak at $N = 11-14$. There are reasons for believing that the latter group of chromosome numbers in the woody dicots is derived phylogenetically from some lower number in the neighborhood of $N = \pm 8$. While we do not know what the original basic number was for the angiosperms, which originated in the early middle Mesozoic era, a number of considerations suggest that the original chromosome number of the group lies most probably in the range $N = 7-9$.

If this estimate is correct, the chromosome numbers from $N = 14$ upwards are of polyploid origin. Actually, $N = 14$ may be attained in some lineages by polyploidy and in others by stepwise changes in the basic chromosome number not involving multiplication of whole sets (viz., aneuploidy); in short, $N = 14$ is probably not always a tetraploid number. On the other hand, some of the lower numbers in our sample from $N = 10$ to $N = 13$ are known or inferred to be of polyploid origin from species with reduced basic numbers of $N = 5$, 6, and 7. There is probably more polyploidy below $N = 14$ in our sample than aneuploidy at and above this level. Let us then consider the species with $N = 14$ or more as polyploids to obtain a conservative estimate of the frequency of polyploidy in the angiosperms.

On this basis, 43 percent of nearly 12,000 species of dicots in Table 17 are polyploid; 58 percent of over 5,000 species of monocots have polyploid numbers; and polyploidy is present in 47 percent of the species of angiosperms in our sample.

In the animal kingdom, by contrast, polyploidy is quite rare. A few

authentic examples exist in various asexually reproducing, parthenogenetic groups of arthropods, namely, in the brine shrimp (*Artemia
salina*, Crustacea), sowbug (Trichoniscus, Isopoda), bagworm moth
(Solenobia, Psychidae, Lepidoptera), weevil (Otiorrhynchus, Curculio-
nidae, Coleoptera), and fly (Ochthiphila, Chamaemyiidae, Diptera);
and in a parthenogenetic group of worms (Lumbricidae).[75]

By far the majority of the natural occurring polyploid species in
plants that have been analyzed taxogenetically have been shown to be
allopolyploids, containing the chromosome sets of different ancestral
species.[76] This fact, taken together with the high frequency of polyploidy in the angiosperms (and in other groups of vascular plants),
clearly indicates that the hybrid origin of species is a common and
widespread method of speciation in the plant kingdom.

REDUCTION OF STERILITY BARRIERS
BY HYBRIDIZATION

Hybridization between reproductively isolated species followed by
repeated backcrossing of the hybrids to one or both parental species, a
process which Anderson has described and designated as introgressive
hybridization, will lead in time to convergences between previously
separate phyletic lines in their morphological, physiological, and
ecological characteristics.[77] Long-continued introgressive hybridization
can also gradually reduce the sterility barrier between the original
species at the same time that their phenotypic characteristics become
more similar. This effect comes about by gamete selection in the
incompletely sterile heterozygotes.[78]

The conditions necessary for the reduction of a sterility barrier by
selection may be enumerated. It is assumed that two species are
prevented from interbreeding freely by a barrier of hybrid sterility
which, however strong, is less than absolute. The type of sterility
barrier involved is gametic sterility. The gamete abortion in the
hybrids could theoretically be either genic or chromosomal in origin,

---

[75] White, 1954; Stalker, 1956; Bungenberg, 1957.
[76] Clausen, Keck, and Hiesey, 1945, ch. 8.
[77] Anderson and Hubricht, 1938; Anderson, 1949, 1953. See the latter two works
for interesting and valuable reviews.
[78] Grant, 1958, 358–59.

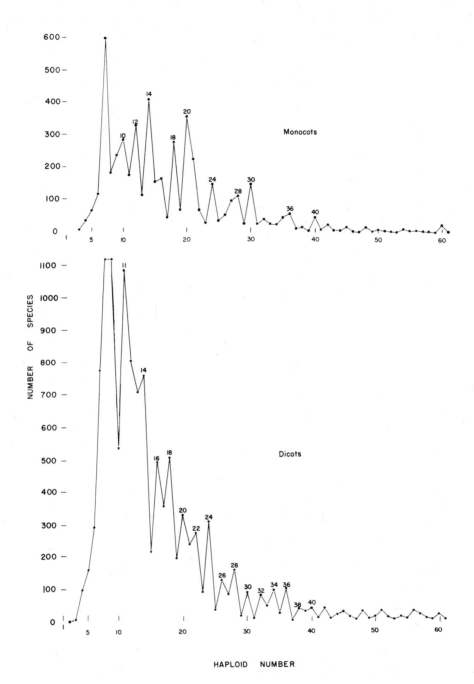

Fig. 87. Frequency distribution of haploid chromosome numbers in about 17,000 species of flowering plants

Frequency distribution in the range from N = 2 to N = 61.

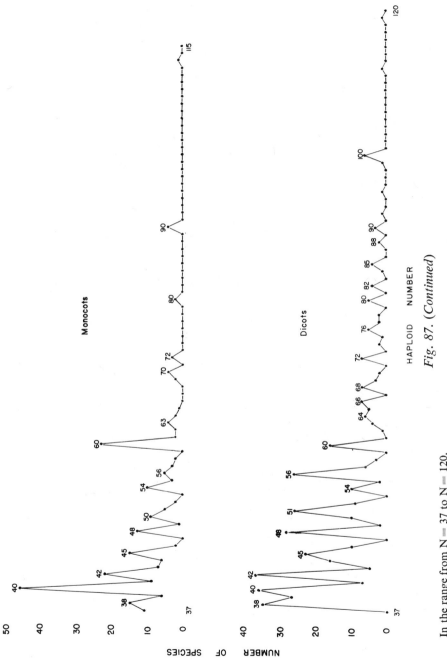

Fig. 87. (Continued)

In the range from N = 37 to N = 120.

but in the known actual cases it happens to be chromosomal gametic sterility, and we will accordingly assume the latter type of sterility for the development of our model. The species and their hybrid derivatives are cross-fertilizing. Introgressive hybridization takes place between the species and is continued over many generations. Certain genes determining morphological and physiological characteristics which are present in the donor species have a positive selective value in the genotype of the recipient species. These morphological and physiological genes, if borne on the same chromosomes as the segmental rearrangements which cause gamete abortion, are separable from the latter by crossing-over. All of the foregoing conditions are known or inferred to exist in numerous groups of higher plants.

Let us take as our starting point the setup described previously and shown in Fig. 84 of two species differing for two independent translocations. The recipient species R has the chromosomal constitution AA BB CC DD; the donor species D has the constitution A_bA_b B_aB_a C_dC_d D_cD_c; and the F_1 hybrid has a gametic fertility of 25 percent.

Now assume that certain genes borne on chromosome B_a in the donor species D are adaptively valuable in the genetic background of species R. The goal of introgressive hybridization is to introduce those genes from the one species into the other in a homozygous condition, that is, to produce a new recombination type of the constitution AA B_aB_aCC DD. The achievement of this goal depends on the functioning, in the F_1 and in the later backcross generations, of the gametes carrying the adaptively valuable genes on chromosome B_a. Thwarting the process of introgression, on the other hand, is the sterility barrier and particularly the linkage of the adaptively valuable genes with the chromosomal sterility factor (t_1) on B_a (see Fig. 84).

The four classes of functional pollen grains produced by the F_1 hybrid are (i) ABCD; (ii) $A_bB_aC_dD_c$; (iii) A_bB_aC D; and (iv) ABC_dD_c (Fig. 84). Backcrossing of the hybrid with the recipient species in the direction Species R ♀ × F_1 ♂ will consist of the union of ABCD ♀ gametes with one or another of the four kinds of pollen. Only classes (ii) and (iii) carry chromosome B_a with its genes derived from the donor species, however, and can serve as a bridge for introgression. But both of these classes of pollen will produce backcross types in the next generation which are heterozygous and semisterile.

This block can be circumvented, however, by crossing-over in the heterozygote B/B_a in the F_1 or any subsequent backcross generation. Crossing-over can occur between chromosomes B and B_a in such a way as to separate the segments of B_a determining adaptive morphological and physiological characters from the translocation segment (t) responsible for pollen-grain abortion. The substitution in chromosome B_a of the non-homologous segment from chromosome B in effect wipes out the set of complementary sterility factors. The chromosome regions of B_a freed from the neighboring sterility factors can then be carried over into the recipient species in a new homozygous combination with chromosome A.[79]

Assume that the process of introgression from Species D into one [branch] of Species R continues over a long period of time, because certain recombination types. . . have a selective advantage in a given environment. The chromosomes contributed by the donor species (D) are purged of their components of each pair of complementary sterility factors as they enter the recipient population of R. This purging takes place through the selective elimination of all the non-crossover gametes, the only functioning gametes being those in which crossing over has occurred between the sterility factors and other regions of the chromosomes. The chromosome pool of the recipient population could eventually become filled with chromosomes like those of the donor species (D) in virtually every respect except the absence of the sterility factors of D and the retention of particular morphological-physiological genes of R. The phenotypic manifestation of these recombination chromosomes would be plants well differentiated taxonomically from the unintrogressed branch of species R.[80]

If, as frequently happens, the introgression of genes affects one branch of the recipient species R but not others, the morphological and physiological differences between the original and the introgressed population systems of R could become very great: as great, indeed, as the differences between R and D. But whereas a barrier of chromosomal sterility separated R and D, the almost equally different unintrogressed and introgressed populations of R are interfertile.[81]

Furthermore, the introgressive population system of species R may retain some chromosome structural arrangements and gamete fertility factors in common with species D. As a result, the meiosis is more normal and the fertility higher in the hybrids of D × introgressive R than in those of D × pure R. And to the extent that introgressive R bridges the chromosomal sterility gap with species D, the former is less

[79] Grant, 1958, 358–59.   [80] Grant, 1958, 358–59.   [81] Grant, 1958, 358–59.

than completely interfertile with pure R.[82] Different hybrid combinations between R and D, in other words, will be expected to form a graded series in their degree of chromosomal fertility, as shown diagrammatically in Fig. 88. Different pure races of R will be most

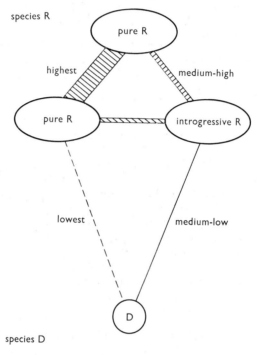

Fig. 88. *Expected relative degree of chromosome pairing and gamete fertility in F_1 hybrids from different interracial and interspecific combinations of species R and D*

Further explanation in text.

interfertile; introgressive R will be somewhat less fertile chromosomally with pure R at the same time that the former retains a trace of chromosome homology and interfertility with D; while the maximum chromosomal sterility will occur in the combination of pure R × D.

Some examples conforming to the above hypothesis have been worked out in the cross-fertilizing Cobwebby Gilias. We saw in Chapters 12

---

[82] Grant and Grant, 1960, 480.

and 13 that *Gilia ochroleuca* and the *Gilia latiflora-leptantha-tenuiflora* group are separated in their sympatric area by a strong barrier of chromosomal sterility, chromosome pairing being much reduced in the sterile hybrids (see Fig. 75 in Chapter 13). However, the two divergent lines of evolution have not remained completely isolated during their history of sympatric contacts, but have hybridized sporadically. This hybridization has led to the formation of some existing races, particularly to the race *vivida* of *Gilia ochroleuca* and to the race *speciosa* among others of *Gilia cana*.[83]

Gilia ochroleuca vivida which inhabits the San Gabriel Mountains is interfertile with other races of *Gilia ochroleuca* and constitutes a geographical race of that species. Yet it possesses a complex of morphological characters which point to a special relationship with *Gilia leptantha* and which could have been acquired most readily by introgression from *G. leptantha* into a preexisting population of *G. ochroleuca* with features like those still preserved in *G. o. bizonata*. In terms of our general symbolism *Gilia ochroleuca bizonata* is equivalent to pure R; *Gilia ochroleuca ochroleuca* is another pure race of R; *G. o. vivida* is equivalent to introgressive R; and *G. leptantha* is equivalent to Species D.

Now the chromosome pairing and fertility relationships between the races of *Gilia ochroleuca* and between these races and *G. leptantha* are very interesting. The F_1 hybrid of *bizonata* × *ochroleuca*, (pure R × pure R), has perfect chromosome pairing (an average of 9.0 bivalents per pollen mother cell) and high fertility (an average of 78 percent good pollen). The F_1 hybrid of *bizonata* × *vivida*, (pure R × introgressive R), exhibits a slight reduction in chromosome pairing and fertility, with an average of 8.9 bivalents per pollen mother cell, most cells having 9 but some having only 8 bivalents, and an average poelln fertility of 32 percent. The interspecific hybrid of *G. o. vivida* × *G. leptantha*, (introgressive R × D), exhibits a medium degree of chromosome pairing with a mean of 6.9 and range of 4 to 9 bivalents per pollen mother cell, and is highly sterile. The interspecific hybrid of *G. o. bizonata* × *G. leptantha*, (pure R × D), finally, has the most reduced chromosome pairing of all, with a mean of 3.0 and range of 0 to 7 bivalents per cell, and it is also highly sterile.[84]

[83] Grant and Grant, 1960, 478 *passim*.   [84] Grant and Grant, 1960.

Gilia cana is a species related to *Gilia ochroleuca* in the Mojave Desert. The western races of *Gilia cana*, for instance *G. c. speciosa*, approach the *Gilia latiflora-leptantha-tenuiflora* group morphologically as well as geographically, and have probably derived their convergent morphological features from the latter by introgression. The artificial hybrids of the introgressive races of *Gilia cana* with the *Gilia tenuiflora* group are fertile or semifertile. This fertility is associated with a relatively high degree of chromosome pairing, the average number of bivalents per cell being 7.9, 8.4, and 8.7 in different hybrid combinations between the two groups. By comparison, *Gilia ochroleuca bizonata* forms on the average only 3 or 4 bivalents per cell in hybrids with the same members of the *Gilia tenuiflora* group. The hybrids in the latter combination are highly sterile.[85]

Gilia cana and *Gilia ochroleuca vivida*, which have independently acquired genetic material from the *Gilia tenuiflora* group, produce in combination with one another a semifertile hybrid with an average of 8 bivalents per cell. The hybrid of *Gilia cana* with *Gilia ochroleuca bizonata*, by contrast, had a mean of 5 bivalents per cell and was completely sterile.[86]

It is interesting that both the morphological gap and the sterility barrier between the *Gilia ochroleuca* and the *G. tenuiflora* groups should be partially bridged in two taxa, *G. o. vivida* and *G. cana*, which are probably of hybrid origin between the species groups. The two ways in which the populations *vivida* and *cana* reveal their relationship to the *G. tenuiflora* group, their morphological resemblances and interfertility, could have a common cause. The chromosomes of *vivida* and *cana* are more similar in structural arrangement to the *G. tenuiflora* genome than are the chromosomes of other, more extreme members of the *G. ochroleuca* group. The introgression from one species group into the other was an influx of chromosome segments which would be expected to affect both the visible traits and the chromosome pairing relationships of the recipient population.

There are theoretical grounds for expecting that long-continued hybridization between intersterile species should lead, under conditions of cross-fertilization, to the elimination of the sterility barrier as a result of the smoothing out of the genomic differences (Grant, 1958). The cross-fertilizing taxa, *G. ochroleuca vivida* and *G. cana*, which in all probability originated from hybrids between intersterile members of the *G. ochroleuca* and *G. tenuiflora* groups, bridge the sterility barrier between the ancestral species groups, as demanded by the hypothesis.[87]

[85] Grant and Grant, 1960.   [86] Grant and Grant, 1960.
[87] Grant and Grant, 1960, 479–80; quoted from *Aliso*.

Related species in many groups of perennial plants, like Pinus, Quercus, Ceanothus, Iris, Aquilegia, are often weakly isolated (see Chapter 17). The weak incompatibility barriers and external isolating mechanisms in such groups are correlated, as we shall see in the later discussion, with the relative infrequency of close sympatric contacts and the high reproductive potential of the plants, both of which factors weaken the intensity of selection for isolation per se. As regards the generally low degree of hybrid sterility in the same plant groups, a somewhat different explanation may be offered. Most of the species in the aforementioned genera and in other perennial groups are cross-fertilizing, and many of these species have been much contaminated by introgressive hybridization during their evolutionary history. If, therefore, sterility barriers ever did exist in the past history in such groups, they would have been lost as new introgressive and interfertile population systems grew up and took the place of the original intersterile population systems (see Fig. 89).[88]

The progeny, if any, of hybridization between intersterile species of self-fertilizing plants are likely to be new species on either the diploid or polyploid level, as we have seen earlier. On the other hand, interspecific hybridization in cross-fertilizing plants frequently takes the form of introgression.[89] And introgression in a cross-fertilizing breeding system is expected to result eventually in the reduction of the original sterility barrier.[90] These expectations are realized in two sections of the genus Gilia in which natural hybridization has occurred among species possessing various breeding systems. In both the Cob-webby Gilias and the Leafy-Stemmed Gilias, hybridization between the autogamous species has led to the formation of many new allopolyploid species. In the same groups of plants a number of cross-fertilizing species or semispecies which are affected by introgression are also separated by reduced sterility barriers (compare Figs. 86 and 89).[91]

MACROMUTATION AND SPECIATION

It is quite possible that the variations necessary for species formation may arise by special bursts of mutation. Mutations with drastic phenotypic effects, or macromutations, would change the adaptive

[88] Grant, 1958, 359.   [89] Baker, 1953; Grant, 1956c.   [90] Grant, 1958.
[91] Grant, 1956c, 1958.

properties of the organisms, and might occasionally change those properties for the better. The possible importance of the mutation process in the origin of species was emphasized by the early Mendelian geneticists, such as Bateson (1894), Castle (1903), De Vries (1906), and

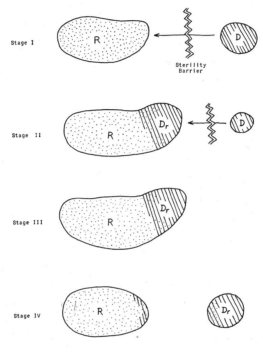

Fig. 89. The replacement of two intersterile species (R *and* D) *by two interfertile species* (R *and* D_r) *as a result of introgressive hybridization in a cross-fertilizing breeding system, followed by replacement of one of the original forms*

From Grant, *Cold Spring Harbor Symposia Quant. Biol.*, 1958, with permission of the Long Island Biological Association.

Goldschmidt (1940). Owing to lack of understanding of natural selection and of the nature of species the early Mendelian geneticists failed to indicate—or even to consider—how the process of speciation, once initiated by mutations, could be carried through to completion. Since we can fill that latter deficiency in theory today, let us consider again here the possible role of macromutations in speciation.

Various cases have been described in plants of macromutations which produce morphological structures characteristic of different genera or even different higher categories. It will be sufficient for our purpose here to consider a few examples of macromutations known in Antirrhinum and other members of the Scrophulariaceae. In the snapdragon, *Antirrhinum majus*, the mutation "transcendens" causes the formation of two instead of the normal four stamens. This mutant type lies outside the normal range of variation of Antirrhinum but within the range of other scrophulariaceous genera like Mohavea and Veronica in its floral structure.[92] Again, the mutant type "hemiradialis" of *Antirrhinum majus*, with five instead of four anthers and a semiradial instead of bilateral symmetry, has the floral pattern of Verbascum.[93] Another mutant of Antirrhinum develops spurs on the flowers, being in this respect like the related genus Linaria.[94] Conversely, there is a mutation ("gratioloides") in *Linaria maroccana* which produces spurless and radially symmetrical, instead of spurred and bilateral, flowers.[95] Finally, a mutant form of Antirrhinum has a single cotyledon in the seedlings, as found normally in the subclass of monocotyledons.[96]

Comparable macromutations are known in Drosophila. The mutant "tetraltera," for example, has the wings reduced to haltere-like organs and is thus essentially wingless.[97]

The phenotypic manifestation of the mutant character is variable in the early generations in all these examples, the mutant alleles evidently possessing low penetrance and expressivity, but selection leads after several generations to a more stabilized and regular development of the mutant phenotype. The effect of selection is probably to build up a set of modifier genes with an enhancing and stabilizing action on the macromutant gene.[98]

As various students have pointed out, a combination of macromutation and selection of modifier genes could lead to rapid and large-scale shifts in phenotypic characteristics.

Goldschmidt has postulated the existence of a special type of macromutation, called systemic mutations, in which the chromosomes are

[92] Stubbe, 1952, 1959.   [93] Stubbe, 1952, 1959.   [94] Schwanitz, 1956.
[95] Schwanitz and Schwanitz, 1955; Schwanitz, 1956.
[96] Schwanitz, 1956.   [97] Goldschmidt, 1952, 1953.
[98] Stubbe and Wettstein, 1941; Stubbe, 1952, 1959; Schwanitz and Schwanitz, 1955; Schwanitz, 1956; Goldschmidt, 1953.

repatterned, the segments being rearranged in new sequences, with profound effects not only on the phenotypic characteristics and adaptive properties of the organism but also on its fertility relationships.[99] There is no doubt that a repatterning of the chromosome set would alter simultaneously the morphological, physiological, and genetical responses of the organism, and would usually alter those responses drastically. The systemic mutant individual would be a monster, and perhaps, as Goldschmidt suggests, it might sometimes be a "hopeful monster."[100] This mutant type is supposed to be the progenitor of a new species originating by a saltation from the ancestral form and isolated from all other organisms by a bridgeless gap.

Chromosome mutations are well known in plant and animal populations, of course, though chromosome mutations involving the very large number of simultaneous rearrangements visualized by Goldschmidt are still a hypothetical quantity. But even assuming that systemic mutations with the properties attributed to them by Goldschmidt do arise, there are serious difficulties in the way of their establishment in populations, which have been faced by Dobzhansky, Mayr,[101] and other evolutionists, and which we should briefly consider here.

In the first place, how often would a completely scrambled genotype be viable? In a complex genotype built up by many generations of selection the viability of any mutation is probably inversely proportional to the radicalness of its phenotypic effects. As Lerner has put it, the hopeful monster had better not be too monstrous, or else it had better not be too hopeful.

As we have seen earlier, macromutations or macrorecombinations involving just one or a few genes in Antirrhinum, Gilia, etc. are generally unstable in development and heredity until a new system of minor modifier genes has been assembled around the major genes by selection. But this latter process, the selection of minor genetic variations, is one which Goldschmidt specifically rejects as a mechanism of evolution at the level of species or higher categories.[102]

A related difficulty concerns the fertility of the systemic mutant. As Dobzhansky points out, the systemic mutation would appear first in the heterozygous condition, unless we can imagine two chromosome

[99] Goldschmidt, 1940.   [100] Goldschmidt, 1940, 390.
[101] Dobzhansky, 1951a, 203; Mayr, 1942, 155.   [102] Goldschmidt, 1940, 8 *passim*.

sets becoming repatterned simultaneously in identical ways, and the structural heterozygote would be highly if not completely sterile.[103] Indeed, accepting Goldschmidt's own premise that all species are separated by "unbridged gaps" due to sterility and other breeding barriers,[104] I think we might have to conclude that the heterozygote for a systemic mutation, being genotypically in effect a species hybrid, could not reproduce at all.

Finally, even if the systemic mutation were viable and fertile, how could it become established in a population? In a large random mating population a mutant type is continually swamped out. However, as Lewis and Raven have pointed out, a systemic mutation might become established by self-fertilization.[105]

The less extreme and drastic type of systemic mutation visualized by Lewis and Raven consists of extensive chromosomal rearrangements with associated genetic changes induced by mutator genes. As these authors point out, spontaneous chromosome breakage, probably genically determined, is known in a wide variety of plants, such as Allium, Bromus, Hyacinthus, Paeonia, Paris, and Tradescantia. Mutator genes arising in a population at any given time could initiate extensive chromosome breakage and segmental rearrangement. Having produced the chromosome mutations, the mutator genes could then drop out of the population again.[106]

The new type of chromosome set would become fixed in the homozygous condition in a few generations of self-fertilization and, if associated with adaptively valuable traits, would soon become characteristic of a successful and expanding derivative population. In this way a very rapid evolutionary change might come about from one segmental arrangement of the chromosomes to another. Associated with the process of chromosome repatterning would be a change in phenotypic characteristics. The new population would be separated from the ancestral one by a chromosomal sterility barrier.[107]

Mutation and hybridization may not be independent of one another as sources of variation. Geneticists have suggested several times that hybridization may stimulate the production of new mutations. The

[103] Dobzhansky, 1951a, 203.   [104] Goldschmidt, 1940, 142–83 *passim*.
[105] Lewis and Raven, 1958.   [106] Lewis and Raven, 1958.
[107] Lewis and Raven, 1958.

most easily recognized example of this is allopolyploidy where hybridization is followed by a mutation involving the entire chromosome set.

Mangelsdorf has compared the mutability of inbred lines of corn (*Zea mays*) carrying particular chromosomes of teosinte (*Zea euchlaena*) with the mutability of pure inbred corn. Whereas inbred strains of pure corn are stable, their derivatives carrying introduced teosinte chromosomes are highly mutable, producing about 1.8 percent of mutant gametes. The mutations in this case probably represent products of crossing-over between structurally differentiated segments in the only partially homozygous parent plants; they are, in other words, recombinations of small segments rather than intragenic changes.[108] A similar mutation-inducing effect of hybridization was observed by Sturtevant in Drosophila.[109]

There is a good possibility that many reported macromutations are essentially macrorecombinations appearing several or many generations after hybridization (see also Chapter 8).

SUMMARY

By way of recapitulation, the various modes of speciation discussed in this chapter may be grouped according to the field and pathway with typical examples of each (see Fig. 88).

1. Allopatric speciation:

   v—s   Organisms with wide outcrossing and diverging geographical races as exemplified by *Parus major* and Diplacus.

   v—i_p Peripheral populations budding off from large widely outcrossing main populations as in New Guinea kingfishers. Also cross-fertilizing but sedentary organisms like cypresses and geese.

2. Neighboringly sympatric speciation:

   v—s   Cross-fertilizing organisms subjected to intense disruptive selection in neighboring habitats, as in race formation in white pines; it is not known whether this process can bring about speciation.

[108] Mangelsdorf, 1958*b*.   [109] Sturtevant, 1939.

v—i$_p$ Insects and parasites restricted to narrow food niches as, for example, the bee Diadasia; hypothetical though not implausible.

h—s   Cross-fertilizing and introgressively hybridizing plants as exemplified by oaks, pines, and Cobwebby Gilias.

3. Biotically sympatric speciation:

v—s   Cross-fertilizing organisms under very intense disruptive selection; hypothetical but doubtful.

v—i or h—i   Flowering plants pollinated assortatively by flower-constant bees or other specialized insects; somewhat hypothetical.

v—i$_f$ Occasionally self-fertilizing organisms, a possible example being Paramecium.

h—i$_f$ Occasionally or regularly self-fertilizing plants which hybridize and give rise to new species on the diploid level.

u—i   Organisms in which macromutations arise and are fixed by inbreeding; hypothetical.

h—u—i   Plant species originating by allopolyploidy. Some reported instances of macromutations may belong in this category too.

SEVENTEEN

Consolidation of the Sympatric Species

SELECTION FOR REPRODUCTIVE ISOLATION

TWO SPECIES which have diverged in their genetic constitutions so that they cross with difficulty to produce inviable or sterile offspring may, on becoming sympatric, hybridize on a limited scale. The production of inviable or sterile progeny from hybrid crosses represents a lowering of the reproductive potential of the species as compared with the formation of viable and fertile offspring by intraspecific matings. The formation of incompatible interspecific unions likewise represents a wastage of gametes. Hybridization, in other words, whether it produces inviable or sterile progeny or no progeny at all, is disadvantageous to the species concerned under these circumstances, whereas the prevention of hybridization is selectively advantageous.

We will make the reasonable supposition that the breeding habits of the species are genetically controlled and are subject to genetic variation. Thus some individuals in each species population might have a stronger ethological preference for their own species and a stronger aversion to interspecific matings than do other individuals of the same population. Or, in a pair of species with partially overlapping breeding seasons, some individuals in each species might be sexually ready at a time when only other members of their species are available as sexual partners, thus very early or very late in the season, whereas other individuals might enter the breeding season simultaneously with members of the foreign species. Or, again, some individuals in a population may be more compatible and others less so with the members of a different species.

The individuals within a species population which can and do cross with the foreign species either produce no offspring, if the species are mutually incompatible, or produce inviable or sterile progeny, if the species are separated by inviability and sterility barriers. In any case the hybridizing individuals contribute few or no offspring to subsequent generations. Meanwhile those individuals in the same population which possess strong inherent tendencies or capacities to mate chiefly or exclusively with conspecific individuals produce many viable and fertile progeny. During the course of generations, therefore, the genetically determined proneness to hybridize is gradually eliminated from the population, while the genetic factors promoting reproductive isolation increase in relative frequency.[1]

Natural selection can thus act to reinforce a partial and incipient degree of reproductive isolation by favoring those individuals within either species which tend to mate predominantly with their own kind, while discriminating against the individuals which are prone to cross with members of the foreign species. As Dobzhansky states, "Once reproductive isolation is initiated, natural selection will tend to strengthen it and eventually to make the isolation complete."[2]

It is of historical interest to note that Darwin in 1868 considered the question of a selective origin of incompatibility barriers, but could not formulate a satisfactory causal explanation in the state of evolutionary theory existing at that time. Darwin writes:

It would profit an incipient species if it were rendered in some slight degree sterile when crossed with its parent-form or with some other variety; for thus fewer bastardised and deteriorated offspring would be produced to commingle their blood with the new species in process of formation.

But, he concludes, the difficulties of this view are so great that "we must give up the belief that natural selection has come into play" in rendering species mutually sterile.[3]

Wallace later, in 1889, presented a model whereby natural selection *could* build up barriers of hybrid sterility and mating behavior between diverging sympatric species. He argued that if the hybrids were adaptively inferior to the races or species, selection would favor sterility and ethological barriers between them.[4] Wallace's treatment of the problem was apparently forgotten, however, and after a lapse of 40 and 50 years several evolutionists developed anew the hypothesis of the

[1] Fisher, 1930, 130–31; Dobzhansky in 1940; see 1951a, 208–9.
[2] Dobzhansky, 1951a, 208.   [3] Darwin, 1868, ch. 19.
[4] Wallace, 1889, 173–80, 217.

selective origin of species barriers, and of course stated it in more precise genetical terms.

The hypothesis that some reproductive isolating mechanisms are adaptations built up by natural selection to protect the genetic constitutions of the species from the deleterious effects of interspecific hybridization was advanced by Fisher in 1930 in relation to the origin of ethological barriers in animals.[5] Huxley in 1943 extended the hypothesis to include the origin of recognition marks, on which ethological isolation is based in animals.[6] Dobzhansky (1940, 1951) further developed the hypothesis and extended it to a wider range of isolating mechanisms in both plants and animals.[7] Since 1950 evidence bearing on the hypothesis of the selective origin of reproductive isolation has been collected by a number of workers.

It follows from the hypothesis of Wallace, Fisher, and Dobzhansky that reproductive isolating mechanisms are of two types with respect to their origin. Some isolating mechanisms arise as a by-product of evolutionary divergence. In addition, certain isolating mechanisms can be developed and strengthened by natural selection to reinforce a partial reproductive isolation already present. The two types of isolating mechanisms are complementary and interdependent in their origin as well as their action. This is because the existence of partial internal barriers—hybrid inviability, sterility, breakdown, and the like—provides a precondition for the building up of other and stronger barriers which prevent hybridization and the wastage of reproductive potential involved therein.[8]

The process of evolutionary divergence can give rise to any isolating mechanism whatever, as we saw in Chapter 15, and is the only known source of hybrid inviability, hybrid sterility, and hybrid breakdown. Selection for reproductive isolation per se is best fitted to develop the ethological barriers, which play such an important part in the isolation of animal species, and various other external barriers operating in the parental generation. The relative importance of the two types of isolating mechanisms as classified according to their mode of origin— the by-product mechanisms and the special reinforcement mechanisms— undoubtedly varies from one group of organisms to another.

[5] Fisher, 1930, 130–31. [6] Huxley, 1943, 45, 288 ff.
[7] Dobzhansky, 1951a, 207–11. [8] Dobzhansky, 1951a, 209–11.

The closely related but intersterile plant species, *Clarkia biloba* and *C. lingulata*, do not grow together in nature. Lewis has found that in experimental plots where they are combined artificially, they do not form stable mixtures, because pollinating insects do not discriminate between them, and sterile hybrids are consequently produced as readily as fertile intraspecific progeny. The species in a mixture which produces the smaller number of flowers loses a higher proportion of its progeny, and eventually dies out. Thus when an experimental population was started with individuals of the two species in the proportion of two-thirds *C. biloba* and one-third *C. lingulata*, though both species grew and flowered well, *C. lingulata* was eliminated from the mixture in four generations.[9] Under such conditions, genetic variations within one species leading to ethological or other hybridization-preventing barriers with the opposite species could be expected to have a selective advantage.

Selection has been shown to be effective in the enhancement of reproductive isolation between *Drosophila pseudoobscura* and *D. persimilis* under experimental conditions. These two species of flies, which form sterile and otherwise inadaptive hybrids, are ethologically isolated in nature. In laboratory cages at certain temperatures, however, the species interbreed fairly freely. Thus mixtures of the two species in several replicated cages set up by Koopman initially produced progeny, of which 22 percent to 49 percent were hybrids. Koopman removed the hybrid individuals from the mixed population during the first and succeeding generations. The spontaneous formation of hybrids fell rapidly during successive generations in response to selection against the hybrid types. In just five generations of selection the ease of hybridization decreased from a level measured by the production of 22 to 49 percent of hybrid offspring to a level measured by the formation of about 5 percent of hybrids. Evidently the genotypes with stronger ethological disinclinations to hybridize under the environmental conditions in the cages were increasing in relative frequency in the two populations during the course of the selection experiment.[10]

Ethological isolation has been built up between different strains of *Drosophila melanogaster* by selection in experiments carried out in a similar fashion.[11]

[9] Lewis, 1962.   [10] Koopman, 1950.
[11] Knight, Robertson, and Waddington, 1956.

In nature also selection can be inferred to have reinforced the reproductive isolation between species in certain instances. The situation most favorable for such inferences is where two species have overlapping distribution ranges, and the allopatric populations of the two species are found to be weakly isolated in the parental generation, whereas the sympatric populations are strongly isolated ethologically. This situation has been reported in several groups of animals.

Two related species of tree frogs, *Microhyla carolinensis* and *M. olivacea*, which occur respectively in the eastern and western United States, overlap in distribution in parts of Texas and Oklahoma. One of the most important barriers to hybridization between these species is their ethological reluctance to form interspecific mating pairs. The nature of the mating call—the duration of the call and frequency of the notes—plays a very important part in the ethological isolation. By means of instruments Blair analyzed and scored the mating call in a series of races of each species. He found that the call notes in races of the two species from their allopatric areas, as for example *M. carolinensis* from Florida and *M. olivacea* from Arizona, are very much alike; but the calls of races of these species from their zone of sympatric overlap and potential hybridization in Texas are completely distinct. It is difficult to explain this distribution of ethological isolating mechanisms within the two species on any other basis than direct selection for reproductive isolation per se.[12]

A similar distribution of weak and strong forms of ethological isolation has been described in several species groups of Drosophila, namely, the *D. pseudoobscura* group in western North America, and the *D. guarani* and *D. willistoni* groups in the American tropics,[13] and again in the butterfly *Erebia tyndarus* and its relatives in Europe.[14]

The effective action of natural selection in the strengthening of reproductive isolation between species presupposes a number of conditions. (1) The species must be sympatric, at least marginally. (2) The products of hybridization must be adaptively inferior. (3) The loss of gametes involved in hybridization must be selectively disadvantageous to the species concerned. These conditions may or may not be fulfilled in the

[12] Blair, 1955.
[13] Dobzhansky, 1951a, 210; Dobzhansky, Ehrman, and Pavlovsky, 1957.
[14] Lorković, 1958.

case of any given pair of species. In the following discussion we will attempt to show by broad comparisons of different major groups of organisms that the presence or absence of hybridization-preventing mechanisms is correlated in a general way with the presence or absence of the foregoing conditions.

THE EFFECT OF HYBRIDIZATION ON SELECTION FOR ISOLATION

One of the conditions which is necessary in order that natural selection can act to strengthen reproductive isolation is this: the interspecific hybridization must in fact lead to the production of F_1 and F_2 progeny with a low adaptive fitness. The genotypes produced by hybridization must be less viable or less fertile than the genotypes produced by crossing within either species population alone. The genes of the separate species, in short, must be incapable of recombining successfully.

This condition is certainly met with in many cases of interspecific hybridization, but it is by no means a universal condition. In many genera of trees, shrubs, and perennial herbs in north temperate floras, as will be shown later, the progeny of many or most interspecific crosses are quite vigorous and fertile. A fair number of interspecific hybrid combinations in mammals and birds are likewise vigorous and fertile.[15]

In those groups of organisms in which interspecific hybridization has no or only slight deleterious effects, selection cannot be expected to act effectively to erect strong barriers to hybridization, and such isolating mechanisms as do exist between the species will be the result primarily of divergence. It is noteworthy that in higher plants a general correlation can be discerned between the presence of hybrid inviability and sterility and the presence of incompatibility barriers which reduce the production of inviable or sterile hybrids.[16] Strong incompatibility barriers generally exist between species of plants which have inviable or sterile hybrids. This combination of isolating mechanisms is typical in such herbaceous genera as Nicotiana, Gossypium, Layia, Clarkia, Gilia, and Nigella. But incompatibility barriers to hybridization are only weakly developed between related species of plants, in genera like

---

[15] Gray, 1954, 1958. [16] Grant, 1958, 353.

Pinus, Quercus, Ceanothus, Aquilegia, and Iris, the interspecific hybrids of which are highly fertile and vigorous.[17]

Some recombination types derived from interspecific hybridization in plants are actually adaptively superior to the parental species. Stebbins and other botanists have described cases of particular genotypes of hybrid origin in grasses and other plant groups which have spread and increased relative to the original species populations.[18]

If interspecific hybridization, so far from being deleterious, has some positively favorable effects, it will be to the advantage of the species concerned to keep open some channels of interspecific gene exchange.[19] Related species and semispecies of plants the world over can and do hybridize in nature, on a limited scale at least, wherever the physical opportunity exists. Perhaps the proneness of plant species to engage in hybridization is due to the absence of strong selective pressures to seal off the channels of interspecific gene flow completely. Or perhaps it is a result of a process of interspecific selection discriminating throughout past ages against species which have become too strongly isolated for their own long-range good, and favoring the species which have retained a limited capacity for hybridization.

We may ask why related species in some groups of organisms are able to exchange genes successfully, whereas related species in other groups produce largely non-functional recombination types. What factors affect the combining ability of the genes carried by different related species? On this question we can only venture some suggestions.

The adaptations of each species depend on combinations of characters which must work harmoniously together. Underlying the character combination is a combination of particular genes, which have been selected for their coordinated action. The adaptive differences between related species are differences in the whole character complex and in the underlying gene combination, as Stebbins has emphasized.[20] Certain genes of one species may not be successful in combination with the genes of another species simply because the phenotypic products of the two sets of genes do not form a harmoniously functioning character complex.

[17] Stebbins, 1950, 234–35; Grant, 1958, 352–55.
[18] Stebbins, 1958, 199. See also Stebbins, 1950, ch. 7, and 1959.
[19] Grant, 1957, 72–74; Stebbins, 1958, 199.
[20] Stebbins, 1950, ch. 5 *passim*.

For example, *Gilia modocensis* and *Gilia malior* are related species of Cobwebby Gilia occurring in arid parts of California and bordering regions. Both species are autogamous or naturally self-pollinating. In the flowers of each species the stamens and style are mutually adjusted in length so that the anthers automatically touch and deposit pollen on the stigma soon after the flower opens. The flowers of *Gilia modocensis* are somewhat larger than those of *G. malior* in all their individual parts. In the F_2, F_3, and F_4 generations derived from the cross between the two species, we find many different recombination types in which the lengths of stamens and style do not correspond properly, so that self-pollination does not take place. Since the flowers are not normally visited by insects, the failure of automatic self-pollination leads to sterility, even though the plants produce functional pollen grains and ovules. Such recombination types have a low adaptive value.

Two species will give rise to inviable or sterile recombination products therefore if, first, they differ genetically with respect to an adaptation and, second, if this adaptation is complex.

In general, the more complex an adaptation is, the more ways there are for something to go wrong with it. For proper functioning, a complex adaptation requires a higher degree of coordination between more component parts than does a simpler adaptation. A novel, untried, and unselected combination of component parts derived from different species has more chances of being poorly coordinated and non-functional in a complex than in a simple adaptation. Gene recombination in interspecific crosses is most apt to produce ill-adapted phenotypes where the species differ in a complex adaptation.

Undoubtedly the individual organism, its mode of growth, its organ systems, its degree of coordination, and basic adaptations are vastly more complex in the higher animals than in the higher plants. Probably the genotype is correspondingly more highly integrated and complex in animals.[21] There is some genetic evidence in favor of this supposition. Thus position effects are much more common in animals like Drosophila than in plants, where they have been looked for but found in only very few instances. In *Drosophila melanogaster* homozygotes for 332 different translocations were obtained, of which 40 percent were viable

[21] Stebbins, 1950, 182.

and 60 percent were lethal[22]; whereas in *Zea mays* 64 translocation types were *all* viable as homozygotes.[23]

It follows that interspecific gene recombination is more likely to upset a finely balanced internal organization in animals than in plants.[24]

One will immediately think of hybrid crosses in complex animals that do not engender conspicuously defective recombination products, as for instance polar bears × brown bears and mallard × pintail ducks. Conversely, there are numerous hybrid crosses in plants which do produce defective recombinations for relatively simple adaptations, as in the floral mechanisms of *Gilia modocensis* and *G. malior*. Such cases emphasize the point that complexity of an adaptation does not in itself guarantee hybrid inviability, sterility, and breakdown in the progeny of species crosses, for the species must also differ genetically in the adaptive characteristics which are being recombined by hybridization. Indeed the latter condition is primary. A marked allelic differentiation between two species with respect to a simple gene system might be expected to yield recombination products that are more defective than those obtained from two species with a slight allelic differentiation in a complex gene system.

Given a substantial degree of differentiation between species in their gene-controlled adaptive properties, however, interspecific hybridization will, on the average, have more deleterious effects in many animal groups than in most plant groups, and is hence more to be guarded against in the animal than in the plant kingdom.[25]

That hybridization *is* more strongly guarded against in higher animals than in plants is shown by a consideration of the relative efficacy of ethological isolating mechanisms in the two kingdoms. The ethological barriers which prevent hybridization are highly effective in most if not all higher animals. Such barriers are absent in many plant groups entirely, are relatively weakly developed in others, and are moderately effective in only a few of the more advanced plant families.

The expectation that natural hybridization is relatively more rare in occurrence and inconsequential in its evolutionary effects in higher animals than in plants is confirmed by the evidence. Several generations

[22] Patterson, Stone, Bedichek, and Suche, 1934.   [23] E. G. Anderson, 1935.
[24] Grant, 1957, 72–74; Stebbins, 1958, 195.
[25] Grant, 1957, 72–74; Stebbins, 1958, 195.

of botanists—Naudin (1863), Kerner (1891), Lotsy (1916), Diels (1921), Anderson (1949, etc.), Stebbins (1950, etc.), and many others—have shown that interspecific hybridization has profoundly affected the course of plant evolution at the species level.[26] Natural hybrids have of course been recorded in various animal groups; furthermore, the natural hybridization has been shown to affect the variation pattern in particular cases in birds, amphibia, fishes, butterflies, etc. Nevertheless, the consensus among students of animal evolution, as stated by Weismann, Fisher, Dobzhansky, Simpson, Mayr, and others, is that natural interspecific hydridization has had relatively unimportant effects in the evolution of higher animals.[27]

THE EFFECT OF FECUNDITY ON SELECTION FOR ISOLATION

Another condition for the reinforcement of incipient reproductive isolation by natural selection is that the loss of gametes and zygotes involved in hybridization must, in fact, be selectively disadvantageous. The selective advantage of a high reproductive potential is, as Stebbins has pointed out, relatively much greater in short-lived animals, from Drosophilas to the smaller mammals, than in most perennial plants, either herbaceous or woody, for most animals can produce young during a lifetime only a few weeks or years long, whereas perennial herbs, shrubs, and trees have scores or hundreds of years in which to produce seeds.[28] Therefore selection for reproductive isolation as a means of conserving the reproductive potential of the species populations will be stronger on the average in short-lived animals than in long-lived plants.[29]

Dobzhansky has noted that by the same reasoning selection for reproductive isolation will be stronger in annual plants, which have but one season in which to produce progeny, than in perennial plants.[30] Incompatibility and other hybridization-preventing barriers are, as expected, more strongly developed between species of annual plants than in long-lived perennial plants.

---

[26] These authors are cited in the bibliography, usually as listed here. I have developed and presented my views on this question in Grant, 1953b, 1957.

[27] Grant, 1957, 70–72, for literature references. Also Weismann, 1913, II, 304.

[28] Stebbins, 1950, 183–86.   [29] Stebbins, 1958, 199–200.   [30] Dobzhansky, 1958.

THE EFFECT OF ECOLOGICAL DEMANDS
ON SELECTION FOR ISOLATION

The development of reproductive isolating mechanisms as a by-product of divergence permits two population systems to coexist without interbreeding. Coexistence has a secular ecological as well as a reproductive aspect, however, and if the reproductively isolated populations establish sympatric contacts, those contacts will lead to ecological interactions. A state of interspecific competition is likely to arise between the populations for limiting resources in the environment.

The outcome of the interspecific competition depends upon its strength of action among other factors, as we attempted to show in Chapter 14. Effective competition between two species may lead to a complete selective replacement in particular areas, habitats, or niches. One species may well prove superior under the environmental conditions prevailing in one area, habitat, or niche, while the other species is superior in other environments under which the issue of the competition is decided. Then the two species will become distributed at equilibrium into separate areas, habitats, or niches.

In organisms with very large ecological demands, the interspecific competition is most likely to result in complete species replacement throughout large areas and hence in a predominantly allopatric distribution of the species. This is the situation we find in many groups of big mammals.

In organisms which have relatively strong but not excessive demands and which enter into a moderately strong interspecific competition, the process of species replacement may run its course separately in different habitats within the same area, leading to a neighboringly sympatric distribution of the different species. This situation is exemplified in many groups of perennial herbaceous plants, woody plants, birds, small mammals, etc., the species of which are restricted to particular ecological zones or habitats.

Biotic sympatry can be attained more easily, and seems to be attained more frequently, in groups of organisms with relatively small ecological demands, for the related species in such groups compete relatively weakly or not at all for essential resources, and can live together in a closer ecological relationship. Different species of small annual plants, of small Diptera, of Paramecium, and so on frequently coexist in similar or even identical niches in the same habitat (see Chapter 14).

Insofar as sympatric contacts are more numerous and varied among organisms with slight ecological demands than among organisms in which heavy ecological demands lead to competitive exclusion, we might predict that interspecific barriers operating in the parental generation would be better developed among small organisms which form flocks of biotically sympatric species than among the larger forms which form predominantly allopatric or neighboringly sympatric species.[31]

Stebbins noted in 1950 that the species in such woody genera as Pinus, Quercus, Larix, Ceanothus, and Platanus, and in some perennial herbaceous genera like Wyethia, are weakly isolated by ecological and seasonal barriers, whereas the species in many herbaceous genera, such as Nicotiana, Gossypium, Layia, Clarkia, and Aegilops, are strongly isolated by incompatibility and sterility barriers.[32] In 1958 Stebbins and I published the results of independent surveys of a larger sample of plant genera with respect to the strength and types of isolating mechanisms, both of which revealed the same general correlation.[33] Incompatibility and sterility barriers are strongly developed between most related species of annual herbs, but are not well developed, by comparison, between many or most related species of perennial herbs, shrubs, or trees (see Table 18).

This correlation can be observed within a single natural group in some cases. Thus in the sunflower genus Helianthus, the annual species are mostly intersterile, whereas the perennial species are more interfertile.[34] The perennial herbaceous species of Polemonium are not strongly isolated by incompatibility or sterility barriers, but the annual species of Gilia in the same family are so isolated to a strong degree.[35]

It will be recalled that the species of woody and perennial herbaceous plants tend to be allopatric or neighboringly sympatric. Among annual plants, by contrast, the species are frequently biotically sympatric in pairs or threes or even in flocks (Chapter 14). The relative infrequency

[31] Also Hutchinson, 1959.   [32] Stebbins, 1950, 234–36.
[33] Stebbins, 1958, 186–88; Grant, 1958, 352–55.
[34] Heiser, Martin, and Smith, 1962.
[35] See Stebbins (1958, 187–88) and Grant (1958, 354) for other examples and references.

of strong incompatibility barriers between related species of perennial herbaceous and woody plants is associated, then, with a relative infrequency of sympatric contacts. The frequent occurrence of strong incompatibility barriers between related species of annual plants, on the other hand, is associated with frequent close sympatric contacts.[36]

Whether the degree of reproductive isolation is proportional to the degree of sympatry also in the animal and protistan kingdoms remains

Table 18. *The distribution of sterility barriers in plant groups with different life forms*[a]

Life form	Number of phylads in which related species are mostly intersterile	Number of phylads in which related species are mostly interfertile
Trees or shrubs	2	19
Perennial herbs	6	23
Annual herbs	19	3

[a] Grant, 1958, 354.

to be seen. It is noteworthy in this connection that in many groups of larger mammals, related species, which are mainly allopatric or neighboringly sympatric in nature, are capable of crossing fairly freely in captivity, as a perusal of Gray's check-list of mammalian hybrids shows. Thus, in the cat family, leopards can cross successfully with lions and cougars, and lions with tigers.[37] Polar bears and brown bears, which are placed in different genera (*Thalarctos maritimus* and *Ursus arctos*), sometimes produce vigorous and fertile hybrids in captivity.[38] The intergeneric cross of the American bison with domestic cattle yields vigorous (but sterile) hybrids,[39] and so on. The contrasting situation is presented by Drosophila, mosquitoes, and Paramecium, which form flocks of biotically sympatric species, and which also possess strongly developed mechanisms for the prevention of hybridization.[40]

The relation between reproductive isolation and ecological coexistence is a reciprocal one. Not only does the formation of reproductive isolating mechanisms permit species to coexist more or less closely,

[36] Also Grant and Grant, 1954, 82–83.
[37] Gray, 1954, 15 ff.   [38] Gray, 1954, 28.   [39] Gray, 1954, 61.
[40] M. Bates, 1949, on mosquitoes; Sonneborn, 1957, as to incompatibility barriers between populations of Paramecium, but not as to the designation of the reproductively isolated populations as species.

but the attainment of sympatry promotes the further development of breeding barriers between species. Reproductive isolation is a cause of sympatry, and sympatry in turn can be a cause of reproductive isolation. And the magnitude of the ecological demands of the organisms is a factor determining how far this circular cause-effect process will go in the formation of species barriers.

Part 5 *Evolution of Major Groups*

EIGHTEEN

Evolutionary Rates and Trends

> Experience shows . . . that there is no way toward understanding of the mechanisms of macroevolutionary changes, which require time on geological scales, other than through understanding of microevolutionary processes observable within the span of a human lifetime, often controlled by man's will, and sometimes reproducible in laboratory experiments.
> TH. DOBZHANSKY[1]

ADAPTIVE RADIATION

A group of related species comprising a genus, tribe, family, or higher taxonomic category generally exhibits a diversity of types correlated with a diversity of ecological niches or habitats within its territory. The different phyletic lines within the natural group are adapted for different ways of life. This diversification of a group of organisms in relation to the ways of making a living and reproducing successfully is a product of an adaptive radiation during the evolutionary history of the group.

The Galapagos finches form a natural subfamily of the finch family (Fringillidae, subfamily Geospizinae) endemic to the Galapagos Archipelago and Cocos Island in the Pacific Ocean off the coast of tropical America. There is every reason to believe that this group of birds has evolved its present diversity of forms on these islands from some ancestral ground finch which colonized the islands from the American mainland in past time.[2]

Lack has shown that many of the differences between the subgenera and genera of Galapagos finches, in the size and shape of the bill, other bodily characters, and habits, are related to the method of food-getting in the various groups.[3]

(1) Thus the species placed in Geospiza subgenus Geospiza (particularly G. *fuliginosa, fortis,* and *magnirostris*) are adapted by their strong

[1] Dobzhansky, 1951a, 16. [2] Lack, 1947, 14, 102–5, 107–14.
[3] Lack, 1947, chs. 6, 11, 16.

519

heavy bills, stout bodies, and hopping and scratching habits for feeding on seeds on the ground. (2) Another branch of the genus Geospiza (*G. scandens* in the subgenus Cactornis) possesses a somewhat narrower and longer, but nevertheless quite stout bill, enabling it to feed both on seeds and on the flowers and fleshy pulp of cactus (Opuntia) (Fig. 90a, b).

(a)

(b)

(c)

(d)

Fig. 90. Four types of Galapagos finches

(a) *Geospiza magni rostris*, seed-eating ground finch
(b) *Geospiza scandens*, cactus-feeding ground finch
(c) *Camarhynchus psittacula*, insectivorous tree finch
(d) *Certhidea olivacea*, warbler finch

Redrawn from originals by J. Gould, *The Zoology of the Voyage of H.M.S. "Beagle,"* 1841, as reproduced in D. Lack, *Darwin's Finches* (New York, Cambridge University Press, 1947).

The genus Camarhynchus deviates from Geospiza in that its members live and feed in trees. (3) In this genus, *Camarhynchus crassirostris* (in the subgenus Platyspiza) feeds on leaves, buds, flowers, and fruits with its parrotlike beak. (4) The species *Camarhynchus psittacula, pauper* and *parvulus* (forming the subgenus Camarhynchus), possessing parrotlike bills combined with agile climbing movements, hunt for and feed primarily on beetles and similar insects among the branches and leaves in the trees (Fig. 90c). (5) The remarkable woodpecker finch, *Camarhynchus pallidus* (subgenus Cactospiza), feeds on arboreal insects in the manner of a woodpecker, climbing up vertical trunks and branches, excavating cracks or holes in the bark, and working the insect out by means of a spine or short twig held in its bill.

(6) The genus Certhidea, as represented by *C. olivacea*, consists of warblerlike birds with light bodies, slender bills, and quick flitting movements. These birds search for small soft insects on leaves, twigs, and sometimes in the air (Fig. 90d). (7) The genus Pinaroloxias on Cocos Island has similar warblerlike characteristics and feeding habits.

The ancestor of the Galapagos finches is inferred to have been a seed-eating ground finch with characteristics like those still found in the *Geospiza fuliginosa* group. From some such ancestral form the cactus finches, insectivorous tree finches, woodpecker finches and warbler finches have developed by a process of adaptive radiation (Fig. 91).[4]

The Hawaiian honeycreepers (Drepaniidae), like the Galapagos finches, have evolved a diversity of forms on a geographically isolated archipelago, the Hawaiian Islands. As in the Galapagos finches, again, the diversification of the drepanids is related to the exploitation of different food sources. But in the Hawaiian honeycreepers the adaptive radiation has proceeded farther and has produced a remarkable range of types.

Within the subfamily Drepaniinae of the Drepaniidae are: (1) nectar-eaters with short thin bills, as for example *Hematione sanguinea;* (2) sickle-billed nectar-feeders like *Drepanis pacifica;* (3) and fruit-eaters like *Ciridops anna* (Fig. 92a–c). The other subfamily, Psittirostrinae, contains (4) thin-billed warblerlike insectivorous birds such as *Loxops coccinea*; (5) birds which feed on insects in the manner of creepers, as

[4] Lack, 1947.

Hemignathus lucidus; (6) and heavy-billed fruit- and seed-eaters like *Psittirostra palmeri* (Fig. 92d–f). The ancestral drepanid was probably a small-billed bird which fed on both nectar and insects, as do the more primitive members of the two subfamilies in the genera Loxops and Hematione today. From some such ancestor arose groups specialized for feeding on nectar in long tubular flowers, for gleaning insects on foliage, for picking insects out of bark, for eating soft fruit, and for cracking hard seeds.[5]

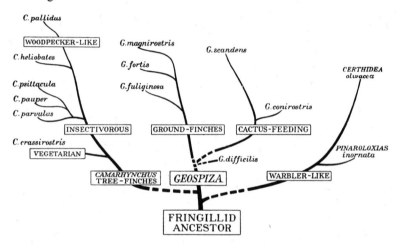

Fig. 91. Probable phylogeny of the Galapagos finches

From D. Lack, *Darwin's Finches* (New York, Cambridge University Press, 1947).

In continental faunas a similar adaptive radiation in relation to feeding habits took place long ago and is seen today in the differences at the family level between tanagers, sunbirds, cedar waxwings, warblers, creepers, and grosbeaks, all of which have counterparts within the relatively small and relatively young family Drepaniidae.[6]

In many groups of flowering plants an adaptive radiation has taken place in relation to methods of pollination. The genus Pedicularis of the family Scrophulariaceae, for example, contains numerous species with an elaborate floral mechanism fitted for pollination by bumblebees. In the basic and original floral mechanism in the genus, the pollen is deposited on the head of the bumblebee. In some derived species like

[5] Amadon, 1950; Baldwin, 1953.   [6] Amadon, 1950.

P. groenlandica the floral mechanism deposits the pollen on the venter of bumblebees. Other species of Pedicularis are specialized for pollination by various other kinds of insects and birds. Thus *P. semibarbata* in California has developed a simpler floral mechanism which is worked by smaller bees, principally Osmia. Another California species, *P. densiflora*, has evolved red tubular flowers and is pollinated by humming-birds. In the Himalaya Mountains the genus has produced some species

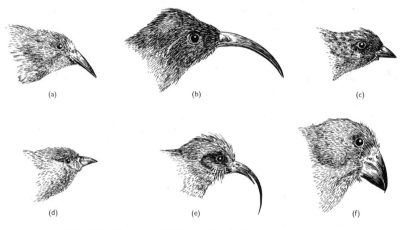

(a)   (b)   (c)

(d)   (e)   (f)

Fig. 92. Types of bills in Hawaiian honeycreepers

(a) Nectar-feeder, *Himatione sanguinea*
(b) Nectar-feeder, *Drepanis pacifica*
(c) Fruit-eater, *Ciridops anna*
(d) Insect-eater, *Loxops coccinea*
(e) Insect-eater, *Hemignathus lucidus*
(f) Fruit- and seed-eater, *Psittirostra palmeri*
Redrawn from Amadon, 1950.

with slender corolla tubes up to 11 cm long, which are probably polli-nated by hawk moths.[7]

In another branch of the family Scrophulariaceae, the tribe Antirrhi-neae, the main lines of an adaptive radiation in floral characters related to mode of pollination are largely commensurate with generic divisions within a tribe. The elaborate flowers in this tribe are generally two-lipped with a pouched lower lip which closes the entrance to the nectar-containing tube. In the genus Antirrhinum the flowers are blue, purple,

[7] Sprague, 1959, 1962.

or whitish, relatively short-tubed, and tightly closed, so that only heavy bees can force their heads into the nectar chamber successfully. In European species of Antirrhinum such as *A. majus* the pollinators are bumblebees,[8] while the California species *A. nuttallianum* is pollinated chiefly by Anthophora. The small California genus Galvesia diverges from the ancestral type of bumblebee flower in possessing trumpet-shaped, odorless, bright red flowers, which are visited and pollinated by hummingbirds. Another divergent type on the California desert is *Mohavea confertiflora*, the unusual flowers of which stand erect on the stems, have an open campanulate form, are pale cream with purple spots, and give off a slightly musty odor. These flowers are pollinated by beetles, principally Melyridae, in favorable years, and otherwise seem to be self-pollinating.

The process of adaptive radiation leads, first, to the origin of diverse specializations within a natural group, one kind of specialist giving rise to various other kinds of specialists. A ground-feeding, seed-eating finch with the characteristics of *Geospiza fuliginosa* produces descendants adapted for other food niches in its territory—hence cactus finches, parrot finches, woodpecker finches, warbler finches, and the like. In later stages of the process, under the pressure of intergroup competition, the specializations within the various phyletic lines may become more narrow as one and the same adaptive zone becomes subdivided. Thus the *Geospiza fuliginosa* phyletic line splits up into derivative species which are specially adapted to utilize different types of hard seeds and fruits on the ground. The small-beaked *Geospiza fuliginosa* itself is specialized for small grass seeds, while the large-beaked *G. magnirostris* is able to crack and eat large hard fruits. This narrowing of the specializations, resulting from interspecific competition, enables the group as a whole to exploit the resources of its environment with maximum effectiveness (see also Chapter 14).

Savile has argued that an adaptive radiation may take place, not only when a migrant reaches an unoccupied territory, but also when it enters a different biotic community in a densely occupied territory. As Savile notes, a species existing in a relatively stable biotic community is subject to constant and strong competition, which promotes specialization and suppresses the development of novel forms. He

[8] Knuth, 1906–9, III, 170 ff.

asks us to assume that another different biotic community, also relatively stable, exists in a neighboring area separated from the first community by a geographical or ecological barrier, like Beringia between Asia and North America, so that individuals belonging to some species in the first community can migrate only rarely into the second community. Then such penetrants as manage to survive in the new community, on encountering new and different biotic conditions, and being released from the particular checks on their variation that operated in the old community, may be able to radiate adaptively. Eventually, some of the new derived forms may be able to return across the same barrier to the original biotic community and radiate again there.[9]

EVOLUTIONARY TRENDS IN THE HORSE FAMILY

Numerous groups of mammals, molluscs, conifers, and other organisms have fossil records extending over long periods of time. When the fossils are arranged according to geological age, progressive evolutionary trends become apparent. A classical example of long-term progressive changes within a natural group is provided by the horse family, in which an almost continuous series of gradations connects the doglike Eohippus of Eocene time with the modern horse.

The evolutionary development of the horse family during the 60-million-year span of the Tertiary period, as unraveled by several generations of paleontologists, is too complex to describe here in all its ramifications.[10] Our object in this section is to consider some of the main trends in horse evolution.

The Equidae along with other ungulates and the carnivores developed out of the ancestral stock of condylarths, a primitive and long extinct order of mammals which lived at the beginning of the Tertiary period. The condylarths were somewhat doglike in size and shape, with a small head and long tail; their feet were padded with five toes, as in dogs; and, judging by their teeth, they lived on an omnivorous diet.

The earliest member of the horse family was Hyracotherium, more

[9] Savile, 1959.

[10] Simpson (1951, 1961) has integrated and summarized the evidence bearing on horse phylogeny in an excellent non-technical book, which has been my main source in preparing the following condensed account. The reader in quest of more details than can be given here is referred to Simpson's book.

commonly known as Eohippus, another small doglike, or rather condylarth-like animal which lived in Europe and North America in Eocene time. Eohippus stood 10 in. high in some species, 20 in. high in others, and had a long tail. The feet were padded, as in condylarths, and most of the body weight was carried on the pads; but unlike the ancestral condylarths, Eohippus had only four toes on the forefeet and three toes on the hind feet, and the toes ended in small separate hoofs. The molar and premolar teeth were short (or low-crowned), capped with enamel (but lacking cement), and possessing relatively simple cusps and low ridges on the biting surface. These and other dental characteristics indicate that Eohippus fed on succulent leaves, small fruits, and soft seeds. In its herbivorous food habits Eohippus was

Fig. 93. Four Tertiary members of the horse family, restored

(a) Hyracotherium, or Eohippus, of Eocene time
(b) Mesohippus, a three-toed browsing horse of Oligocene age
(c) Merychippus, a Miocene three-toed grazing horse
(d) Pliochippus, the earliest one-toed horse in the Pliocene epoch

Reproduced from paintings in the collection of the American Museum of Natural History, by courtesy of the Museum.

more specialized than the earlier omnivorous condylarths. Eohippus was a forest-inhabiting browsing animal (Fig. 93a).[11]

The genus Equus, which arose in the late Pliocene, and is represented in the recent fauna by the true horse (*E. caballus*), two species of donkey,

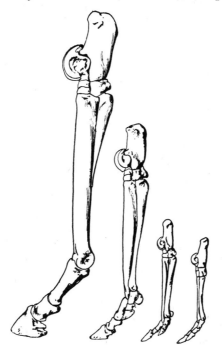

Fig. 94. Stages in the evolution of horse feet

Forefeet are shown in side view to the same scale. From right to left: Hyracotherium Mesohippus, Merychippus, Equus.

From G. G. Simpson, *Horses* (New York, Oxford University Press, 1951).

and three species of zebra, differs from Eohippus in every part of the body. Equus is large and heavy and, unlike Eohippus, has a short tail, which is tipped with long hairs. The feet are singled-toed and the toes terminate in large hoofs, so that the animal stands on its tiptoes. The bones of the toe are connected by powerful elastic ligaments which give the foot a spring action when the animal is running (Figs. 94, 95). The proportions of the skull are very different from those in Eohippus, the

[11] Simpson, 1951, 1961, ch. 13.

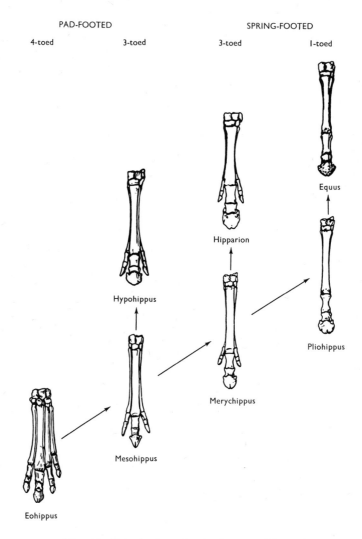

Fig. 95. Principal mechanical types of horsefeet

Forefeet are shown in front view, not to scale.

From G. G. Simpson, *Horses* (New York, Oxford University Press, 1951).

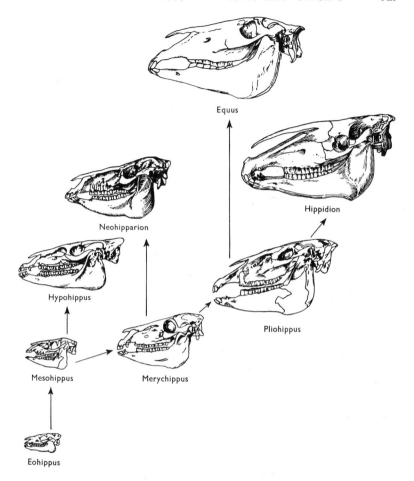

Fig. 96. Stages in the evolution of horse skulls

Drawn to the same scale.

Redrawn from G. G. Simpson, *Horses* (New York, Oxford University Press, 1951).

eyes being set to the rear and the muzzle greatly elongated (compare the extreme types depicted in Fig. 96). Inside the horse skull is an elaborately developed and quite intelligent brain.[12]

Equus lives on a diet of grass. Now grass is a harsh material, owing to the presence of silica inside the plant cells and the normal covering

[12] Simpson, 1951, 1961, chs. 22, 24.

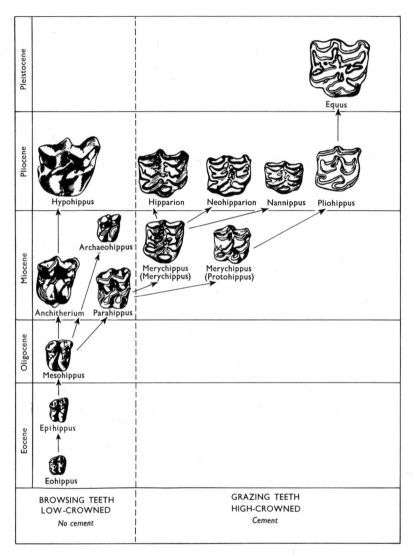

Fig. 97. Stages in the evolution of horse teeth

Grinding surfaces of upper molars are shown to the same scale.
From G. G. Simpson, *Horses* (New York, Oxford University Press, 1951).

of dust and grit on the outside of the plant. A diet of grass consequently wears down teeth. In Equus, however, the chisel-like front teeth are well fitted for nipping off grass, and the rear cheek teeth are specially adapted for grinding this hard food material.[13]

The grinding capacity of the cheek teeth is increased in Equus in various ways. In the first place, the teeth are large in size. Secondly, and of greater importance, the grinding surface of the cheek teeth consists of a series of elaborately contorted and sharply developed ridges or crests, which maximize the effective grinding area on a tooth of given size. Whereas the molars of Eohippus had simple cusps and low crests, those of Equus are provided with a complicated pattern of high crests (compare the extreme types shown in Fig. 97). Thirdly, the grinding teeth are rendered more durable in Equus by the development of cement in the pockets between the enamel crests. This cement was not present in the primitive browsing teeth of Eohippus.[14]

Fourthly, the length of the cheek teeth, or height of the crown, is greatly increased in Equus, as compared with the low-crowned molar teeth of Eohippus, thus allowing the teeth to wear down without wearing out during a long lifetime. And, fifthly, the number of grinding cheek teeth is increased in Equus to six on each side of each jaw. This increase in number of grinding teeth came about by the transformation of premolars into teeth which can function as molars. Eohippus had three molars and four smaller premolars on each side of each jaw; three of the premolars are represented in Equus by large crested teeth similar in size, sculpturing, and function to the true molars.[15]

Equus then differs from Eohippus, among other characters, in size, strength, intelligence, length of muzzle, length of tail, foot mechanism, food habits, and dentition. These character differences are related to differences in the way of life of the two animals. Eohippus was a forest-inhabiting browsing animal, Equus a grazing animal of dry grassy plains.[16]

The differences between Eohippus and the modern horse, enormous as they are, are bridged by an almost continuous series of transitional forms. Let us now look at three types of Tertiary horses representing intermediate stages in horse evolution.

[13] Simpson, 1951, 1961, ch. 23.   [14] Simpson, 1951, 1961, ch. 23.
[15] Simpson, 1951, 1961, ch. 23.   [16] Simpson, 1951, 1961.

Mesohippus, which lived in North America in the Oligocene epoch, was small in size, though somewhat larger than Eohippus on the average. The feet were padded as in Eohippus but had only three functional toes (Figs. 94, 95). The head of Mesohippus was horselike in appearance with the eye set back and the muzzle elongated (Fig. 96). The brain of Mesohippus shows a marked development, as compared with that of Eohippus, of the parts associated with intelligence in modern mammals. The cheek teeth were low-crowned and lacking in cement. The development of crests was greater than in Eohippus but less complex than in later horses (Fig. 97). Three of the premolars had become molarlike in pattern and size, as in later horses, to form a set of six grinding teeth on each jaw. Mesohippus was a three-toed browsing horse which probably inhabited woods or river banks at the edge of woods (Fig. 93b).[17]

The American Miocene horse Merychippus was a large animal, attaining a height of 40 in. Its feet still possessed three toes ending in separate hoofs, but in the more advanced species of this genus the pads were absent. The body weight was carried mainly on the large hoof of the central toe, which possessed a spring mechanism as in Equus, and the smaller but functional side toes may have absorbed some of the shock from the central toe in running (Figs. 94, 95). The skull of Merychippus was very equine, with a long muzzle, and contained a well developed and probably intelligent brain (Fig. 96). The cheek teeth of Merychippus differed from those of Mesohippus and remotely approached those of Equus in possessing high crowns, a complex pattern of crests, and cement, thus fitting this horse for a diet of grass (Fig. 97). Merychippus was a group of three-toed grazing animals which probably lived in open grassland savannah (Fig. 93c).[18]

Pliohippus was a one-toed grazing horse which roamed the North American plains in the Pliocene epoch. Some primitive species of Pliohippus had minute vestigial side toes, but these vestiges were evidently absent in the more advanced species. Pliohippus arose from Merychippus and gave rise at the end of the Pliocene to Equus (Figs. 93d, 95–97).[19]

The five stages in horse evolution considered so far—Eohippus,

[17] Simpson, 1951, 1961, ch. 15.   [18] Simpson, 1951, 1961, ch. 16.
[19] Simpson, 1951, 1961, ch. 18.

Mesohippus, Merychippus, Pliohippus, and Equus—are nodal points in a continuous sequence. Figure 98 places these five ancient and modern genera in the perspective of the phylogeny of the whole horse family.

The evolutionary trends in body size can be seen in Figs. 94, 96, and 98, and are summarized diagrammatically in Fig. 99. A prevailing trend in the evolution of the horse, as in many other land vertebrates during the Tertiary, is increase in size. As indicated in Fig. 99, the small primitive members of the Equidae remained small throughout the Eocene, but a rapid increase in body size took place in the Oligocene. This trend continued, though at a somewhat slower rate, during the Miocene and Pliocene, several phyletic lines reaching the maximum size in Pliocene time. However, a reverse trend toward smaller size has also occurred in various phyletic lines at various times, including Equus during the Pleistocene epoch (Fig. 99).[20]

The evolution of the foot mechanism is summarized in diagrammatic form in Fig. 99. The primitive condition of padded feet with four toes on the front feet and three toes on the hind feet prevailed throughout the Eocene. Between the Eocene and Oligocene a change took place (in Mesohippus) to padded feet with three, more compact, less spreading toes. This condition persisted in at least some phyletic lines until the Pliocene. But in the Miocene another phyletic line underwent a rapid change from three-toed padded to three-toed spring feet, as in Merychippus, where the middle toe carries most of the body weight and the smaller side toes act as shock absorbers for the main hoof. The side toes decreased in size during the Miocene, and from late Miocene to early Pliocene another rapid change occurred in one phyletic line from three-toed to one-toed springing feet, as found in Pliohippus. This change involved the progressive reduction and eventual disappearance of the side toes.[21]

The changes in the proportion of the skull, as shown in Fig. 96, involve a progressive elongation of the muzzle and deepening of the jaw. The primitive and stupid brain of the Eocene horses gave way to a more complex and probably more intelligent brain in the Oligocene Mesohippus. Increases in brain size and structures associated with intelligence

[20] Simpson, 1953a, 265.   [21] Simpson, 1951, 1961, ch. 24.

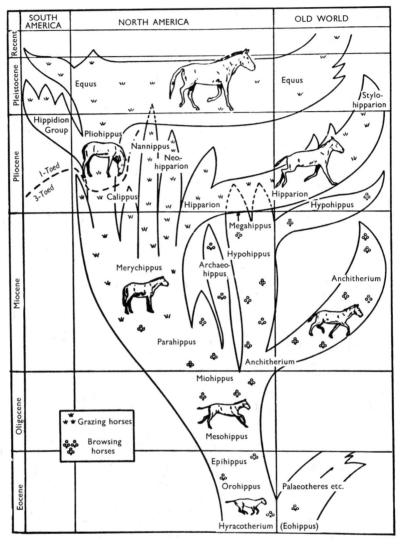

Fig. 98. Phylogeny of the horse family, simplified

The animals shown to the same scale.

From G. G. Simpson, *Horses* (New York, Oxford University Press, 1951).

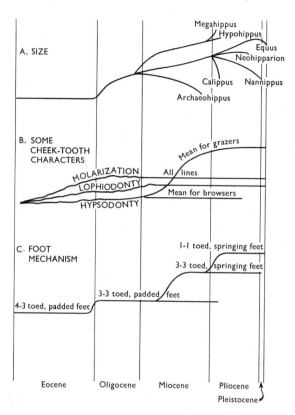

Fig. 99. Relative rates and directions of change in three characters in the evolution of the horse

Hypsodonty refers to height of crown, lophiodonty to the development of crests on the grinding surface, and molarization to the transformation of premolars into molarlike teeth.

From G. G. Simpson, *The Major Features of Evolution* (New York, Columbia University Press, 1953).

then took place steadily until the Miocene, when another rapid advance in the brain was made in Merychippus. After Merychippus, the brain continued to evolve steadily in the phyletic line leading to Equus.[22]

The various changes in the complex of tooth characters that differentiate Equus from Eohippus did not all take place at the same time. One

[22] Simpson, 1951, 1961, ch. 22.

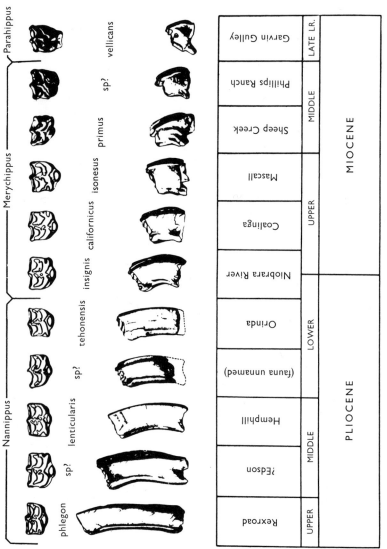

Fig. 100. Stages in a trend toward increased height of crown of cheek teeth in a series of Miocene and Pliocene horses

Stirton, 1947.

of the first developments was the transformation of three premolars into molarlike teeth. This change, known as molarization, proceeded all through the Eocene and became completed by early Oligocene time (Fig. 99). The development of more elaborate crests on the grinding surface of the cheek teeth proceeded slowly through the Eocene and Oligocene, underwent a more rapid change in the Miocene with the appearance of grazing horses, and tapered off thereafter (Fig. 97). Cement as a reinforcing material for grinding teeth also entered the picture in the Miocene. In the same epoch the tooth crown rapidly became higher in the grazing horses but not in the browsers, and this trend continued into the Pliocene (see curve for hypsodonty in Fig. 99).[23] Figure 100 shows the progressively increasing height of crown in a series of Miocene and Pliocene horses.[24]

When the Tertiary horses are represented by a good fossil record in a region in which they were evolving actively, the transition from one genus to the next successive one is gradual, and the dividing lines between the genera are consequently arbitrary. The successive genera Mesohippus-Miohippus-Parahippus in the Oligocene and early Miocene of North America, for example, intergrade perfectly.[25] The intergradation between the Miocene Merychippus and the Pliocene Nannipus is shown graphically in Fig. 100.[26]

THE CAUSES OF EVOLUTIONARY TRENDS
IN THE HORSE FAMILY

One of the older explanations of long-continued progressive trends, in the Equidae and in other groups of organisms, held that evolutionary changes were oriented in a given direction by some force within the organisms themselves. This theory of orthogenesis has been presented in different versions according to the nature of the supposed inner directing force. Lamarck simply stated, as a law of nature, that the guiding principle in evolution is the animal's need.[27] Various later authors referred to a purposeful urge in living organisms, an *élan vital*, or the like, impelling them to evolve in a given direction. Some recent

---

[23] Simpson, 1951, 1961, ch. 23.   [24] Stirton, 1947.
[25] Simpson, 1951, 1961, chs. 15–16.   [26] Stirton, 1947.
[27] Lamarck, 1815–22. See also Chapter 6 of this book for a discussion of another aspect of Lamarck's theory of evolution.

attempts to place the theory of orthogenesis on a more tangible, material foundation have attributed the alleged inner orientation of evolution to an alleged inner orientation of the mutation process.[28]

There is not a shred of evidence for the theory of orthogenesis in any of its versions. The postulation of a purposive urge within organisms which can guide their evolution does not even merit consideration as a scientific explanation. As Huxley aptly remarks[29]:

> To say that an adaptive trend towards a particular specialization or towards all-round biological efficiency is explained by an *élan vital* is like saying that the movement of a railway train is "explained" by an *élan locomotif* of the engine.

The idea that evolutionary trends could be determined by a sustained tendency of genes to mutate progressively in certain directions is nothing more than a suggestion unsupported by, or rather in contradiction with, the facts of mutation genetics (see Chapter 8).

The facts of horse phylogeny, when considered as they really are and not as they are supposed to be by orthogenecists, do not suggest an internally directed evolution process. There was, as Simpson points out, no single trend that occurred constantly throughout the evolutionary history of the horse family.[30]

Indeed, the evolution of the horse family was highly branched. One wave of speciation occurred in the Eocene in the European branch of Eohippus; a second major period of divergence took place in the American browsing horses derived from Miohippus in the early Miocene; and a third period of speciation occurred in the American grazing horses descended from Merychippus in the late Miocene (see Fig. 98).[31] In this complex history of branchings there is no single straight line starting with Eohippus and culminating in the one-toed grazing horses. In fact, as Simpson notes, the closest thing to a straight line in horse phylogeny led from Eohippus to the three-toed browsers, Anchitherium, Hypohippus, and Megahippus, which became extinct in the Pliocene, while the one-toed grazers represent a side branch, a new direction in horse evolution (Fig. 98).[32]

[28] I.e., Cronquist, 1951; Werth, 1956, 158–70.
[29] Huxley, 1943, 458, quoted from *Evolution: The Modern Synthesis* (London, George Allen and Unwin, Ltd).
[30] Simpson, 1953a, 264; 1961, 167, 244, 261.
[31] Simpson, 1953a, 260; 1961, 166–67, 183–84, 279.
[32] Simpson, 1953a. 260; 1961, 170, 261.

Simpson sums the case up as follows[33]:

The evolution of the horse family was definitely not orthogenetic. The preceding chapters have made this abundantly clear. There was, for instance, no constant and over-all increase in size. Most recent horses are larger than most ancient horses, but when the history is examined in detail it shows that there have been long periods when no increase in size occurred and also several branches of horse evolution in which the animals became markedly smaller, not larger. The feet did not steadily change from four toes to three and then to one. In the last chapter it was shown that foot evolution followed different lines in different branches of the family and that toe reduction, for instance, from three to one was not a continuation of a previous trend, an orthogenetic process, but was a change in evolutionary direction, a new development that occurred at just one time and in only one group of horses. And so it goes for all the changes that have occurred in the history of the family; not one of them shows the constant, guided change in a single direction that is demanded by the theory of orthogenesis. In reality, the horse family goes far to disprove that theory, and when other supposed examples of orthogenesis are examined with similar care they, too, are found to be opposed to the reality of orthogenesis in any strict sense.

There is, on the other hand, a body of evidence in the fossil record of the horse family which points to an evolutionary process caused by the selection of sporadic variations. An example is provided by one of the characters of the grinding surface of the cheek teeth. One of the crests on these teeth has a short projection called a crochet. This crochet is prominent in Equus but absent in most Oligocene horses. In some populations of Mesohippus in the early Oligocene most individuals lack the crochet, but occasional individuals have a tiny crochet. The crochet-containing individuals are otherwise indistinguishable from the non-crochet individuals, with which they are associated in the same fossil population, and the former evidently represent a new mutant type. Later in the Oligocene some early species of the derivative genus Miochippus still remain polymorphic for the presence or absence of the crochet. But in some early Miocene species of Parahippus, derived from and intergrading through time with Miohippus, the crochet is established uniformly in whole horse populations. The crochet was small in Parahippus but became larger later in the line leading to Equus.[34]

The appearance of cement in the cheek teeth of Parahippus and its immediate successor Merychippus behaves in the same way. When

[33] Simpson, 1951, 206, quoted from *Horses* (New York, Oxford University Press).
[34] Simpson, 1953a, 105–6.

first encountered in the fossil record the cement is a polymorphic character found in the teeth of some individuals but not others of the same population. Later and by gradual stages the presence of cement comes to characterize whole populations. Still later the cement increases in average thickness up to a certain optimal level, after which the trait shows no further trend.[35]

Downs has made a detailed study of the variation in tooth characters of two populations of Merychippus which lived one or two million years apart in western North America. The populations are *Merychippus seversus* from the Mascall formation of Oregon, and *M. californicus* from Coalinga, California. Typical teeth of the adjacent Mascall and Coalinga formations of Upper Miocene age, are shown in Fig. 100 (where *M. seversus* is labeled by an older name, *M. isonesus*). Twenty-six different tooth characters all showed some variation within each population. For about 70 percent of these characters there are significant differences between the two populations, indicating a micro-evolutionary shift from the earlier Mascall to the later Coalinga population. But these differences are not clearcut, in that the range of variation overlaps. Thus, as shown in Fig. 101, the average height of the crown is greater in the Coalinga than in the earlier Mascall formation, but some individual specimens in the Mascall sample have crowns as high as some Coalinga specimens. Some characters form combinations which usually shift together; yet individual specimens are found with recombinations of these characters.[36]

Some at least of the selective factors which have guided the evolution of horses can be reconstructed with reasonable certainty. The development of intelligence is clearly adaptive in an animal which has to cope with conditions in a varied terrain, and elude increasingly intelligent predators. Size increase in the horse family, as in other land animals, can be accounted for by selection for greater strength in combat, for better protection from enemies, and for the improved thermodynamic efficiency of larger warm-blooded bodies. Increase in skull size is clearly related to increase in brain size and intelligence, and may well be dependent in part on increased body size.[37]

Now a larger animal requires a disproportionately greater amount of food and places disproportionately greater demands on the teeth which

[35] Simpson, 1953a, 106–7.   [36] Downs, 1961.   [37] Rensch, 1959, 164–65.

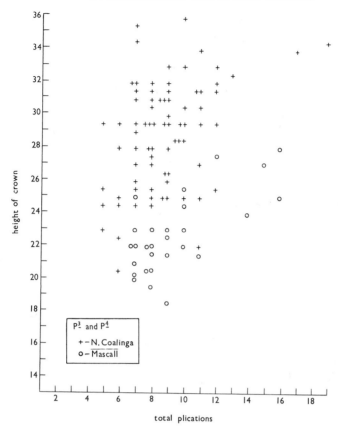

Fig. 101. The range of variation for two characters of the premolar teeth within and between two Upper Miocene populations of Merychippus Downs, 1961.

have to prepare this food than does a smaller animal. As Simpson notes, one animal twice as tall as another requires about eight times as much food for maintenance during a given period of time. Furthermore, the larger animal is apt to grow more slowly and live longer, and therefore needs longer-lasting teeth. The trend toward larger body size in the horses would thus, in itself and without considering the type of food consumed, have placed a selective premium on a more efficient and durable tooth mechanism.[38]

[38] Simpson, 1961, 242.

One of the main defenses of the horse against its predators is its fleetness of foot. Eohippus was built for fast running. Its larger descendents, to maintain the same speed under the handicap of a heavier weight, had to develop an ever more efficient foot mechanism. The light Eohippus ran on four flexible toes (on its forefeet), whereas the heavier Mesohippus ran on three compact toes. The still heavier Merychippus increased the effectiveness of its feet and legs by raising itself up and running on one central tiptoe, with supplementary assistance from the side toes. In Pliohippus, which was larger still, the thrust in running came from one hoof which had a stronger spring mechanism than that of Merychippus.[39]

The selective value of greater size and intelligence and of improved tooth and foot mechanisms was probably enormously enhanced by environmental changes during the Miocene. The early Tertiary forest was beginning to break up and grassy savannahs and prairies were developing in the Miocene, and in the Pliocene the area of grassland continued to increase. The horses which moved into the open savannahs and steppes, and took to a diet of grass, would have been subjected to very intense selective pressures for strength, alertness, speed, and durable grinding teeth.[40]

EVOLUTIONARY RATES

Evolutionary trends frequently proceed at different rates in different phyletic lines. This conclusion is supported by the comparative morphology and ecology of Recent species belonging to the same major group and living at the same time. In the first section in this chapter we have considered several examples in both birds and flowering plants (and many more could be cited) where the same family, tribe, or genus contains some species with many primitive characters and other species with many advanced features. The inference is clear that some phyletic lines within a large natural group have changed greatly, and other phyletic lines have changed only slightly, from the same common ancestor during the same period of time.

If further proof of this conclusion is needed, it is abundantly provided by the geological record. During the same 60-million-year period in the Tertiary there was a succession of eight or nine genera of horses, but

[39] Simpson, 1951, 1961, ch. 24.   [40] Simpson, 1951, 1961, ch. 23.

only small changes in the lineage leading to the modern opossum (Didelphis). The opossum, indeed, has changed little since the late Cretaceous some 80 million years ago.[41]

Among other classical examples of evolutionary conservatism are those so-called living fossils which have come down more or less unchanged to Recent times from past geological periods. There has been little change in the crocodiles since the early Cretaceous; in the lizard, Sphenodon, since the Jurassic; in the horseshoe crab, Limulus, since the Triassic; and, in a coelacanth fish, since the Devonian.[42] The Recent marine brachiopod, Lingula, is only specifically different from forms that lived in the Ordovician period of early Paleozoic time.[43] This sturdy little creature, as Haldane says, "has watched the legions of evolution thunder by for [over four] hundred million years without changing its shell form to any serious extent."[44]

Evolutionary rates have differed, not only between separate phyletic lines living through the same period, but also within one and the same phyletic line at different times in its history. The morphological difference between modern opossums and Cretaceous ones, separated by 80 million years of history, is slight. If evolutionary changes in this same phyletic line took place at a comparable rate prior to the Cretaceous, the transition from reptile to opossum would have required a time span many hundreds of millions years long stretching back long before the Cambrian period. The conclusion is inescapable that the phyletic line to which the opossum belongs must have evolved rapidly prior to late Cretaceous time but slowed down thereafter.[45]

The length of recorded history of the orders of mammals ranges from 25 million years in the long extinct condylarths to 70 million years in the marsupials and insectivores, and averages 45 million years for 16 orders combined. The unrecorded previous period when the orders were originating is estimated to be only 10 to 15 million years long in the case of most (11) of these orders, and is estimated to average 19 million years for all 16 orders. Therefore the basic differentiation of each mammalian order from the ancestral stock must have taken place in a much shorter time period than the later expansion and diversification of the same order. Now the structural and functional changes

[41] Simpson, 1949, 99.   [42] Simpson, 1944, 125; 1949, 47, 101.
[43] Simpson, 1953a, 36.   [44] Haldane, 1932, 167.   [45] Simpson, 1944, 119.

involved in the basic differentiation of some orders of mammals are comparable in magnitude to all the subsequent changes, and in other orders the early changes were much more profound than the later ones. Great changes thus took place in a relatively short time in the early history of each mammalian order. It follows that the rate of evolutionary change in each order was usually higher, and in some cases many times more rapid, during the period of origination than during its later history.[46]

In groups possessing a nearly continuous fossil record, the evidence for different evolutionary rates in the same line of descent may be even more direct. In the Equidae as discussed earlier in this chapter, for example, the Eocene was a period of slow change in height of crown and practically no change in body size and foot mechanism, whereas the Miocene witnessed very rapid changes in all of these characters (see Fig. 99).[47]

As a first approximation to a classification of evolutionary rates, Simpson distinguishes three main conditions. There are phyletic lines which show (1) a slow rate of change or complete stagnation; (2) a medium rate of evolution; or (3) exceptionally rapid change.

Simpson has further given reasons for expecting that the frequency distribution of the different evolutionary rates as found in the various phyletic lines belonging to the same large group, say a class, will generally be leptokurtotic, with a prominent peak sloping off on both sides, and asymmetrical, the peak being skewed to one side of center. The peaked region on the frequency distribution curve, representing the phyletic lines with evolutionary rates that are near the mode, provides a standard of reference for the group in question. This part of the frequency distribution, and by extension the evolutionary rates that are standard or medium in the group, is called horotelic. The sectors of the distribution curve corresponding to the exceptionally slow and the exceptionally rapid rates in the same group are designated as bradytelic and tachytelic, respectively. The bradytelic rates, ranging from slow evolutionary changes to complete stagnation, will usually be found in a fairly large number of phyletic lines within the class. The tachytelic range on the distribution curve is more poorly represented.

[46] Simpson, 1944, 120.   [47] Simpson, 1953a, 265; 1961, 281.

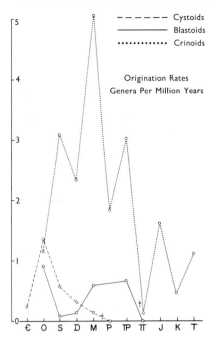

Fig. 102. Rate of evolution, as measured by the number of genera originating per million years, in three groups of echinoderms

The time scale runs from Cambrian to Tertiary.

From G. G. Simpson, *The Meaning of Evolution* (New Haven, Yale University Press, 1949).

In other words, a large group is likely to include a few lineages, but only a few lineages, exhibiting rapid or tachytelic rates.[48]

The quantification of the rate of change beyond this qualitative division is beset with numerous technical difficulties, which we will not attempt to discuss.[49] One measure of evolutionary rates is the number of new genera originating per million years. As shown in Fig. 102, when three groups of echinoderms are compared with respect to origination of genera per unit of time, marked differences in evolutionary rate become evident between the groups and from period to period within the same group.[50]

[48] Simpson, 1944, ch. 4.   [49] See Simpson, 1953a, chs. 1, 2, 10.
[50] Simpson, 1949, ch. 8.

Tachytelic changes as exhibited in the fossil record take place in time spans measured in millions of years. In numerous genera of living plants the experimental evidence indicates that new species can arise and become stabilized genetically in just a few generations, which would be equivalent to a few years in annual plants; and it seems altogether probable that larger changes such as the divergence of a new section or a new minor genus in such plants and no doubt also in some animals could be carried out in one or a few centuries. The very rapid rates of evolution inferred from the fossil record and those inferred from the study of certain modern groups are orders of magnitude apart and need to be distinguished. The evolutionary rates in the fast end of the tachytelic range, where changes are rapid on a time scale measured in centuries, might be called explosive, to distinguish them from the tachytelic rates that are rapid in terms of geological time.

ENVIRONMENTAL FACTORS AFFECTING EVOLUTIONARY RATES

Differences in evolutionary rates can be attributed, broadly, to two sets of conditions: the constancy or changeability of the environment, and the inherent characteristics of the organism which determine its reaction to the environment.

A population living in a stable environment, once it has reached the highest adaptive level possible, can be maintained at the same adaptive level by stabilizing selection for long periods of time; whereas a population living in a changing environment is exposed to progressive selection, and may respond to this mode of selection by changing in phase with its environment.

These relationships are borne out by the environmental distribution of low-rate and high-rate groups. As Simpson states[51]:

Bradytelic groups do not occur in impermanent physiographic surroundings such as lakes or, to less degree, deserts or in extreme climatic zones such as the far Arctic. They are usually to be found in the sea, its strand, the shifting but essentially permanent major rivers, or the more slowly shifting and almost equally permanent great forest belts. They are especially common in the most nearly permanent of climatic zones, the subtropical, whence they ebb and flow into the tropical or into the temperate zone as occasion arises.

[51] Simpson, 1944, 141, quoted from *Tempo and Mode in Evolution* (New York, Columbia University Press).

The opossum, for example, is broadly and almost perfectly adapted to live in a forest environment which has been continuously available for millions of years. Almost any genetic variation arising in an opossum population—and such variations do arise fairly frequently in opossums[52]—will be disadvantageous in the ancestral environment and will be eliminated by stabilizing selection, leaving the population unchanged. Or if a genetic variation arises which, though disadvantageous in the ancestral environment, is adaptively valuable in some other neighboring environment, a population characterized by the new trait may branch off and occupy the new habitat, still leaving the ancestral stock unchanged. The opossums have in fact repeatedly given rise to divergent branch groups while remaining bradytelic themselves.[53]

In the horse family, as we saw previously, a period of relative stability in the forest environment during the Eocene was associated with slow evolution, while rapid evolution took place concurrently with rapid environmental changes in the Miocene. Indeed, the tachytelic rates in the Miocene horses were confined to just those phyletic lines which occupied the new grassland frontier, while the woods-inhabiting browsers remained conservative in the Miocene as in earlier ages.

Botanists have long recognized that the rich forest flora of the southeastern United States, in regions unaffected by marine inundations or ice ages, has lived under relatively constant environmental conditions since the Eocene, and includes many conservative plant groups.[54] The western United States, on the other hand, which has been subjected to mountain-building, volcanic activity, drought, pluvial cycles, and other environmental disturbances throughout much of the Tertiary and the Pleistocene, is inhabited by numerous tachytelic plant groups. Not all genera in the western flora are tachytelic by any means, however, for scattered pockets where ancient humid environmental conditions are preserved harbor conservative plant types like Sequoia and its associates. Rapid evolutionary rates in the western flora, as Stebbins has noted, are best exhibited by genera like Ceanothus, Arctostaphylos, Lupinus, Crepis, and many others which occur in the various new arid environments.[55]

---

[52] Simpson, 1944, 39–40. [53] Simpson, 1944, 140–41.
[54] See for example Wulff, 1943, chs. 9, 10; and Sharp, 1955.
[55] Stebbins, 1949.

In one and the same community the physical factors of the environment may be relatively stable, while the biotic factors continue to change in consequence of the unceasing process of mutual adjustment between species. Evolution in a phyletic group then slows down in characteristics related to the physical factors, but remains active in characters concerned with the biotic factors. Thus in tropical forests, the large orchid family is relatively uniform in the vegetative organs, but contains an enormous diversity of flower forms related to different methods of pollination by insects and birds.[56]

We have so far considered the environment as being stable or changing, and the organism as being responsive to the state of the environment. There is another possibility. The organism may take the initiative, so to speak, by invading an environment which has long been present, as when aquatic plants and animals colonized the land, or mesic organisms the desert. The net result as far as the pioneering population is concerned is the same as a rapid environmental change. The population is exposed to intense selective pressures favoring a rapid shift to a new adaptive state.[57]

INTERNAL BIOLOGICAL FACTORS AFFECTING EVOLUTIONARY RATES

In an environment changing at a given rate, the response of the organism, its rate of evolutionary change, will be determined by various inherent characteristics, which may well differ from one group to another. In this section we will consider two biological factors, length of generations and population size and structure, as principally affecting evolutionary rates. We will also consider briefly the possible relation between the position of an organism in the food chain and its rate of evolution.

Evolutionary rates are measured in chronological time, whereas the evolutionary forces of selection and drift operate from generation to generation. Now the length of generation varies enormously in different higher organisms, from a few weeks in a Drosophila or a year in annual herbaceous plants, to many years in a horse or centuries in some

[56] At this writing, a review of pollination mechanisms in the orchid family, containing numerous original observations from the tropics, is in preparation by L. van der Pijl and C. Dodson.

[57] Simpson, 1944, chs. 3, 7; Mayr, 1960.

perennial herbs and trees. It might be expected, therefore, that plants or animals with frequent short generations would evolve more rapidly in periods of environmental change than organisms with a long period between generations.

There is little empirical evidence from the fossil record to confirm this expectation. In the mammals, the rapidly reproducing opossum is distinctly bradytelic, while the slow-breeding elephants have evolved rapidly. As between the horses Neohipparion and Hypohippus with similar generation lengths, the former evolved more rapidly than the latter.[58] The carnivorous mammals with relatively long breeding cycles have evolved more rapidly on the average than the clams and mussels (Pelecypoda) with shorter life cycles, and some carnivores have evolved many times faster than some groups of pelecypods.[59]

This annomaly stimulates us to reexamine our concept of "rapid" rates as applied to environmental changes. The development of grass-land in the Miocene or the rise of a mountain range in the Pliocene undoubtedly represents a rapid environmental change when considered in terms of geological time. An outpouring of lava, a landslide, a forest fire, and most human activities also bring about environmental changes which have to be described as "rapid." But geologically rapid changes like those mentioned above are measured in millions of years, while the catastrophic environmental changes take place in years or even days. We need to distinguish between the geologically rapid changes and the short-term catastrophic changes when discussing rapid evolutionary rates.

The length of generation in horses or elephants is no limiting factor at all in the evolutionary response of the population to environmental changes proceeding gradually through one million years of time. The slow-breeding horses or elephants will keep pace with their environment, as it changes millenium by millenium, and an annual plant or Drosophila can do no more. But catastrophic environmental changes may release in the fast-breeding organisms a potentiality for explosive evolutionary rates which is never expressed in times of geologically rapid change. And, since evolutionary rates are measured by the amount of structural and genetic change per unit of chronological time, the organisms with a short life cycle are capable of developing bursts of evolutionary speed

[58] Simpson, 1944, 63.   [59] Simpson, 1944, 128.

which are impossible in forms with a long life cycle. Confronted with the same sudden environmental changes, a population of small plants or insects might be able to develop new adaptive properties at an explosive rate, where a population of large mammals or trees could only migrate or become extinct.

The conclusion that length of generations is not a significant factor in evolutionary rates is based on paleontological evidence referring to non-explosive rates, whether tachytelic or otherwise, and to geologically rapid environmental changes. In this context this conclusion is sound. We also have reason to believe that, in relation to catastrophic environmental changes, the difference between a short and a long breeding cycle may be crucial in determining whether rapid evolution is or is not possible.

The size and structure of the breeding population is another biological factor, and an extremely important one, affecting the evolutionary rate in a changing environment.

It seems to be generally agreed among evolutionists that a small population which has been isolated for a long time will ordinarily lack the genetic variability necessary for shifting adaptively in response to rapid or even slow environmental changes.[60] In such a population most polymorphic genes quickly become fixed in a monomorphic condition by the combined action of selection and drift, and some polymorphic genes could theoretically be fixed by drift alone (see Chapter 11). As a result, the population is adapted to its present condition with regard to most of its traits, and may in addition possess some inadaptive characters. If the environment to which the population is adapted remains stable, the small isolated population may persist indefinitely, but if the environment changes very much the population is likely to become extinct. Mayr notes that of all the species of birds known to have become extinct during the last 200 years (a period of rapid environmental change caused chiefly by man), 97 percent were inhabitants of small or well-isolated islands.[61]

The conditions for evolutionary response to environmental changes are more favorable in a large, continuous, freely interbreeding population, insofar as a store of unfixed genetic variability is usually present. However, wide outcrossing tends to have a leveling effect in such

[60] Wright, 1931, 147–49.   [61] Mayr, 1942, 224.

populations. An adaptive genotype that is formed by the sexual process and favored by selection in one part of the population will be broken up and swamped out by interbreeding with other neighboring parts of the population.[62] This factor of free interbreeding checks the establishment of deviations from the established norm in a large continuous population and gives such a population a somewhat conservative character in evolution.[63]

The resistance to change of a large continuous outcrossing population may be increased by an additional factor. Under conditions of continual gene flow within a population, selection is apt to favor those alleles which have good combining ability as heterozygotes. The population comes to exploit the advantages of heterozygosity, particularly heterosis and homeostasis, in achieving adaptive fitness. But a system of permanent heterozygosity has its corresponding disadvantages. Stabilizing selection in favor of buffered heterozygotes containing coadapted alleles and chromosomes will, by the same token, eliminate from the population any genes which do not fit into the coadapted system, but which instead produce phenotypes deviating radically from the population norm.[64]

The large outcrossing population composed of heterotic and homeostatic heterozygotes may possess maximum fitness for its present conditions and may be buffered against ordinary fluctuations in the environment. Under conditions of gradual environmental change, either slow or rapid in geological terms, the population is capable of making the necessary adaptive shifts. The Miocene grazing horses considered earlier provide an example of tachytelic evolution in large populations living under geologically rapid environmental changes. But the large continuous population is unable to respond adaptively to catastrophic changes in the environment; a victim of its own conservatism, it cannot develop the explosive rates of evolution called for in times of sudden environmental change.

There is, however, a way out of this impasse. Mayr has argued that the isolation of a few individuals from the stream of gene flow in the large population, as when a new small isolated daughter colony is founded, will bring novel genetic variations to expression. Those alleles and chromosome types which did not fit into the coadapted

[62] Wright, 1943.   [63] Mayr, 1954.   [64] Mayr, 1954.

system in the ancestral large population, and which were suppressed by stabilizing selection on that account, or else swamped out by inter-breeding, can now become homozygous and produce their radically divergent phenotypic effects in the small inbred daughter colony. To be sure, most such deviants from the norm will prove non-adaptive in any available environment, but in a large number of trials some new phenotypes of superior adaptive value could be expected to arise.[65] (A discussion of this hypothesis is given also in Chapter 16.)

In this way, although the large ancestral population cannot itself sustain explosive evolutionary rates, or keep pace with catastrophic environmental changes, a small isolated colony derived from it can. The founder population, indeed, may evolve rapidly during a brief period without any major changes in the secular environment at all. The condition of inbreeding in a group of individuals which was formerly permanently heterozygous, and the isolation of a few particular alleles or chromosome types from a formerly polymorphic and co-adapted system, produce a radical change in the "genetic environment" which is enough in itself to bring about a more or less radical shift in phenotypic traits.[66]

Another type of population structure which is capable of a wide range of evolutionary rates, including explosive ones, is that of a series of semi-isolated colonies.[67] In each colony, owing to its relatively small size, the action of selection is boosted by the cooperative action of drift, so that adaptive traits and probably also some non-adaptive ones are quickly established. Since the action of selection and drift is likely to follow a different course in separate colonies, the latter become racially different. A good example of numerous local races in a species with a colonial population structure is provided by *Gilia achilleaefolia* (Fig. 54 in Chapter 11).

Although the genetic variability within any given colony tends to become fixed monomorphically in a few generations, the occasional dispersal of an individual or gamete from neighboring, racially different colonies provides a recurrent source of new variations. The population system as a whole thus possesses a large and decentralized store of variability. Now the immigrant genotypes in a colony may increase the variability of that colony directly, by the simple addition of new genetic

[65] Mayr, 1954.   [66] Mayr, 1954.   [67] Wright, 1931, 150–51; 1960, 463–64.

variants or, more importantly, they may increase the variability by intercrossing with the indigenous individuals and engendering new recombination types. Some of the variability introduced into a colony by occasional intercolonial migration, then, is free and immediate, but even more of it is potential,[68] to be realized in later generations.

The colonial population system has at its disposal various modes of evolutionary response to changes in the environment: (1) some original genotypes within a colony may rise in frequency relative to other original genotypes; (2) or some whole colonies may expand while other colonies decline; (3) or, finally, certain recombination products of intercolonial hybridization may prove successful and spread relative to any of the original types.

Simple Darwinian selection and intercolonial selection [modes (1) and (2) above] will ordinarily prove adequate for the purpose of maintaining a state of adaptedness in an environment which is changing slowly or rapidly on the geological scale. In relation to catastrophic environmental changes, modes (2) or (3) or the combination of both may come into play. The sudden environmental change may find some local race of the species preadapted for the altered conditions [mode (2)]. A more likely result of catastrophic changes, however, is vast expansions, contractions, and migrational displacements in the various colonies, leading to an increased amount of intercolonial hybridization [the combination of mode (2) and mode (3)]. Among the new genotypes produced by intercolonial hybridization may be some which are adapted to the new conditions. In this way, the great potential variability of the colonial population system can be converted into free, expressed variability in times of environmental change which call for a burst of explosive evolution.

The partial isolation of genetically different individuals, which in cross-fertilizing plants and animals results from the spatial separation of the colonies, is accomplished in another way in some self-fertilizing plants. A single population of a self-fertilizing plant may consist of several or many genetically different biotypes growing together and reproducing as a series of isolated pure lines. Occasionally, however, intercrossing between biotypes takes place and leads to the formation of new gene recombinations.[69]

[68] To use Mather's (1943) terminology.   [69] Stebbins, 1957a.

A modified form of semi-isolation is found in many plant genera consisting of species or semispecies which are reproductively isolated but can and do hybridize under conditions of environmental disturbance. That some of the products of this interspecific hybridization are adaptive is indicated by their establishment as new widespread races or species in the new environments (see also Chapters 8, 15, and 16 for further discussion with examples). Hybridization between reproductively isolated species or semispecies may well surpass hybridization between spatially isolated colonies, in plants at least, as a process making possible explosive evolutionary changes.

Schmalhausen has suggested that the effectiveness of natural selection and hence the rate of change due to selection is related to the position of an organism in the food pyramid.[70] Let us compare organisms low in the food pyramid, such as plankton, diatoms, bacteria, worms, lower crustaceans, and many insects, with predatory animals at the top of the pyramid.

Organisms standing low in the hierarchy of nutrition frequently have no special means of defense against predators. Their populations are then subject to large-scale decimations by the predators. The death or survival of any given individual in such decimations, moreover, is largely a matter of chance, rather than of differences in adaptive fitness, and the process of elimination is consequently non-selective to a considerable extent.[71] This type of elimination favors increase in fecundity, since the most fecund individuals will by chance have the greatest number of survivors, but does not lead to progressive evolutionary changes in other traits. The defenseless types of organisms are in fact conservative.[72]

Predatory animals which are well protected themselves, on the other hand, will tend to increase in numbers and then to compete for a limited food supply. The ensuing elimination of individuals is apt to have a large selective component. In proportion to the magnitude of this selective component in the elimination process, some progressive evolutionary changes will occur at relatively rapid rates. The animals evolve more or less rapidly toward more specialized modes of food-getting, improved sensory organs, higher intelligence, more economical

[70] Schmalhausen, 1949, 69–72, 92–93, 254–70.
[71] Schmalhausen, 1949, 69–70.   [72] Schmalhausen, 1949, 257–59.

metabolism, and greater longevity along with lower fecundity. Rapid evolutionary rates are often found in such types of organisms, as exemplified by the carnivorous mammals.[73]

Undoubtedly the different biological factors affecting evolutionary rates can and do work hand in hand. Thus organisms low in the food chain may exhibit slow evolutionary rates partly because of the small selective component in elimination, and partly because of their usually large populations. Carnivorous animals, on the other hand, are not only higher in the food pyramid than their prey, but also and for the same reason are less numerous in individuals. Relatively rapid evolution may be promoted in the carnivorous animals, therefore, by the characteristics of their populations as well as by their mode of securing food.

QUANTUM EVOLUTION

The sequence of evolutionary rates in the history of many paleontologically well-known groups consists of an early, relatively short phase of tachytelic or explosive evolution followed by a later phase, often lasting for millions of years, of horotelic or bradytelic evolution.

The group in its later phase of expansion and stabilization must have existed in large populations to judge from the abundance of fossil remains. The same group is typically represented only poorly or not at all in the fossil record during the earlier period of origin and divergence. The fossil representation of the group in its early phase is less than would be expected if it had large populations at that time, but can be accounted for on the assumption that its populations were small then.[74]

Now the phase of large populations was also the phase of evolutionary conservatism, whereas the rapid evolutionary changes in the group took place when its populations were presumably small. These conclusions, based on direct and indirect paleontological evidence, are seen to agree with the requirements of population genetics theory, which points to small populations under certain conditions as the most likely fields of explosive evolution.[75]

Simpson, who presented the foregoing line of reasoning in 1944, suggests that the origin of new genera, families, and orders, with new

[73] Schmalhausen, 1949, 265–70.   [74] Simpson, 1944, ch. 3.
[75] Simpson, 1944, ch. 3.

and distinctive traits, takes place by the transition of a population from the ancestral way of life to a new adaptive mode. This transition involves the crossing of an adaptive no man's land lying between the ancestral and the new environment in which the population is necessarily poorly adapted. The shift must therefore occur rapidly and in small populations. Such rapid shifts from an ancestral adaptive mode to a new and quite different adaptive mode, via small populations which are temporarily out of equilibrium with their conditions, he terms quantum evolution. Quantum evolution is believed to be the normal process by which new major groups come into being.[76]

Although the hypothesis of quantum evolution as applied to the origin of families and orders in past eras is not susceptible to experimental verification, the various elements of this hypothesis can be tested in living populations. We have considered in this chapter and in Chapter 16 various processes that can bring about major structural changes in a relatively few generations.

Macromutations, or macrorecombinations resulting from hybridization, or new homozygotes segregated from large polymorphic populations may engender radical changes in structural type and adaptive properties (see Chapter 16). Under catastrophic environmental changes, or upon the invasion of new habitats, some of these novel variations may prove to have high selective value. Given a population structure which permits inbreeding, as in small daughter colonies derived from a large ancestral population, or a system of semi-isolated colonies, or a group of hybridizing species capable of inbreeding, the new adaptive traits may then be established quickly by the combined action of selection and drift.

The profound and rapid changes in the morphological and ecological characters of a population, which are subsumed under the term quantum evolution, may take place in a single unbranched phyletic line; or, and perhaps more frequently, such changes may involve the divergence of a branch line. The immediate product of quantum evolution in the latter case, is a new race or species which deviates markedly from the ancestral stock.

Strongly divergent races and species are known in many genera of plants and animals; it will suffice to recall the example of the New

[76] Simpson, 1944, ch. 7.

Guinea kingfishers presented in Chapters 15 and 16. If the divergent population bears fruit in a macroevolutionary sense, if in the course of time it gives rise to one or more series of derivative species possessing its basic morphological and ecological characters in various modified forms, so that its descendants assume the taxonomic proportions of a distinct genus, tribe, family, or order, the process of quantum evolution can be regarded as completed.

QUANTUM EVOLUTION IN THE PHLOX FAMILY

The Phlox family (Polemoniaceae) has its main center of distribution in North America, where 290 species belonging to 13 genera and grouped into two tribes occur in a variety of habitats. The phylogenetic relationships of these North American genera and tribes can be reconstructed with reasonable certainty from several lines of indirect evidence. The direct evidence of fossils is lacking, unfortunately, but is compensated for to a large extent by a rich representation of living species, including the connecting links between genera in some cases.

From the comparative morphology of the genera we can infer certain phylogenetic relationships and morphological trends. The geographical distribution and ecology of these genera, when correlated with the known geological history of the plant communities with which they are associated today,[77] indicate the probable migrational history and ecological trends in the group. Combining the various lines of evidence, we obtain an internally consistent picture of the evolutionary development of the Phlox family in North America from the late Tertiary to the present.[78] It is a picture of progressive trends from tropical woody plants to desert, arctic, or alpine herbs, effected in part at least by occasional quantum shifts from one ecological zone to another more extreme one.[79]

The most primitive characters in the Phlox family are found in several tropical American genera such as Cobaea, Bonplandia, Cantua, and others. No one of these tropical genera has a monopoly of the primitive features; each of them possesses also some advanced characters; and the distribution of primitive and advanced characters varies from genus to genus. This pattern of morphological variation, together with the

---

[77] As worked out by Axelrod and others; see Axelrod, 1950a, 1950b.
[78] Grant, 1959, I, chs. 7–9.   [79] Grant, 1959, I, 220–22.

fact that the tropical genera are mostly small in number of species and only remotely related to one another, suggests that they are the remnants of an ancient group that has suffered much extinction. There is no single genus living that we can point to as a representative of the common ancestor of the family as a whole, or of the temperate North American genera in particular; nor is there any reason to expect that ancestral form to be alive today.[80]

Making as few assumptions as possible, it is reasonable to suppose that the now extinct ancestral form was an evergreen woody plant, perhaps a shrub or small tree, living in some facies of the old tropical forest that covered a wide area from South America to southern North America during the early Tertiary. The divergence of the two north temperate tribes, the Gilia and the Polemonium tribes, from their ancestor evidently took place within the northern arm of the old tropical forest.[81]

Let us consider some of the probable steps involved in the evolutionary development of the Gilia tribe.[82]

During the Late Oligocene and Early Miocene the northern arm of this neotropical flora developed under the influence of more arid climatic conditions into a xeric type of vegetation. The woodland vegetation composed of subtropical trees, shrubs and grasses adapted to a warm semi-humid semi-arid climate is known as the Madro-Tertiary flora [cf. Axelrod 1950a, 1950b]. By Middle Miocene time it spread as far north as southern California.

The Bonplandia tribe, which is phylogenetically a derivative of some tropical Polemoniaceae having many of the characteristics of the modern Cobaeas, occurs today in the subtropical forests and xeric woodlands of Mexico and Central America. It is probable that the divergence of this tribe from the ancestral Polemoniaceae involved a shift in ecological zone from the neotropical rainforest to the more xeric Madro-Tertiary flora, i.e., quantum evolution in the sense of Simpson. . . .

Through Miocene and Pliocene time the climate continued to become increasingly arid. Keeping pace with these changes, the more humid phase of the Madro-Tertiary flora gradually gave way to a more arid phase in western North America. Many evergreen oaks, thorn forest species, and subtropical trees and shrubs, which had been represented in western North America during the Miocene, retreated to Mexico by Middle Pliocene time, while other elements of the Madro-Tertiary flora, such as pinyon pine, juniper, live-oak, chaparral species, and plains grasses, remained in

[80] Grant, 1959, I, 197–98.   [81] Grant, 1959, I, 197–98, 220–22.
[82] Grant, 1959, I, 220–22, quoted from *Natural History of the Phlox Family* (The Hague, Martinus Nijhoff).

possession of the western American territory. In the Middle Pliocene the northern branch of the Madro-Tertiary flora broke up into segregates: pinyon-juniper woodland in the areas with cold winters; live-oak woodland in areas with mild but moist winters and dry summers; short-grass plains in the interior areas with low rainfall; and chaparral in warm dry zones.[83]

The climatic and floristic associations of the earliest Gilieae are plainly suggested by their tendency at the present time to reach their best development in warm arid regions. When we examine the vegetation types with which the Gilia tribe is associated we find that nearly every North American species in the tribe occurs in one or more of the present northern segregates and remnants of the old Madro-Tertiary flora: the pinyon-juniper woodland, live-oak woodland, chaparral, and short-grass plains. The Gilieae are found in other vegetation types too, in the deserts, alpine zones, and coniferous forests, but the association with the Madro-Tertiary flora is like a common denominator running through the whole tribe. This association involves the more primitive groups in each genus and section, whereas the associations with other vegetation types can be viewed as later invasions by more specialized species and races in the different lineages.

No single existing member of the Gilia tribe exemplifies the ancestral condition of the tribe, though primitive characters are found within the genera Gilia, Ipomopsis, Eriastrum and Leptodactylon. By pooling the primitive features still preserved and extrapolating from them we concluded that the ancestor of the Gilia tribe was a small alternate-leaved xeric shrub related to the modern Loeselias. The most probable inference from the considerations presented above is that this unknown ancestor of the Gilia tribe diverged from a Loeselia-like stock within the Madro-Tertiary flora. This divergence represented another shift in ecological zone, in this case from the subtropical to the arid phase of the Madro-Tertiary flora.

In the segregation of the Madro-Tertiary flora into its present-day communities, which occurred under the influence of increasing aridity and a climatic differentiation in the Middle Pliocene, different phylads within the Gilia tribe retained associations with different segregate communities. The genera Eriastrum, Navarretia, Linanthus, and the section Gilia of Gilia remained with the California live-oak woodland and digger-pine savanna. Ipomopsis and Gilia sect. Giliandra associated themselves chiefly with the pinyon-juniper woodland, and Gilia sect. Giliastrum with the short-grass plains and xeric subtropical woodland. The expansion of arid habitats now permitted these groups to undergo a rich speciation and development in their respective regions.

In the Late Pliocene the continuing trend toward lower rainfall in western North America, accompanied by the rise of a series of mountain ranges along the Pacific coast, led to the appearance and expansion of deserts in the interior. Large areas in the southwest situated in the

[83] This summary statement is based on Axelrod, 1950a, 1950b *passim*.

rain-shadow of the rising Pacific mountain ranges, which had supported Madro-Tertiary woodland and grassland communities earlier in the Pliocene, now became deserts.[84]

The appearance of the desert habitats made possible a new wave of adaptive evolution in the Gilia tribe. This tribe has responded to its opportunities by producing numerous species of characteristic desert annuals belonging to 14 genera or sections. Most of these groups occur also in semiarid communities bordering the desert, but one of them, the small genus, Langloisia, is endemic to the desert.[85]

Among the members of the Gilia tribe on the Mojave Desert are the related genera, Gilia, Eriastrum, and Langloisia. The relative ages of these genera can be inferred from their taxonomic structure, and their relative degree of advancement from the evidence of comparative morphology. The genus Gilia is morphologically and ecologically heterogeneous, with five different sections and about 60 species, and is geographically widespread, occurring in a variety of desert and non-desert habitats. Eriastrum, a small and homogeneous genus of 14 species in a single series, occurs in a narrower range of arid habitats, in and near the desert, and over a smaller geographical area; while Langloisia is a minor genus of four species endemic in the desert. The evidence of comparative morphology and cytology indicates that Eriastrum is more advanced than Gilia, and that Langloisia in turn is more advanced than Eriastrum. The whole pattern of evidence suggests that Gilia is an old and basic stock, that Eriastrum has been evolving for a shorter time than Gilia, and that Langloisia is a relatively recent group derived from Eriastrum. Now Gilia, Eriastrum, and Langloisia exhibit a series of progressively increasing specializations for the desert environment.

The desert species of Gilia, like their close relatives in the bordering woodland and savannah communities, germinate with the winter rains and bloom in the spring months before the summer heat and drought set in. Their relatively soft-tissued leaves and stems are only moderately resistant to desiccation, and they survive only as long as moisture is available in the upper layer of the soil. These plants grow most abundantly in the western part of the Mojave Desert where the rainfall is greatest.

[84] Axelrod, 1950a, 1950b.   [85] Grant, 1959, I, 225.

In Eriastrum, which also occurs in woodland and savannah as well as desert communities, the plants have wiry stems and small tough leaves often covered with wooly hairs. These plants grow and bloom in the heat and drought of early summer after the tender and more mesic Gilias have gone to seed. Langloisia, finally, is a genus of small, tufted, bristly-leaved plants, which also bloom in late spring and early summer. Langloisia is confined to the desert, and grows abundantly in the very dry eastern parts of the Mojave Desert.[86]

A prevailing trend in the unrelated genus Clarkia (Onagraceae) in the California flora is likewise in the direction of adaptation for more arid habitats. Lewis has described this trend as a series of speciational shifts. There is good genetic evidence to indicate that the relatively mesic species, *Clarkia biloba*, in the northern and central Sierra Nevada, has given rise to the more xeric daughter species, *C. lingulata*, on its southern periphery.[87] Similar changes if repeated in series would account for the whole trend in Clarkia from some mesic Oenothera-like ancestor to the xeric species of Clarkia that now grow in the dry foothills of southern California.[88]

The climatic changes during the latter part of the Tertiary and the Pleistocene involved not only an increase in aridity, but also an increase in cold, culminating in the ice age. The formerly widespread tropical climate characterized by an abundance of both moisture and warmth gave way over large areas in North America variously to hot dry, cold moist, and cold dry climates. The North American Polemoniaceae underwent adaptive changes corresponding to all these climatic changes. Some tropical or subtropical stock with requirements for both mild temperatures and abundant moisture, gave rise to the Gilia tribe, which, with reduced moisture requirements, was to achieve considerable evolutionary success in the various hot dry areas and some cold dry ones.[89]

Meanwhile the other temperate North American tribe, the Polemonium tribe, also derived from some tropical or subtropical ancestor, and at some point in the past from an ancestor shared in common with the Gilia tribe, was sacrificing the ancestral requirements for abundant warmth but retaining the ancestral requirements for moisture, and was

---

[86] Grant, 1959, I, 79, 196–97, 258.   [87] Lewis and Roberts, 1956.
[88] Lewis, 1962.   [89] Grant, 1959, I, 224.

thereby fitting into the new cold moist climates.[90] Some members of the Polemonium tribe then went on to colonize such cold dry habitats as the Great Basin desert and the rocky talus slopes on high mountain peaks.

As in the Gilia tribe, the ecological evolution of the Polemonium tribe apparently took place as a series of adaptive shifts. Let us quote again from an earlier account.[91]

As we survey the present floristic associations of the Polemonium tribe we cannot help but note that this tribe is only a casual member of the present remnants of the Madro-Tertiary flora. Its basic associations seem to be with the various modern derivatives of the temperate Arcto-Tertiary flora which covered the northern half of North America in Tertiary time. Thus Polemonium and Phlox both occur in the eastern deciduous forest, the western coniferous forest, the temperature forest of Mexico, and the arctic tundra. Collomia is present in the Pacific coniferous forest and Gymnosteris in the sagebrush plains. These are all segregates of the old Arcto-Tertiary flora. The associations of the Polemonieae with Madro-Tertiary floras can be regarded as later invasions by advanced members belonging to several of the genera.

The Arcto-Tertiary forest was in contact with the neotropical forest during the Eocene and has been in contact with the Madro-Tertiary flora since the late Oligocene. The Polemonium tribe is believed to be derived [through the genus Polemonium] from a Cobaea-like or Bonplandia-like ancestor existing in the neotropical forest or the humid phase of the Madro-Tertiary forest. Its divergence from the tropical or subtropical ancestor must, consequently, have entailed an ecological shift from tropical or subtropical to temperate conditions.

We find the Polemonium tribe growing today in eastern North America, on the highest western mountain peaks, in Eurasia, and in the tundra around the pole. Obviously the ecological shift which accompanied the rise of the Polemonium tribe was not a sudden transition from a tropical rainforest to an arctic tundra environment. It is more likely that the evolutionary transition from one ecological zone to another occurred at a point where the gap was narrowest. A shift from a cloud-forest environment to a warm temperate environment seems feasible. In this connection it is worth noting that Cobaea, representing a tropical form, today occurs within the territory of Polemonium, which represents a primitive type of Polemonieae. The montane rainforest zone of Cobaea and the temperate forest zone of Polemonium occur in juxtaposition in the Mexican highland. A shift from one zone to the other under conditions such as these would suffice to start the derivative group on a new course of evolution, leading eventually to the occupation of more extreme environments.

[90] Grant, 1959, I, 224.
[91] Grant, 1959, I, 222, quoted from *Natural History of the Phlox Family* (The Hague, Martinus Nijhoff).

An interesting series of ecological stages is exhibited by different species and sections of the single genus Polemonium. The perennial herb, *Polemonium carneum*, with various primitive morphological characters, grows in moist places on the cool but equable coast of the Pacific northwest. The related *Polemonium caeruleum* group occurs in temperate forests where long cold winters prevail. *Polemonium pulcherrimum*, which shows a number of advanced characters, lives at and above treeline in the western American mountains. Still higher in these ranges, on the high peaks and alpine tundra, where even the summers are stormy, are the derived species of alpine plants belonging to the section Melliosma. And on the sagebrush plains of the Great Basin there occurs a reduced species of annual herb placed in the monotypic section Polemoniastrum. This series suggests that, by the process of speciation, a group originally adapted for a moist equable climate can give rise to derivative groups adapted for the conditions on high mountain peaks or cold deserts.

THE ROUTES TO ADAPTATION

The result of evolution, or, to employ a metaphor useful for visualization, the goal of evolution, is a state of adaptation. A population or species existing at any given moment of time, and adapted to its present conditions, may be able to evolve adaptations fitting it for one or more different environments. The descendants of the ancestral population, in other words, may progress through one or more evolutionary trends. To continue the figure of speech, the ancestral population can, potentially, move in one or more phyletic lines toward as many new adaptive goals.

The number of such potentially attainable adaptive goals varies from zero to some number in the billions according to the degree of specialization already built into the population at its starting point and the amount of competition it faces from superior competitor populations.

Some primitive single-celled form living when the earth was still young could and did give rise to many billions of descendant protistan, fungus, plant, and animal species adapted to many billions of environmental niches or habitats. The first land plants and land animals were already specialized in ways which restricted them thereafter (with minor exceptions) to the sedentary food-manufacturing and the motile food-getting modes of life, respectively; yet the earliest land plants and animals separately found a very wide range of terrestrial and aerial habitats open to explore and colonize. At the opposite extreme, a highly specialized population living in a biotic community which has become saturated with species may be able to reach few or no new adaptive goals.

Let us assume a typical intermediate situation in which several adaptive goals, differing in the amount of structural and physiological change required to reach them, are open to colonization by an original population. The ancestral stock gives rise to some descendant species which attain adaptive goals not very different from the starting point, and to other descendant species which come to occupy markedly different adaptive zones. Each evolving descendant population, in progressing through its particular evolutionary trend, toward its particular adaptive goal, must follow a route commensurate with the magnitude of the evolutionary change required. If for any reason the population is incapable of taking a suitable route, of course, it will not reach the adaptive goal.

The new gene combinations underlying new adaptive characteristics can be established in a large population by selection alone. Their establishment in this case is either expensive in terms of the number of genetic deaths per generation, perhaps too expensive for the population to bear and still maintain its position in the biotic community, or else slow, as we saw in Chapter 9. Because of this and other factors (discussed in Chapter 16 and earlier in this chapter), a large population is likely to be conservative in evolution. Yet conservatives do change in time.

Given an adaptive goal not too different from the ancestral condition, and plenty of time in which to reach it, so that a new gene combination which is similar to an ancestral one can be built up slowly and gradually by progressive selection, the large population may be quite capable of reaching its goal. In its progression toward the new adaptive state, moreover, the large population follows the course of long-continued responses to steady selective pressures in a single phyletic line. In other words, it takes the route of phyletic evolution in the strict sense.

If the increments of structural change are plotted against time on a graph, the evolutionary trend describes a course which, if not perfectly rectilinear, is at least generally straight or gently curving.[92] An example of such a progressive evolutionary trend is found in the development of the grazing horses from the three-toed savannah-inhabiting Merychippus to the one-toed steppe-inhabiting Equus.

The conservative properties of large populations, while permitting them to change gradually and steadily, strongly limit their ability to make such rapid shifts or abrupt changes in the direction of evolution as may be required in order to attain a new and profoundly different set of characteristics. There are, however, alternative routes by which an adaptive goal very different from the ancestral condition can be reached.

New gene combinations, which are established only slowly when controlled by selection alone in a large population, can be fixed quickly by the combined action of selection and drift in a small population (Chapter 11). Consequently, the variations present in the large population, even though it is conservative itself, can be released and utilized for the rapid development of radically different adaptations by the budding off of small daughter colonies in which these variations are fixed (Chapter 16). Such daughter colonies inevitably diverge from the ancestral stock as new races or species. These may later become large and conservative again.

The route from an existing state of adaptedness to a very different adaptive goal, therefore, may involve a quantum shift followed by a period of slower adjustment and stabilization. This route begins as a major change of direction, as a departure from the ancestral condition, and continues as a slower and more gradual series of phyletic changes. The course of the phyletic line, obtained by plotting structural change against time, contains a sharp curve at the point where the group branches off from its more conservative ancestors, but is smooth where the group becomes conservative again. The whole evolutionary trend, from the point of origin of the group to its later stabilization in its new adaptive zone, is a composite of quantum shifts and slower phyletic changes.

Now the quantum shift might take place in a single phyletic line which passes through a bottleneck of small size and rapid change at

[92] Simpson, 1944, 152–53.

some point or points in its history. The quantum shift may also take place frequently by the budding off of a divergent daughter population from a conservative ancestral species, which itself remains unchanged. Speciation, in other words, may be involved in many cases of quantum evolution. This leads to an interesting deduction.

It has generally been assumed that progressive changes within a population and speciational divergences between populations are distinct modes of evolution. This time-honored assumption goes back to Darwin who drew the distinction between "descent with modifications" and "origin of species." Among modern authors, Simpson contrasts phyletic evolution with speciation or splitting; Rensch refers to anagenesis or progressive evolution versus kladogenesis or phylogenetic branching; in Chapter 15 we found it necessary to distinguish between phyletic evolution and evolutionary divergence, and so on (see Fig. 76 in Chapter 15).[93] It is indeed useful for analytical purposes to separate these two directions of change in the evolutionary process. It is also widely recognized—by Simpson, Rensch, and other authors— that this distinction becomes rather arbitrary when applied to actual cases of evolution, in which both modes of change are likely to be occurring simultaneously.

I wish now to suggest that the two modes, intrapopulational phyletic changes and interpopulational divergence, not only take place simultaneously within an evolving group of populations, but may be so intimately combined as to be inseparable or indistinguishable in some phyletic lines which reach and successfully colonize new adaptive zones. It is contended that evolutionary trends, in some cases, are the resultant of a stepwise succession of divergences, or a series of branch lines each of which diverges from its ancestral stock in the same general direction (Fig. 103).

Granting that a very different adaptive goal can be attained successfully and effectively by the formation of a divergent daughter species, it then seems to follow that a still more distant adaptive goal, or one that is constantly receding owing to continual rapid change in the environment, might call for the repeated formation *of a series of* divergent daughter species. Each new species in such a series can branch off from its more conservative ancestor in the direction of the adaptive

[93] Simpson, 1944, ch. 7, 1953a, ch. 12; Rensch, 1959, 97, and earlier works.

goal, and each species in its turn, even after it has expanded and grown conservative itself, can give rise to a new generation of daughter species diverging toward the same goal. The evolutionary trend in a very rapidly changing environment or toward a remote adaptive goal is thus considered to be a succession of quantum speciational changes, as is inferred to have taken place in the adaptation of the Polemoniaceae to

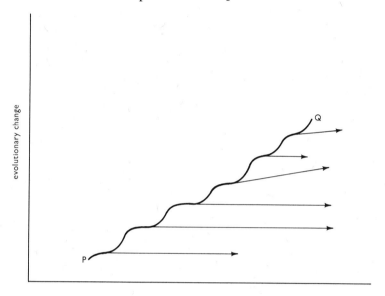

Fig. 103. *An evolutionary trend seen as a series of branch lines* (P — Q) *resulting from a succession of speciational divergences*

increasingly arid and cold conditions in North America since the Miocene. The phyletic line representing such an evolutionary progression follows the steplike path shown in Fig. 103.

The environment in all its diverse facets can be represented as a varied terrain, an adaptive landscape, consisting of level plains, hills, and towering mountain ranges.[94] Assume that an ancestral population

[94] The useful analogy of an adaptive landscape has been employed by Wright, Dobzhansky, Simpson, and other evolutionists. In borrowing the analogy from these authors, I am also changing the symbolism in several important respects to suit the purposes of the present discussion. The meaning attached here to the landscape features—the plains, peaks, etc.—is explained in the text.

exists on the plain in the center of the landscape at a point P. The plain slopes downward gradually to the east, while to the west foothills and mountains rise up more or less abruptly. The difference in elevation between the original station P and various adaptive goals on the eastern plain or in the western mountains is taken to represent the amount of structural and physiological change required in evolving from P to a new adaptive state. And the route to any station or goal on the adaptive landscape is representative of an evolutionary trend.

It is obvious that the same kind of easy straight route, as would suffice to take the population from P to a station only slightly different in elevation on the eastern plain, might be totally inadequate for the purpose of reaching an adaptive goal in the mountains to the west. We will consider three types of routes corresponding to as many types of adaptive goals. A goal on the plain can be reached by a route which is mostly straight but sometimes gently curving. The route from P to a station in the foothills reaches its destination by curving and changing direction occasionally. To climb a steep slope, finally, it may be necessary to follow a series of switchbacks.

The straight route across the plain symbolizes an evolutionary trend effected by normal phyletic evolution. The combination of curves and straight paths into the foothills represents an evolutionary trend which is carried out by a combination of quantum and phyletic evolution. The series of switchbacks up the steep slope suggests an evolutionary trend composed of a succession of speciations. The diverse lineages descended from the original ancestral population, and progressing toward their respective adaptive goals, follow routes appropriate to the location of the various goals on the adaptive landscape.

NINETEEN

Conclusions

AS THE ARGUMENT presented in this book is long, it may be worthwhile to attempt to summarize the central thesis in abstract form here.

1. The adaptations fitting organisms to live successfully in their environment generally consist, in the first place, of characters which are genetically and developmentally complex in themselves and, in the second place, of coordinated combinations of such characters.

2. These complex adaptations are determined by correspondingly complex combinations of gene alleles, or in other words, by the joint actions of particular alleles at numerous gene loci.

3. New adaptive allele combinations can be built up out of raw genetic variations in a large outcrossing population by the action of natural selection alone. But the cost to the large outcrossing population of selection for a new and complex combination of gene alleles, that is, the total number of genetic deaths involved in the replacement of the ancestral allele combination by the new one, is very great.

4. A population living in the midst of competitor populations can suffer only so many genetic deaths in any generation and still maintain its position in the biotic community. Under normal conditions of strong interspecific competition, therefore, the costly process of remolding a complex gene combination by selection alone in a large outcrossing population can take place slowly, but not rapidly, if the population is to survive, and frequently not rapidly enough to meet the requirements of adaptation to a changing environment.

5. Occasionally in its history a population may exist in an open habitat in which it is relatively free from the pressure of strong interspecific competition. The pioneering population living under these exceptional conditions can tolerate a large number of genetic deaths

569

concentrated in a few generations, and hence can evolve rapidly to a new adaptive state under the control of selection alone.

6. Inbreeding combined with selection of the inbred products provides a means of fixing a new allele combination in a population quickly and with a reduced number of genetic deaths.

7. Many actual situations are such that the two foregoing factors, inbreeding and freedom from strong interspecific competition, will be present together. Then they can work hand in hand to permit selection to bring about rapid and far-reaching changes in the genetic constitution of a population.

8. The pathway to the fixation of a new allele combination by the combined action of inbreeding and selection leads concomitantly to the formation of a new and divergent species in many cases.

9. It follows that long-continued evolutionary trends in some phyletic lines which undergo rapid and continual formation of new adaptive allele combinations may be expected to follow a course of repeated speciational branchings. The phyletic line in such cases progresses through a steplike succession of divergences, in which each daughter species, with its particular new adaptive allele combination, branches off from a genetically different and more conservative ancestral population, and later gives rise in its turn to another new daughter species which diverges again in the same general direction.

Bibliography

Ainsworth, G. C., and G. R. Bisby. 1954. A Dictionary of the Fungi. 4th ed. Kew, England.

Allee, W. C., A. E. Emerson, O. Park, T. Park, and K. P. Schmidt. 1949. Principles of Animal Ecology. Philadelphia and London.

Amadon, D. 1950. The Hawaiian honeycreepers (Aves, Drepaniidae). Bull. Am. Museum Nat. Hist. 95 (4).

Anderson, E. 1948. Hybridization of the habitat. Evolution 2: 1–9.

—— 1949. Introgressive Hybridization. New York.

—— 1953. Introgressive hybridization. Biol. Revs. 28: 280–307.

Anderson, E., and B. R. Anderson. 1954. Introgression of *Salvia apiana* and *Salvia mellifera*. Ann. Missouri Botan. Garden 41: 329–38.

Anderson, E., and L. Hubricht. 1938. Hybridization in Tradescantia. III. The evidence for introgressive hybridization. Am. J. Botany 25: 396–402.

Anderson, E., and G. L. Stebbins. 1954. Hybridization as an evolutionary stimulus. Evolution 8: 378–88.

Anderson, E. G. 1935. Chromosomal interchanges in maize. Genetics 20: 70–83.

Andrewartha, H. G., and L. C. Birch. 1954. The Distribution and Abundance of Animals. Chicago.

Andrews, H. N. 1959. Evolutionary trends in early vascular plants. Cold Spring Harbor Symposia Quant. Biol. 24: 217–34.

Anonymous. 1956. Genetics and cytology of cotton, 1948–1955. Southern Coöp. Ser. Bull. (U.S. Dept. Agriculture) 47.

Arnold, C. G. 1958. Selektive Befruchtung. Ergeb. Biol. 20: 67–96.

Arnon, D. I. 1960. The role of light in photosynthesis. Sci. American, November, 1960.

Avery, A. G., S. Satina, and J. Rietsema. 1959. Blakeslee: The Genus Datura. New York.

Avery, O. T., C. M. Macleod, and M. McCarty. 1944. Studies on the chemical nature of the substance inducing transformation of pneumococcal types. J. Exptl. Med. 79: 137–58.

Axelrod, D. I. 1950a. Classification of the Madro-Tertiary flora. Carnegie Inst. Wash. Publ. No. 590: 1–22.

—— 1950b. Evolution of desert vegetation in western North America. Carnegie Inst. Wash. Publ. No. 590: 215–306.

Baker, H. G. 1953. Race formation and reproductive method in flowering plants. Symposia Soc. Exptl. Biol. No. 7: 114–45.

Baldwin, J. M. 1896. A new factor in evolution. Am. Naturalist 30: 441–51, 536–53.

Baldwin, P. H. 1953. Annual cycle, environment and evolution in the Hawaiian honeycreepers (Aves: Drepaniidae). Univ. Calif. (Berkeley) Publs. Zoöl. 52: 285–398.

Bateman, A. J. 1949. Pollinating agents and population genetics. Hereditas, suppl. vol. 1949: 532–33.

—— 1950. Is gene dispersion normal? Heredity 4: 353–63·.

—— 1951. The taxonomic discrimination of bees. Heredity 5: 271–78.

Bateman, K. G. 1956. Experiments on genetic assimilation. (Abstract.) Heredity 10: 281.

—— 1959. The genetic assimilation of four venation phenocopies. J. Genet. 56: 443–74.

Bates, H. W. 1862. Contributions to an insect fauna of the Amazon Valley. Lepidoptera: Heliconidae. Trans. Linnean Soc. London (Zoology) 23: 495–566.

Bates, M. 1949. The Natural History of Mosquitoes. New York.

Bateson, Wm. 1894. Materials for the Study of Variation. London.

Beadle, G. W. 1955. The gene: carrier of heredity, controller of function and agent of evolution. Nieuwland Lectures (Notre Dame) 7: 1–24.

—— 1957. The physical and chemical basis of inheritance. Eugene, Oregon.

—— 1960. Physiological aspects of genetics. Ann. Rev. Physiol. 22: 45–74.

—— 1962. "Structure of the Genetic Material and the Concept of the Gene," in W. H. Johnson and Wm. C. Steere, eds., This Is Life. New York.

Beamish, K. I. 1955. Seed failure following hybridization between the hexaploid *Solanum demissum* and four diploid Solanum species. Am. J. Botany 42: 297–304.

Beardmore, J. A., Th. Dobzhansky, and O. Pavlovsky. 1960. An attempt to compare the fitness of polymorphic and monomorphic experimental populations of *Drosophila pseudoobscura*. Heredity 14: 19–33.

Beaudry, J. R. 1960. The species concept: its evolution and present status. Revue can. biol. 19: 219–40.

Beeks, R. M. 1961. Variation and hybridization in southern California populations of Diplacus (Scrophulariaceae). Ph.D. thesis, Claremont University College, Claremont, California.

—— 1962. Variation and hybridization in southern California populations of Diplacus (Scrophulariaceae). Aliso 5: 83–122.

Belt, T. 1874. The Naturalist in Nicaragua. London.

Bemis, W. P. 1959. Selective fertilization in lima beans. Genetics 44: 555–62.

Benzer, S. 1955. Fine structure of a genetic region in bacteriophage. Proc. Natl. Acad. Sci. U.S. 41: 344–54.

Bertalanffy, L. von. 1952. Problems of Life. Translation. London.

Black, G. A., Th. Dobzhansky, and C. Pavan. 1950. Some attempts to estimate species diversity and population density of trees in Amazonian forests. Botan. Gaz. 111: 413–25.

Blair, W. F. 1953. Population dynamics of rodents and other small mammals. Advances in Genet. 5: 1–41.

—— 1955. Mating call and stage of speciation in the *Microhyla olivacea-M. carolinensis* complex. Evolution 9: 469–80.

Blum, H. F. 1955. Time's Arrow and Evolution. 2d ed. Princeton, New Jersey.

Bonnett, O. T. 1954. The inflorescences of maize. Science 120: 77–87.

Bridges, C. B., and K. S. Brehme. 1944. The mutants of *Drosophila melanogaster*. Carnegie Inst. Wash. Publ. No. 552.

Briggs, M. H. 1959. Dating the origin of life on earth. Evolution 13: 416–18.

Brink, R. A. 1952. "Inbreeding and Crossbreeding in Seed Development," in J. Gowen, ed., Heterosis. Ames, Iowa.

Brink, R. A., and D. C. Cooper. 1947. The endosperm in seed development. Botan. Rev. 13: 423–541.

Brock, R. D. 1954. Spontaneous chromosome breakage in Lilium endosperm. Ann. Botany (London) 17: 7–14.

—— 1955. Chromosome balance and endosperm failure in hyacinths. Heredity 9: 199–222.

Brower, J. V. Z. 1958a. Experimental studies of mimicry in some North American butterflies. I. The monarch, *Danaus plexippus*, and viceroy, *Limenitis archippus archippus*. Evolution 12: 32–47.

—— 1958b. Experimental studies of mimicry in some North American butterflies. II. *Battus philenor* and *Papilio troilus*, *P. polyxenes* and *P. glaucus*. Evolution 12: 123–36.

—— 1960. Experimental studies of mimicry. IV. The reactions of starlings to different proportions of models and mimics. Am. Naturalist 64: 271–82.

Brower, L. P., J. V. Z. Brower, and P. W. Westcott. 1960. Experimental studies of mimicry. V. The reactions of toads (*Bufo terrestris*) to bumblebees (*Bombus americanorum*) and their robberfly mimics (*Mallophora bomboides*), with a discussion of aggressive mimicry. Am. Naturalist 64: 343–55.

Brücher, H. 1943. Experimentelle Untersuchungen über den Selektionswert künstlich erzeugter Mutanten von *Antirrhinum majus*. Z. Botan. 39: 1–47.

Brues, C. T., A. L. Melander, and F. M. Carpenter. 1954. Classification of Insects. 2d ed. Bull. Museum Comp. Zool., Harvard, Vol. 108.

Buchsbaum, R. 1948. Animals Without Backbones. Chicago.

Buell, K. M. 1953. Developmental morphology in Dianthus. III. Seed failure following interspecific crosses. Am. J. Botany 40: 116–23.

Buffon, C. de. 1749. Histoire Naturelle. Vol. 2. Paris.

Bunge, M. 1959. Causality; The Place of the Causal Principle in Modern Science. Cambridge, Massachusetts.

Bungenberg de Jong, C. M. 1957. Polyploidy in animals. Bibliographia Genet. 17: 111–228.

Burma, B. H. 1954. Reality, existence, and classification: a discussion of the species problem. Madroño 12: 193–209.

Cain, A. J. 1954. Animal Species and their Evolution. London.

Calvin, M. 1956. Chemical evolution and the origin of life. Am. Scientist 44: 248–63.

—— 1959. Evolution of enzymes and the photosynthetic apparatus. Science 130: 1170–74.

Carson, H. L. 1959. Genetic conditions which promote or retard the formation of species. Cold Spring Harbor Symposia Quant. Biol. 24: 87–105.

Castle, W. E. 1903. Mendel's Principle of Heredity. Cambridge.

Chapman, F. 1929. My Tropical Air Castle; Nature Studies in Panama. New York.

Clark, E., L. R. Aronson, and M. Gordon. 1954. Mating behavior patterns in two sympatric species of xiphophorin fishes; their inheritance and significance in sexual isolation. Bull. Am. Museum Nat. Hist. 103: 135–226.

Clarke, C. A., and P. M. Sheppard. 1959–60. The genetics of *Papilio dardanus*, Brown. I–III. Genetics 44: 1347–58, 45: 439–57, 683–98.

—— 1960a. Evolution of mimicry in the butterfly, *Papilio dardanus*. Heredity 14: 163–73.

—— 1960b. Super-genes and mimicry. Heredity 14: 175–85.

—— 1962. Disruptive selection and its effect on a metrical character in the butterfly *Papilio dardanus*. Evolution 16: 214–26.

Clausen, J. 1951. Stages in the Evolution of Plant Species. Ithaca, New York.

Clausen, J., and Wm. M. Hiesey. 1958. Experimental studies on the nature of species. IV. Genetic structure of ecological races. Carnegie Inst. Wash. Publ. No. 615.

Clausen, J., D. D. Keck, and Wm. M. Hiesey. 1939. The concept of species based on experiment. Am. J. Botany 26: 103–6.

—— 1940. Experimental studies on the nature of species. I. Effect of varied environments on western North American plants. Carnegie Inst. Wash. Publ. No. 520.

—— 1943. Experimental taxonomy. Carnegie Inst. Wash. Yearbook 42: 91–100.

—— 1945. Experimental studies on the nature of species. II. Plant evolution through amphiploidy and autoploidy with examples from the Madiinae. Carnegie Inst. Wash. Publ. No. 564.

Clausen, R. E., and D. R. Cameron. 1944. Inheritance in *Nicotiana tabacum*. XVIII. Monosomic analysis. Genetics 29: 447–77.

Cleveland, L. R. 1949. The whole life cycle of chromosomes and their coiling systems. Trans. Am. Phil. Soc. 39: 1–100.

Colwell, R. N. 1951. The use of radioactive isotopes in determining spore distribution patterns. Am. J. Botany 38: 511–23.

Commoner, B. 1961. In defense of biology. Science 133: 1745–48.

Cooper, D. M., and Th. Dobzhansky. 1956. Studies on the ecology of Drosophila in the Yosemite region of California. I. The occurrence of species of Drosophila in different life zones and at different seasons. Ecology 37: 526–33.

Cordeiro, A. R., and Th. Dobzhansky. 1954. Combining ability of certain chromosomes in *Drosophila willistoni* and invalidation of the "wild-type" concept. Am. Naturalist 88: 75–86.

Cott, H. B. 1940. Adaptive Coloration in Animals. London.

Crick, F. H. C. 1954. The structure of the hereditary material. Sci. American, October, 1954.

Cronquist, A. 1951. Orthogenesis in evolution. Research Studies State Coll. Wash. 19: 3–18.

Crosby, J. L. 1956. A suggestion concerning the possible role of plasmagenes in the inheritance of acquired adaptations. J. Genet. 54: 1–8.

Crow, J. F. 1960. Genetics Notes. 4th ed. Minneapolis, Minnesota.

Cunha, A. B. da. 1951. Modification of the adaptive values of chromosomal types in *Drosophila pseudoobscura* by nutritional variables. Evolution 5: 395–404.

—— 1955. Chromosomal polymorphism in the Diptera. Advances in Genet. 7: 93–138.

—— 1960. Chromosomal variation and adaptation in insects. Ann. Rev. Entomol. 5: 85–110.

Cunha, A. B. da, and Th. Dobzhansky. 1954. A further study of chromosomal polymorphism in *Drosophila willistoni* in its relation to the environment. Evolution 8: 119–34.

Cunha, A. B. da, Th. Dobzhansky, O. Pavlovsky, and B. Spassky. 1959. Genetics of natural populations. XXVIII. Supplementary data on the chromosomal polymorphism in *Drosophila willistoni* in its relation to the environment. Evolution 13: 389–404.

Cunha, A. B. da, A. M. E. Shehata, and W. de Oliveira. 1957. A study of the diets and nutritional preferences of tropical species of Drosophila. Ecology 38: 98–106.

Darlington, C. D. 1932, 1937a. Recent Advances in Cytology. 1st and 2nd ed. London.

—— 1937b. The early hybridisers and the origins of genetics. Herbertia 4: 63–69.

—— 1939. The Evolution of Genetic Systems. 1st ed. Cambridge.

—— 1953. The Facts of Life. London.

—— 1956*a*. Natural populations and the breakdown of classical genetics. Proc. Royal Soc. London B 145: 350–64.

—— 1956*b*. Chromosome Botany. London.

Darlington, C. D., and E. K. Janaki-Ammal. 1945. Chromosome Atlas of Cultivated Plants. London.

Darlington, C. D., and K. Mather. 1949. The Elements of Genetics. London.

Darlington, C. D., and A. P. Wylie. 1955. Chromosome Atlas of Flowering Plants. London.

Darwin, Ch. 1859, 1872. On the Origin of Species by means of Natural Selection. 1st and 6th ed. London.

—— 1868. The Variation of Animals and Plants under Domestication. 2 vols. London.

—— 1871. The Descent of Man and Selection in Relation to Sex. London.

—— 1876. The Effects of Cross and Self-fertilisation in the Vegetable Kingdom. London.

Davidson, J. F. 1954. A dephlogisticated species concept. Madroño 12: 246–51.

Demerec, M. 1929. Cross sterility in maize. Z. induktive Abstammungs-u. Vererbungslehre 50: 281–91.

DeVries, H. 1906. Species and Varieties; their Origin by Mutation. London.

Diels, L. 1921. Die Methoden der Phytographie und der Systematik der Pflanzen. Abderhaldens Handbuch der biologischen Arbeitsmethoden, Abt. 11, Teil 1, Heft 2. Berlin and Vienna.

Diesselhorst, G. 1951. Erkennen des Geschlechts und Paarbildung bei der Goldammer (*Emberiza. c. citrinella* L.). Ornithol. Ber. 3: 69–112.

Dillon, L. S. 1962. Comparative cytology and the evolution of life. Evolution 16: 102–17.

Dobzhansky, Th. 1937, 1951*a*. Genetics and the Origin of Species. 1st and 3rd ed. New York.

—— 1943. Genetics of natural populations. IX. Temporal changes in the composition of populations of *Drosophila pseudoobscura*. Genetics 28: 162–86.

—— 1947*a*. Genetics of natural populations. XIV. A response of certain gene arrangements in the third chromosome of *Drosophila pseudoobscura* to natural selection. Genetics 32: 142–60.

—— 1947*b*. A directional change in the genetic constitution of a natural population of *Drosophila pseudoobscura*. Heredity 1: 53–64.

—— 1947*c*. Effectiveness of intraspecific and interspecific matings in *Drosophila pseudoobscura* and *Drosophila persimilis*. Am. Naturalist 81: 66–71.

—— 1948*a*. Genetics of natural populations. XVI. Altitudinal and seasonal changes produced by natural selection in certain populations of *Drosophila pseudoobscura* and *Drosophila persimilis*. Genetics 33: 158–76.

—— 1948*b*. Genetics of natural populations. XVIII. Experiments on chromosomes of *Drosophila pseudoobscura* from different geographic regions. Genetics 33: 588–602.

—— 1949. Observations and experiments on natural selection in Drosophila. Hereditas, suppl. vol. 1949: 210–24.

—— 1950. Mendelian populations and their evolution. Am. Naturalist 84: 401–18.

—— 1951*a*. (See Dobzhansky, 1937 and 1951*a*, above.)

—— 1951*b*. Experiments on sexual isolation in Drosophila. X. Reproductive isolation between *Drosophila pseudoobscura* and *Drosophila persimilis* under natural and under laboratory conditions. Proc. Natl. Acad. Sci. U.S. 37: 792–96.

—— 1954. Evolution as a creative process. Caryologia, suppl. vol. 1954: 435–49.

—— 1955*a*. Evolution, Genetics, and Man. New York.

—— 1955*b*. A review of some fundamental concepts and problems of population genetics. Cold Spring Harbor Symposia Quant. Biol. 20: 1–15.

—— 1955*c*. "The Genetic Basis of Systematic Categories," in, Biological Systematics. Biology Colloquium, Oregon State College, Corvallis, Oregon.

—— 1958. "Species after Darwin," in, S. A. Barnett, ed., A Century of Darwin. London.

—— 1960. Evolutionism and man's hope. The Sewanee Review, April, 1960.

Dobzhansky, Th., and A. B. da Cunha. 1955. Differentiation of nutritional preferences in Brazilian species of Drosophila. Ecology 36: 34–39.

Dobzhansky, Th., L. Ehrman, and O. Pavlovsky. 1957. *Drosophila insularis*, a new sibling species of the *willistoni* group. Studies in Genet. Drosophila Univ. Texas 9: 39–47.

Dobzhansky, Th., and C. Epling. 1944. Contributions to the genetics, taxonomy, and ecology of *Drosophila pseudoobscura* and its relatives. Carnegie Inst. Wash. Publ. No. 554.

Dobzhansky, Th., and H. Levene. 1948. Genetics of natural populations. XVII. Proof of operation of natural selection in wild populations of *Drosophila pseudoobscura*. Genetics 33: 537–47.

—— 1951. Development of heterosis through natural selection in experimental populations of *Drosophila pseudoobscura*. Am. Naturalist 85: 247–64.

Dobzhansky, Th., and O. Pavlovsky. 1953. Indeterminant outcome of certain experiments on Drosophila populations. Evolution 7: 198–210.

—— 1957. An experimental study of interaction between genetic drift and natural selection. Evolution 11: 311–19.

Dobzhansky, Th., and B. Spassky. 1947. Evolutionary changes in laboratory cultures of *Drosophila pseudoobscura*. Evolution 1: 191–216.

—— 1953. Genetics of natural populations. XXI. Concealed variability in two sympatric species of Drosophila. Genetics 38: 471–84.

Dobzhansky, Th., B. Spassky, and N. Spassky. 1952. A comparative study of mutation rates in two ecologically diverse species of Drosophila. Genetics 37: 650–64.

Dobzhansky, Th., and N. Spassky. 1954. Environmental modification of heterosis in *Drosophila pseudoobscura*. Proc. Natl. Acad. Sci. U.S. 40: 407–15.

Dobzhansky, Th., and A. H. Sturtevant, 1938. Inversions in the chromosomes of *Drosophila pseudoobscura*. Genetics 23: 28–64.

Dobzhansky, Th., and S. Wright. 1947. Genetics of natural populations. XV. Rate of diffusion of a mutant gene through a population of *Drosophila pseudoobscura*. Genetics 32: 303–24.

Dodson, C. H., and G. P. Frymire. 1961. Preliminary studies in the genus Stanhopea (Orchidaceae). Ann. Missouri Botan. Garden 48: 137–72.

Downs, T. 1961. A study of variation and evolution in Miocene Merychippus. Contrib. in Science, Los Angeles County Museum, No. 45.

Dronamraju, K. R. 1960. Selective visits of butterflies to flowers: a possible factor in sympatric speciation. Nature 186: 178.

Dunbar, M. J. 1960. The evolution of stability in marine environments. Natural selection at the level of the ecosystem. Am. Naturalist 94: 129–36.

Ehrlich, P. R. 1958. The comparative morphology, phylogeny and higher classification of the butterflies (Lepidoptera: Papilionoidea). Univ. Kansas Sci. Bull. 39: 305–70.

—— 1961. Intrinsic barriers to dispersal in checkerspot butterfly. Science 134: 108–9.

Einstein, A., and L. Infeld. 1938. The Evolution of Physics. New York.

Epling, C. 1947a. Actual and potential gene flow in natural populations. Am. Naturalist 81: 104–15.

—— 1947b. Natural hybridization of *Salvia apiana* and *S. mellifera*. Evolution 1: 69–78.

Epling, C., D. F. Mitchell, and R. H. T. Mattoni. 1953. On the role of inversions in wild populations of *Drosophila pseudoobscura*. Evolution 7: 342–65.

Fernald, M. L. 1950. Gray's Manual of Botany. New York.

Finkner, M. D. 1954. The effect of dual pollinations in upland cotton stocks differing in genotype. Agron. J. 46: 124–28.

Fischer, A. G. 1960. Latitudinal variations in organic diversity. Evolution 14: 64–81.

Fisher, R. A. 1930, 1958. The Genetical Theory of Natural Selection. 1st ed., Oxford; 2d ed., New York.

Fisher, R. A., and E. B. Ford. 1947. The spread of a gene in natural conditions in a colony of the moth *Panaxia dominula* L. Heredity 1: 143–74.

Ford, E. B. 1955. Moths. London.

Fowler, W. A. 1956. The origin of the elements. Sci. American, September, 1956.

Fraenkel-Conrat, H. 1956. Rebuilding a virus. Sci. American, June, 1956.

Frank, P. 1957. Philosophy of Science. Englewood Cliffs, New Jersey.

Gaffron, H. 1960. "The Origin of Life," in S. Tax, ed., Evolution After Darwin. Chicago.

Gajewski, W. 1953. Some observations on disturbances of floral development in Geum species and hybrids. (In Polish with English summary.) Acta Soc. Botan. Polon. 22: 587–604.

Gamov, G. 1955a. The Creation of the Universe. New York.

—— 1955b. "Modern Cosmology," in The New Astronomy. (Scientific American, ed.) New York.

Gates, R. R. 1951. The taxonomic units in relation to cytogenetics and gene ecology. Am. Naturalist, 85: 31–50.

Gause, G. F. 1934. The Struggle for Existence. Baltimore.

Gerassimova, H. 1939. Chromosome alterations as a factor of divergence of forms. I. New experimentally produced strains of *C. tectorum* which are physiologically isolated from the original forms owing to reciprocal translocation. Compt. rend. acad. sci. U.R.S.S. 25: 148–54.

Gering, R. L. 1953. Structure and function of the genitalia in some American agelenid spiders. Smithsonian Inst. Publs. Misc. Collections 121.

Gershenson, S. 1945. Evolutionary studies on the distribution and dynamics of melanism in the hamster (*Cricetus cricetus* L.). Genetics 30: 207–51.

Gerstel, D. U. 1954. A new lethal combination in interspecific cotton hybrids. Genetics 39: 628–39.

Gilmour, J. S. L. 1961. "Taxonomy," in A. M. Macleod and L. S. Cobley, eds., Contemporary Botanical Thought. Chicago.

Glass, B. 1954. Genetic changes in human populations, especially those due to gene flow and genetic drift. Advances in Genet. 6: 95–139.

Glass, B., M. S. Sacks, E. F. Jahn, and C. Hess. 1952. Genetic drift in a religious isolate: an analysis of the causes of variation in blood groups and other gene frequencies in a small population. Am. Naturalist 86: 145–59.

Gleason, H. A. 1926. The individualistic concept of the plant association. Bull. Torrey Botan. Club 53: 7–26.

Goetsch, W. 1957. The Ants. Translation. Ann Arbor, Michigan.

Goldschmidt, R. B. 1940. The Material Basis of Evolution. New Haven, Connecticut.

—— 1952. Homoeotic mutants and evolution. Acta Biotheoret. 10: 87–104.

—— 1953. Experiments with a homoeotic mutant, bearing on evolution. J. Exptl. Zool. 123: 79–114.

—— 1955. Theoretical Genetics. Berkeley, California.

Grant, V. 1949. Pollination systems as isolating mechanisms in flowering plants. Evolution 3: 82–97.

—— 1950a. The flower constancy of bees. Bot. Rev. 16: 379–98.

—— 1950b. Genetic and taxonomic studies in Gilia. I. *Gilia capitata*. Aliso 2: 239–316.

—— 1952a. Genetic and taxonomic studies in Gilia. III. The *Gilia tricolor* complex. Aliso 2: 375–88.

—— 1952b. Isolation and hybridization between *Aquilegia formosa* and *A. pubescens*. Aliso 2: 341–60.

—— 1953a. "Pollination," in Encyclopedia Americana, Vol. 22.

—— 1953b. The role of hybridization in the evolution of the Leafy-stemmed Gilias. Evolution 7: 51–64.

—— 1954a. Genetic and taxonomic studies in Gilia. IV. *Gilia achilleaefolia*. Aliso 3: 1–18.

—— 1954b. Genetic and taxonomic studies in Gilia. VI. Interspecific relationships in the Leafy-stemmed Gilias. Aliso 3: 35–49.

—— 1956a. Chromosome repatterning and adaptation. Advances in Genet. 8: 89–107.

—— 1956b. The genetic structure of races and species in Gilia. Advances in Genet. 8: 55–87.

—— 1956c. The influence of breeding habit on the outcome of natural hybridization in plants. Am. Naturalist 90: 319–22.

—— 1956d. The development of a theory of heredity. Am. Scientist 44: 158–79.

—— 1957. "The Plant Species in Theory and Practice," in E. Mayr, ed., The Species Problem. Washington, D.C.

—— 1958. The regulation of recombination in plants. Cold Spring Harbor Symposia Quant. Biol. 23: 337–63.

—— 1959. Natural History of the Phlox Family. The Hague.

Grant, A., and V. Grant. 1956. Genetic and taxonomic studies in Gilia. VIII. The Cobwebby Gilias. Aliso 3: 203–87.

Grant, V., and A. Grant. 1954. Genetic and taxonomic studies in Gilia. VII. The Woodland Gilias. Aliso 3: 59–91.

—— 1960. Genetic and taxonomic studies in Gilia. XI. Fertility relationships of the diploid Cobwebby Gilias. Aliso 4: 435–81.

Gray, A. P. 1954. Mammalian Hybrids. Commonwealth Agricultural Bureaux, Farnham Royal, England.

—— 1958. Bird Hybrids. Commonwealth Agricultural Bureaux, Farnham Royal, England.

Green, M. M. 1957. Reverse mutation in Drosophila and the status of the particulate gene. Genet. 29: 1–38.

Greenshields, J. E. R. 1954. Embryology of interspecific crosses in Melilotus. Can. J. Botany 32: 447–65.

Gregory, J. W. 1928. "The Nature of Species," in F. Mason, ed., Creation by Evolution. New York.

Grüneberg, H. 1947. Animal Genetics and Medicine. New York and London.

Gustafsson, Å. 1948. Polyploidy, life-form and vegetative reproduction. Hereditas 34: 1–22.

—— 1951. Mutations, environment and evolution. Cold Spring Harbor Symposia Quant. Biol. 16: 263–81.

Hairston, N. G. 1958. Observations on the ecology of Paramecium, with comments on the species problem. Evolution 12: 440–50.

Hairston, N. G., F. E. Smith, and L. B. Slobodkin. 1960. Community structure, population control, and competition. Am. Naturalist 64: 421–25.

Håkansson, A. 1952. Seed development after 2X, 4X crosses in *Galeopsis pubescens*. Hereditas 38: 425–48.

—— 1956. Seed development of *Brassica oleracea* and *B. rapa* after certain reciprocal pollinations. Hereditas 42: 373–96.

Haldane, J. B. S. 1932. The Causes of Evolution. London.

—— 1954. The Biochemistry of Genetics. London.

—— 1957. The cost of natural selection. J. Genet. 55: 511–24.

Haldane, J. S. 1931. The Philosophical Basis of Biology. London.

—— 1935. The Philosophy of a Biologist. Oxford.

Harper, J. L. 1958. Famous plants. The buttercup. New Biology 26: 1–19.

Harper, J. L., J. N. Clatworthy, I. H. McNaughton, and G. R. Sagar. 1961. The evolution and ecology of closely related species living in the same area. Evolution 15: 209–27.

Hartman, W. D. 1957. Ecological niche differentiation in the boring sponges (Clionidae). Evolution 11: 294–97.

Haskins, C. P., and E. F. Haskins. 1949. The role of sexual selection as an isolating mechanism in three species of poeciliid fishes. Evolution 3: 160–69.

Hayase, H. 1950. The pollen tube growth in interspecific crosses. Japan. J. Genet. 25: 181–90.

Heiser, C. B., W. C. Martin, and D. M. Smith. 1962. Species crosses in Helianthus. I. Diploid species. Brittonia 14: 137–47.

Heslop-Harrison, J. 1958. "Ecological Variation and Ethological Isolation," in O. Hedberg, ed., Systematics of Today. Uppsala, Sweden.

Hiraizumi, Y., L. Sandler, and J. F. Crow. 1960. Meiotic drive in natural populations of *Drosophila melanogaster*. III. Populational implications of the segregation-distorter locus. Evolution 14: 433–44.

Holm, R. W. 1950. The American species of Sarcostemma R. Br. (Asclepiadaceae). Ann. Missouri Botan. Garden 37: 477–560.

Holtum, R. E. 1953. Evolutionary trends in an equatorial climate. Symposia Soc. Exptl. Biol. No. 7: 159–73.

Horowitz, N. H. 1945. On the evolution of biochemical syntheses. Proc. Natl. Acad. Sci. U.S. 31: 153–57.

Hovanitz, Wm. 1953. Polymorphism and evolution. Symposia Soc. Exptl. Biol. 7: 238–53.

Hoyle, F. 1956. The steady-state universe. Sci. American, September, 1956.

—— 1961. The age of the galaxy. Am. Scientist 49: 188–91.

Humboldt, F. H. A. von. 1850. Views of Nature. Translation. London.

Hutchinson, G. E. 1957. Concluding remarks. Cold Spring Harbor Symposia Quant. Biol. 22: 415–27.

—— 1959. Homage to Santa Rosalia or why are there so many kinds of animals? Am. Naturalist 93: 145–59.

Hutchinson, G. E., and R. MacArthur. 1959. A theoretical ecological model of size distributions among species of animals. Am. Naturalist 93: 117–26.

Huxley, J. S. 1938. "The Present Standing of the Theory of Sexual Selection," in G. R. de Beer, ed., Evolution. Oxford.

—— 1943. Evolution; The Modern Synthesis. New York and London.

—— 1945. Chance and anti-chance in evolution. New Republic, December 10, 1945.

Ives, P. T. 1950. The importance of mutation rate genes in evolution. Evolution 4: 236–52.

Johannsen, W. 1911. The genotype conception of heredity. Am. Naturalist 45: 129–59.

Johnston, E. F., R. Bogart, and A. W. Lindquist. 1954. The resistance to DDT by houseflies. J. Heredity 45: 177–82.

Jones, G. N. 1941. How many species of plants are there? Science 94: 234.

Jones, W. T. 1952. A History of Western Philosophy. New York.

Kant, I. 1775. Von den verschiedenen Rassen des Menschen. Gesammelte Schriften, Vol. 2. Berlin.

Kearney, T. H., and G. J. Harrison. 1932. Pollen antagonism in cotton. J. Agr. Research 44: 191–226.

Kerner, A. 1894–1895. The Natural History of Plants. 2 vols. Translation. London.

Kerr, W. E., and S. Wright. 1954. Experimental studies of the distribution of gene frequencies in very small populations of *Drosophila melanogaster*. I. Forked. Evolution 8: 172–77.

Kettlewell, H. B. D. 1956. Further selection experiments on industrial melanism in the Lepidoptera. Heredity 10: 287–301.

Khoshoo, T. N. 1958. Biosystematics of *Sisymbrium irio* complex. I. Modifications in phenotype. Caryologia 11: 109–32.

Khoshoo, T. N., and V. B. Sharma. 1959. Biosystematics of *Sisymbrium irio* complex. VI. Reciprocal pollinations and seed failure. Caryologia 12: 71–97.

King, W. V., and J. B. Gahan. 1949. Failure of DDT to control house flies. J. Econ. Entomol. 42: 405–9.

Kiss, A., and T. Rajhathy. 1956. Untersuchungen über die Kreuzbarkeit innerhalb des Subtribus Triticinae. Züchter 26: 127–36.

Klopfer, P. H., and R. H. MacArthur. 1961. On the causes of tropical species diversity: niche overlap. Am. Naturalist 95: 223–26.

Knight, G. R., A. Robertson, and C. H. Waddington. 1956. Selection for sexual isolation within a species. Evolution 10: 14–22.

Knopf, A. 1957. Measuring geologic time. Sci. Monthly 85: 225–36.

Knuth, P. 1906–9. Handbook of Flower Pollination. 3 vols. Translation. Oxford.

Koopman, K. F. 1950. Natural selection for reproductive isolation between *Drosophila pseudoobscura* and *Drosophila persimilis*. Evolution 4: 135–48.

Kruckeberg, A. R. 1957. Variation in fertility of hybrids between isolated populations of the serpentine species, *Streptanthus glandulosus* Hook. Evolution 11: 185–211.

Kulp, J. L. 1961. Geologic time scale. Science 133: 1105–14.

Lack, D. 1947. Darwin's Finches. Cambridge.

Lamarck, J. B. P. de. 1809. Philosophie Zoologique. Paris.

—— 1815–22. Histoire Naturelle des Animaux sans Vertèbres. Paris.

Lammerts, W. E. 1932. Inheritance of monosomics in *Nicotiana rustica*. Genetics 17: 689–96.

Lamotte, M. 1951. Recherches sur la structure génétique des populations naturelles de *Cepaea nemoralis* (L.). Bull. biol. France et Belg., suppl. 35: 1–238.

—— 1959. Polymorphism of natural populations of *Cepaea nemoralis*. Cold Spring Harbor Symposia Quant. Biol. 24: 65–86.

Lamprecht, H. 1954. Selektive Befruchtung im Lichte des Verhaltens interspezifischer Gene in Linien und Kreuzungen. Agr. Hort. Genet. 12: 1–37.

Latimer, H. 1958. A study of the breeding barrier between *Gilia australis* and

Gilia splendens. Ph.D. thesis, Claremont University College, Claremont, California.

Laughlin, W. S. 1950. Blood groups, morphology and population size of the Eskimos. Cold Spring Harbor Symposia Quant. Biol. 15: 165–73.

Lea, D. E. 1955. Actions of Radiations on Living Cells. 2d ed. Cambridge.

Leng, E. R. 1960. Long-term selection of corn for oil and protein content. Mimeographed annual report, Illinois Agr. Exptl. Station, Urbana, Illinois.

Lenz, L. W. 1956. Development of the embryo sac, endosperm and embryo in *Iris munzii* and the hybrid *I. munzii* × *sibirica* "Caesar's Brother." Aliso 3: 329–43.

Lenz, L. W., and D. E. Wimber. 1959. "Hybridization and Inheritance in Orchids," in C. L. Withner, ed., The Orchids. New York.

Lerner, I. M. 1954. Genetic Homeostasis. London.

—— 1958. The Genetic Basis of Selection. New York.

Levene, H., and Th. Dobzhansky. 1959. Possible genetic difference between the head louse and the body louse (*Pediculus humanus* L.). Am. Naturalist 93: 347–53.

Levine, R. P. 1952. Adaptive responses of some third chromosome types of *Drosophila pseudoobscura*. Evolution 6: 216–33.

Lewis, D. 1954. Annual Report of the Department of Genetics. Ann. Rep. John Innes Hort. Inst. 45: 12–17.

Lewis, H. 1953. The mechanism of evolution in the genus Clarkia. Evolution 7: 1–20.

—— 1962. Catastrophic selection as a factor in speciation. Evolution 16: 257–71.

Lewis, H., and M. Lewis. 1955. The genus Clarkia. Univ. Calif. (Berkeley) Publs. Botany 20: 241–392.

Lewis, H., and P. H. Raven. 1958. Rapid evolution in Clarkia. Evolution 12: 319–36.

Lewis, H., and M. R. Roberts. 1956. The origin of *Clarkia lingulata*. Evolution 10: 126–38.

Lindley, J. 1831. An Introduction to the Natural System of Botany. New York.

Linsley, E. G., T. Eisner, and A. B. Klots. 1961. Mimetic assemblages of sibling species of lycid beetles. Evolution 15: 15–29.

Linsley, E. G., and J. W. MacSwain. 1958. The significance of floral constancy among bees of the genus Diadasia (Hymenoptera, Anthophoridae). Evolution 12: 219–23.

Lorković, Z. 1958. Some peculiarities of spatially and sexually restricted gene exchange in the *Erebia tyndarus* group. Cold Spring Harbor Symposia Quant. Biol. 23: 319–25.

Lotsy, J. P. 1916a. *Antirrhinum rhinanthoides* mihi, une novelle espèce Linnéenne, obtenue expérimentalement. Arch. néerl. sci.IIIB 3: 195–204.

—— 1916*b*. Evolution by Means of Hybridization. The Hague.

Lull, R. S. 1947. Organic Evolution. 2d ed. New York.

Lundman, B. 1948. Geography of human blood groups (A, B, O system). Evolution 2: 231–37.

MacArthur, R. 1955. Fluctuations of animal populations, and a measure of community stability. Ecology 36: 533–36.

Maheshwari, P. 1957. Hormones in reproduction. Indian J. Genet. and Plant Breeding 17: 386–97.

Mangelsdorf, P. C. 1958*a*. Reconstructing the ancestor of corn. Proc. Am. Phil. Soc. 102: 454–63.

—— 1958*b*. The mutagenic effect of hybridizing maize and teosinte. Cold Spring Harbor Symposia Quant. Biol. 23: 409–21.

Mangelsdorf, P. C., and D. F. Jones. 1926. The expression of Mendelian factors in the gametophyte of maize. Genetics 11: 423–55.

Mangelsdorf, P. C., and R. G. Reeves. 1959. The origin of corn. Botan. Museum Leaflets (Harvard) 18: 329–440.

Manning, A. 1957. Some evolutionary aspects of the flower constancy of bees. Proc. Roy. Phys. Soc. 25: 67–71.

Mason, H. L. 1947. Evolution of certain floristic associations in western North America. Ecol. Monographs 17: 201–10.

—— 1950. Taxonomy, systematic botany and biosystematics. Madroño 10: 193–208.

Mather, K. 1943. Polygenic inheritance and natural selection. Biol. Revs. 18: 32–64.

—— 1947. Species crosses in Antirrhinum. I. Genetic isolation of the species *majus*, *glutinosum* and *orontium*. Heredity 1: 175–86.

—— 1953. The genetical structure of populations. Symposia Soc. Exptl. Biol. No. 7: 66–95.

—— 1955. Polymorphism as an outcome of disruptive selection. Evolution 9: 52–61.

Mayr, E. 1940. Speciation phenomena in birds. Am. Naturalist 74: 249–78.

—— 1942. Systematics and the Origin of Species. New York.

—— 1947. Ecological factors in speciation. Evolution 1: 263–88.

—— 1948. The bearing of the new systematics on genetical problems. The nature of species. Advances in Genet. 2: 205–37.

—— 1950. The role of the antennae in the mating behavior of female Drosophila. Evolution 4: 149–54.

—— 1954. "Change of Genetic Environment and Evolution," in J. Huxley, ed., Evolution as a Process. London.

—— 1955*a*. "The Species As a Systematic and As a Biological Problem," in Biological Systematics. Biology Colloquium, Oregon State College, Corvallis, Oregon.

—— 1955*b*. "Systematics and Modes of Speciation," in Biological Systematics. Biology Colloquium, Oregon State College, Corvallis, Oregon.

—— 1955*c*. Karl Jordan's contribution to current concepts in systematics and evolution. Trans. Royal Entomol. Soc. London 1955: 45–66.

—— 1957*a*. "Species Concepts and Definitions," in E. Mayr, ed., The Species Problem. Washington, D.C.

—— 1957*b*. "Difficulties and Importance of the Biological Species Concept," in E. Mayr, ed., The Species Problem. Washington, D.C.

—— 1959. Where are we? Cold Spring Harbor Symposia Quant. Biol. 24: 1–14.

—— 1960. "The Emergence of Evolutionary Novelties," in S. Tax, ed., Evolution After Darwin. Chicago.

Mayr, E., E. G. Linsley, and R. L. Usinger. 1953. Methods and Principles of Systematic Zoology. New York.

McNaughton, I. H., and J. L. Harper. 1960*a*. The comparative biology of closely related species living in the same area. I. External breeding-barriers between Papaver species. New Phytologist 59: 15–26.

—— 1960*b*. The comparative biology of closely related species living in the same area. II. Aberrant morphology and a virus-like syndrome in hybrids between *Papaver rhoeas* L. and *P. dubium* L. New Phytologist 59: 27–41.

McVeigh, I., and C. J. Hobdy. 1952. Development of resistance by *Micrococcus pyogenes* var. *aureus* to antibiotics: morphological and physiological changes. Am. J. Botany 39: 352–59.

Mendel, G. 1866. Experiments in plant hybridization. Translation, reprinted in various texts. Brünn, Czechoslovakia.

Merrell, D. J. 1954. Sexual isolation between *Drosophila persimilis* and *Drosophila pseudoobscura*. Am. Naturalist 88: 93–99.

Michaelis, P. 1954. Cytoplasmic inheritance in Epilobium and its theoretical significance. Advances in Genet. 6: 287–401.

Moos, J. R. 1955. Comparative physiology of some chromosomal types in *Drosophila pseudoobscura*. Evolution 9: 141–51.

Morgan, T. H., C. B. Bridges, and A. H. Sturtevant. 1925. The Genetics of Drosophila. Bibliographia Genet. Vol. 2.

Mourant, A. E. 1954. The Distribution of the Human Blood Groups. Oxford.

—— 1959. Human blood groups and natural selection. Cold Spring Harbor Symposia Quant. Biol. 24: 57–63.

Muller, C. H. 1952. Ecological control of hybridization in Quercus: a factor in the mechanism of evolution. Evolution 6: 147–61.

Muller, H. J. 1929. The gene as the basis of life. Proc. Intern. Congr. Plant Sci. 1st Congr. Ithaca 1926 1: 897–921.

Müntzing, A. 1929. Cases of partial sterility in crosses within a Linnean species. Hereditas 12: 297–319.

—— 1930. Outlines to a genetic monograph of the genus Galeopsis. Hereditas 13: 185–341.

—— 1935. The evolutionary significance of autopolyploidy. Hereditas 21: 263–378.

—— 1938. Sterility and chromosome pairing in intraspecific Galeopsis hybrids. Hereditas 24: 117–88.

Naudin, C. 1863. Nouvelles recherches sur l'hybridité dans les végétaux. Ann. sci. nat. sér. Bot. (Paris) 19: 180–203.

Naylor, A. F. 1962. Mating systems which could increase heterozygosity for a pair of alleles. Am. Naturalist 96: 51–60.

Needham, J. 1936. Order and Life. New Haven, Connecticut.

Newton, I. 1687. Mathematical Principles of Natural Philosophy. Translation, 1729. Reprinted, 1934, Berkeley, California.

Nishiyama, I. 1928. On hybrids between *Triticum spelta* and two dwarf wheat plants with 40 somatic chromosomes. Botan. Mag. Tokyo 42: 154–77.

Nobs, M. A. 1954. Genetic studies on Mimulus. Carnegie Inst. Wash. Yearbook 53: 157–59.

Oparin, A. I. 1938. The Origin of Life. 2d ed. Translation. New York.

—— 1957. Origin of Life on the Earth. 3rd ed. Translation, New York.

Owen, D. F. 1961. Industrial melanism in North American moths. Am. Naturalist 95: 227–33.

Pätau, K. 1939. Die mathematische Analyse der Evolutionsvorgänge. Z. induktive Abstammungs-u. Vererbungslehre 76: 220–28.

Patterson, J. T., and W. S. Stone. 1952. Evolution in the Genus Drosophila. New York.

Patterson, J. T., W. Stone, S. Bedichek, and M. Suche. 1934. The production of translocations in Drosophila. Am. Naturalist 68: 359–69.

Penrose, L. S. 1959. Self-reproducing machines. Sci. American, June, 1959.

Pontecorvo, G. 1958. Trends in Genetic Analysis. New York.

Poulton, E. B. 1890. The Colours of Animals. London and New York.

Prout, T. 1954. Genetic drift in irradiated experimental populations of *Drosophila melanogaster*. Genetics 39: 529–45.

Rae, A. L. 1956. The genetics of the sheep. Advances in Genet. 8: 189–265.

Raisz, E. 1940. "Map of the Landforms of the United States," in W. W. Atwood, Physiographic Provinces of North America. Boston.

Ramsbottom, J. 1938. Linnaeus and the species concept. Proc. Linnean Soc. London 150: 192–219.

Randolph, L. F. 1955. Cytogenetic aspects of the origin and evolutionary history of corn. Agronomy 5: 16–61.

Ray, P. M. and Chisaki, H. F. 1957. Studies on Amsinckia. III. Aneuploid diversification in the Muricatae. Am. J. Botany 44: 545–54.

Rendel, J. M. 1951. Mating of ebony, vestigial and wild type *Drosophila melanogaster* in light and dark. Evolution 5: 226–30.

Rensch, B. 1959. Evolution Above the Species Level. Translation. New York.

—— 1960. "The Laws of Evolution," in S. Tax, ed., Evolution After Darwin. Chicago.

Rhoades, M. 1941. The genetic control of mutability in maize. Cold Spring Harbor Symposia Quant. Biol. 9: 138–44.

Richards, P. W. 1952. The Tropical Rain Forest; An Ecological Study. Cambridge.

Rick, C. M., and P. Q. Smith. 1953. Novel variation in tomato species hybrids. Am. Naturalist 87: 359–73.

Rizki, M. T. M. 1951. Morphological differences between two sibling species, *Drosophila pseudoobscura* and *Drosophila persimilis*. Proc. Natl. Acad. Sci. U.S. 37: 156–59.

Roberts, M. R., and H. Lewis. 1955. Subspeciation in *Clarkia biloba*. Evolution 9: 445–54.

Robertson, C. 1887. Insect relations of certain asclepiads. Botan. Gaz. 12: 207–16, 244–50.

Roman, H. 1948. Selective fertilization in maize. Genetics 33: 122.

Romanes, G. J. 1896. Darwin, and After Darwin. 2 vols. 2d ed. Chicago.

Ross, H. H. 1956. A Textbook of Entomology. 2d ed. New York.

—— 1957. Principles of natural coexistence indicated by leafhopper populations. Evolution 11: 113–29.

Sagan, C. 1957. Radiation and the origin of the gene. Evolution 11: 40–55.

Sanderson, I. T. 1955. Living Mammals of the World. New York.

Sandler, L., and E. Novitski. 1957. Meiotic drive as an evolutionary force. Am. Naturalist 91: 105–10.

Sauer, J. D. 1951. Studies of variation in the weed genus Phytolacca. II. Latitudinally adapted variants within a North American species. Evolution 5: 273–79.

Savage, J. Evolution. New York. In press.

Savile, D. B. O. 1959. Limited penetration of barriers as a factor in evolution. Evolution 13: 333–43.

—— 1960. Limitations of the competitive exclusion principle. Science 132: 1761.

Sax, K. 1931. Crossing-over and mutation. Proc. Natl. Acad. Sci. U.S. 17: 601–3.

Schmalhausen, I. I. 1949. Factors of Evolution. Translation. Philadelphia.

Schrödinger, E. 1935. Science and the Human Temperament. Translation. New York.

Schwanitz, F. 1956. Grossmutationen. Umschau 56: 45–48.

—— 1957. Spornbildung bei einem Bastard zwischen drei Digitalis-Arten. Biol. Zentr. 76: 226–31.

Schwanitz, F., and H. Schwanitz. 1955. Eine Grossmutation bei *Linaria maroccana* L.: mut. *gratioloides*. Beitr. Biol. Pflanz. 31: 473–97.

Sears, E. R. 1944. Cytogenetic studies with polyploid species of wheat. II. Additional chromosomal aberrations in *Triticum vulgare*. Genetics 29: 232–46.

Sharp, A. J. 1955. Elements in the Tennessee flora with tropical relationships. J. Tenn. Acad. Sci. 30: 53–56.

Sheppard, P. M. 1959. Natural Selection and Heredity. New York.

Sibley, C. G. 1954. Hybridization in the red-eyed towhees of Mexico. Evolution 8: 252–90.

Simpson, G. G. 1944. Tempo and Mode in Evolution. New York.

—— 1949. The Meaning of Evolution. New Haven, Connecticut.

—— 1951, 1961. Horses; The Story of the Horse Family in the Modern World and through Sixty Million Years of Evolution. 1st and 2d eds. Oxford and New York.

—— 1952. How Many Species? Evolution 6: 342.

—— 1953a. The Major Features of Evolution. New York.

—— 1953b. The Baldwin effect. Evolution 7: 110–17.

Simpson, G. G., C. S. Pittendrigh, and L. H. Tiffany. 1957. Life: An Introduction to Biology. New York.

Sinnott, E. W., L. C. Dunn, and Th. Dobzhansky. 1958. Principles of Genetics. 5th ed. New York.

Sinskaja, E. N. 1931. The study of species in their dynamics and interrelation with different types of vegetation. Bull. Appl. Botany Genet. and Plant Breeding Leningrad 25: 1–97.

Sirks, M. J. 1952. Variability in the concept of species. Acta Biotheoret. 10: 11–22.

Skalinska, M. 1958. Seed development after crosses of Aquilegia with Isopyrum. Studies in Plant Physiol., Prague, 213–21.

Slobodkin, L. B. 1961. Growth and Regulation of Animal Populations. New York.

Smith, F. H., and Q. D. Clarkson. 1956. Cytological studies of interspecific hybridization in Iris, subsection Californicae. Am. J. Botany 43: 582–88.

Smith, G. M. 1938, 1955. Cryptogamic Botany. 2 vols. 1st and 2d eds. New York.

Smith, H. H. 1953. Studies on gene systems that differentiate species. Erfelijkh. i. Prakt. 14: 16–17.

—— 1954. Development of morphologically distinct and genetically isolated populations by interspecific hybridization and selection. Caryologia 6, suppl. vol.: 867–70.

Sonneborn, T. M. 1957. "Breeding Systems, Reproductive Methods, and

Species Problems in Protozoa," in E. Mayr, ed., The Species Problem. Washington D.C.

Spector, W. S., ed. 1956. Handbook of Biological Data. Philadelphia and London.

Spencer, H. 1884. First Principles. 5th ed. London.

Spieth, H. T. 1952. Mating behavior within the genus Drosophila (Diptera). Bull. Am. Museum Nat. Hist. 99: 397–474.

Sprague, E. F. 1959. Ecological life history of California species of Pedicularis. Ph.D. thesis, Claremont University College, Claremont, California.

—— 1962. Pollination and evolution in Pedicularis (Scrophulariaceae). Aliso 5: 181–209.

Squillace, A. E., and R. T. Bingham. 1958. Localized ecotypic variation in western white pine. Forest Sci. 4: 20–34.

Srb, A. M., and R. D. Owen. 1952. General Genetics. San Francisco.

Stalker, H. D. 1956. A case of polyploidy in Diptera. Proc. Natl. Acad. Sci. U.S. 42: 194–99.

—— 1961. The genetic systems modifying meiotic drive in Drosophila paramelanica. Genetics 46: 177–202.

Stebbins, G. L. 1938. Cytological characteristics associated with the different growth habits in the dicotyledons. Am. J. Botany 25: 189–98.

—— 1949. "Rates of Evolution in Plants," in G. Jepsen et al., eds., Genetics, Paleontology, and Evolution. Princeton, New Jersey.

—— 1950. Variation and Evolution in Plants. New York.

—— 1957a. Self fertilization and population variability in the higher plants. Am. Naturalist 91: 337–54.

—— 1957b. The hybrid origin of microspecies in the Elymus glaucus complex. Cytologia, suppl. vol.: 336–40.

—— 1958. The inviability, weakness, and sterility of interspecific hybrids. Advances in Genet. 9: 147–215.

—— 1959. The role of hybridization in evolution. Proc. Am. Phil. Soc. 103: 231–51.

—— 1960. "The Comparative Evolution of Genetic Systems," in S. Tax, ed., Evolution after Darwin. Chicago.

Stebbins, G. L., and L. Ferlan. 1956. Population variability, hybridization, and introgression in some species of Ophrys. Evolution 10: 32–46.

Stebbins, G. L., and A. Vaarama. 1954. Artificial and natural hybrids in the Gramineae, tribe Hordeae. VII. Hybrids and allopolyploids between Elymus glaucus and Sitanion spp. Genetics 39: 378–95.

Stebbins, R. C. 1957. Intraspecific sympatry in the lungless salamander Ensatina eschscholtzi. Evolution 11: 265–70.

Stent, G. S. 1955. "The Reproduction of Viruses," in The Physics and Chemistry of Life (Scientific American, ed.). New York.

Stephens, S. G. 1946. The genetics of "corky." I. The New World alleles and their possible role as an interspecific isolating mechanism. J. Genet. 47: 150–61.

—— 1950. The genetics of "corky." II. Further studies on its genetic basis in relation to the general problem of interspecific isolating mechanisms. J. Genet. 50: 9–20.

—— 1956. The composition of an open pollinated segregating cotton population. Am. Naturalist 90: 25–39.

Stern, C. 1949, 1960. Principles of Human Genetics. 1st and 2d eds. San Francisco.

—— 1958. Selection for subthreshold differences and the origin of pseudo-exogenous adaptations. Am. Naturalist 92: 313–16.

—— 1959. Variation and hereditary transmission. Proc. Am. Phil. Soc. 103: 183–89.

Stirton, R. A. 1947. Observations on evolutionary rates in hypsodonty. Evolution 1: 32–41.

Straw, R. M. 1955. Hybridization, homogamy and sympatric speciation. Evolution 9: 441–44.

—— 1956. Floral isolation in Penstemon. Am. Naturalist 90: 47–53.

Streams, F. A., and D. Pimentel. 1961. Effects of immigration on the evolution of populations. Am. Naturalist 95: 201–10.

Streisinger, G. 1948. Experiments on sexual isolation in Drosophila. IX. Behavior of males with etherized females. Evolution 2: 187–88.

Stubbe, H. 1952. Über einige theoretische und praktische Fragen der Mutationsforschung. Abhandl. sächs. Akad. Wiss. Leipzig, Math. naturw. Kl. 47: 3–23.

—— 1959. Considerations on the genetical and evolutionary aspects of some mutants of Hordeum, Glycine, Lycopersicon and Antirrhinum. Cold Spring Harbor Symposia Quant. Biol. 24: 31–40.

—— 1960. Mutanten der Wildtomate *Lycopersicon pimpinellifolium* (Jusl.) Mill. Die Kulturpflanze (Berlin) 8: 110–37.

Stubbe, H., and F. von Wettstein. 1941. Über die Bedeutung von Kleinund Grossmutationen in der Evolution. Biol. Zentr. 61: 265–97.

Sturtevant, A. H. 1939. High mutation frequency induced by hybridization. Proc. Natl. Acad. Sci. U.S. 25: 308–10.

Sturtevant, A. H., and Th. Dobzhansky. 1936. Geographical distribution and cytology of "sex-ratio" in *Drosophila pseudoobscura* and related species. Genetics 21: 473–90.

Sumner, F. B. 1935. Evidence for the protective value of changeable coloration in fishes. Am. Naturalist 69: 245–66.

Swanson, C. P. 1957. Cytology and Cytogenetics. Englewood Cliffs, New Jersey.

Symposium. 1942. Levels of integration in biological and social systems. Biological Symposia, Vol. 8. R. Redfield, ed.

Symposium. 1956. Genetic mechanisms: structure and function. Cold Spring Harbor Symposia Quant. Biol., Vol. 21.

Symposium. 1959. The Origin of Life on the Earth. A. Oparin, ed. Moscow.

Ter-Avanesjan, D. V. 1949. The role of the number of pollen grains per flower in fertilization in plants. (In Russian.) Bull. Appl. Botany Genet. Plant Breeding Leningrad 28: 119–33. See Plant Breeding Abstracts 25 (1955): 491.

Thoday, J. M., and T. B. Boam. 1959. Effects of disruptive selection. II. Polymorphism and divergence without isolation. Heredity 13: 205–18.

Timofeeff-Ressovsky, N. W. 1934. The experimental production of mutations. Biol. Revs. 9: 411–57.

—— 1940. "Mutations and Geographical Variation," in J. Huxley, ed., The New Systematics. Oxford.

Trimen, R. 1869. On some remarkable mimetic analogies among African butterflies. Trans, Linnean Soc., Zool. 26: 497–522.

Turesson, G. 1922. The genotypical response of the plant species to its habitat. Hereditas 3: 211–350.

Tyler, S. A., and E. S. Barghoorn. 1954. Occurrence of structurally preserved plants in pre-Cambrian rocks of the Canadian shield. Science 119: 606–8.

Urey, H. C. 1952. The Planets. New Haven, Connecticut.

—— 1960. The origin and nature of the moon. Endeavour 19: 87–99.

Valentine, D. H. 1955. Studies in British primulas. IV. Hybridization between *Primula vulgaris* Huds. and *P. veris* L. New Phytologist 54: 70–80.

—— 1961. "Evolution in the Genus Primula," in P. J. Wanstall, ed., A Darwin Centenary. Arbroath, Scotland.

Valentine, D. H., and A. Löve. 1958. Taxonomic and biosystematic categories. Brittonia 10: 153–66.

Viosca, P. 1935. The irises of southeastern Louisiana. Bull. Am. Iris. Soc. 57: 3–56.

Volpe, E. P. 1954. Hybrid inviability between *Rana pipiens* from Wisconsin and Mexico. Tulane Studies in Zool. 1: 111–23.

—— 1955. Intensity of reproductive isolation between sympatric and allopatric populations of *Bufo americanus* and *Bufo fowleri*. Am. Naturalist 89: 303–17.

—— 1959. Hybridization of *Bufo valliceps* with *Bufo americanus* and *Bufo terrestris*. Texas J. Sci. 11: 335–42.

—— 1960. Interaction of mutant genes in the leopard frog. J. Heredity 51: 151–55.

Waddington, C. H. 1953. Genetic assimilation of an acquired character. Evolution 7: 118–26.

—— 1956. Genetic assimilation of the bithorax phenotype. Evolution 10: 1–13.

—— 1957. The Strategy of the Genes. London.

—— 1960. "Evolutionary adaptation," in S. Tax, ed., Evolution after Darwin. Chicago.

Waddington, C. H., B. Woolf, and M. M. Perry. 1953. Environmental selection by Drosophila mutants. Evolution 8: 89–96.

Wald, G. 1954. The origin of life. Sci. American, August, 1954.

Wallace, A. R. 1878. Tropical Nature. London.

—— 1889. Darwinism; An Exposition of the Theory of Natural Selection. London.

Wallace, B., and Th. Dobzhansky. 1959. Radiation, Genes, and Man. New York.

Warburton, F. E. 1956. Genetic assimilation: adaptation versus adaptability. Evolution 10: 337–39.

Watson, J. D., and F. H. C. Crick. 1953. Molecular structure of nucleic acids. Nature 171: 737–38.

Weaver, J. B. 1957. Embryological studies following interspecific crosses in Gossypium. I. G. hirsutum × G. arboreum. Am. J. Botany 44: 209–14.

—— 1958. Embryological studies following interspecific crosses in Gossypium. II. G. arboreum × G. hirsutum. Am. J. Botany 45: 10–16.

Weismann, A. 1889–92. Essays upon Heredity and Kindred Biological Problems. 2 vols. Translation. Oxford.

—— 1892. The Germ-plasm; A Theory of Heredity. Translation. New York.

—— 1902. On Germinal Selection as a Source of Definite Variation. 2d ed. Translation. London.

—— 1904. The Evolution Theory. 2 vols. Translation. London.

—— 1913. Vorträge über Deszendenztheorie. 3rd ed. Jena.

Werth, E. 1956. Bau und Leben der Blumen. Die blütenbiologischen Bautypen in Entwicklung und Anpassung. Stuttgart.

Wettstein, W., and M. Onno. 1948. Blütenbiologische Beobachtungen an Koniferen und bei Tilia. Oester. Botan. Z. 95: 475–78.

Wherry, E. T. 1936. Reflections on the origin of life. Proc. Penna. Acad. Sci. 10: 12–15.

Whewell, Wm. 1859. History of the Inductive Sciences. 2 vols. 3rd ed. New York.

White, M. J. D. 1954. Animal Cytology and Evolution. 2d ed. Cambridge.

—— 1958. Restrictions on recombination in grasshopper populations and species. Cold Spring Harbor Symposia Quant. Biol. 23: 307–17.

—— 1959. Speciation in animals. Australian J. Sci. 22: 32–39.

Whittaker, R. H. 1959. On the broad classification of organisms. Quart. Rev. Biol. 34: 210–26.

Winter, F. L. 1929. The mean and variability as affected by continuous selection for composition in corn. J. Agr. Research 39: 451–76.

Wolf, C. B. 1948. Taxonomic and distributional studies of the New World cypresses. Aliso 1: 1–250.

Woodger, J. H. 1929. Biological Principles. London and New York.

Woodworth, C. M., E. R. Leng, and R. W. Jugenheimer. 1952. Fifty generations of selection for protein and oil in corn. Agron. J. 44: 60–65.

Wright, J. W. 1955. Species crossability in spruce in relation to distribution and taxonomy. Forest Science 1: 319–49.

Wright, S. 1931. Evolution in Mendelian populations. Genetics 16: 97–159.

—— 1943. Isolation by distance. Genetics 28: 114–38.

—— 1948. On the roles of directed and random changes in gene frequency in the genetics of populations. Evolution 2: 279–94.

—— 1956. Modes of selection. Am. Naturalist 90: 5–24.

—— 1960. "Physiological Genetics, Ecology of Populations, and Natural Selection," in S. Tax, ed., Evolution after Darwin. Chicago.

Wright, S., and Th. Dobzhansky. 1946. Genetics of natural populations. XII. Experimental reproduction of some of the changes caused by natural selection in certain populations of *Drosophila pseudoobscura*. Genetics 31: 125–56.

Wulff, E. V. 1943. An Introduction to Historical Plant Geography. Translation. Waltham, Massachusetts.

Zinder, N. D. 1958. "Transduction" in bacteria. Sci. American, November, 1958.

Zirkle, C. 1946. The early history of the idea of the inheritance of acquired characters and of pangenesis. Trans. Am. Phil. Soc. 35: 91–151.

—— 1960. "Evolution after Darwin" (review). Science 131: 1519–21.

Author Index

Subject Index

Achatinella, 437
Adaptability, defined and characterized, 118, 138
Adaptation, defined and characterized, 93–95, 114–17, 270, 519
Adaptive radiation, 519 ff.
Aegilops, 513
Agelenopsia, 360–61
Allele frequency, 143 ff., 196 ff., 221, 233, 272 ff.
Allele pairs, formula for number of, 142, 178
Allen's rule, 435
Allopolyploidy, 481
Amauris, 106
Antirrhinum, 161–63, 365, 470, 497, 523–24
Ants, 74–75, 108
Apis, 104, 260, 365
Aquilegia, 357 ff, 376, 379, 391, 416, 508
Archaeopteryx, 29–30
Artemia, 487
Artificial selection, 224 ff.; *see also* Natural selection, experiments
Asclepias, 359
Assortative mating, 261, 465–66
Atmosphere of primitive earth, 52–53, 60, 62
Atriplex, 436

Baldwin effect, 136 ff.
Barley, 160
Battus, 111–12, 114
Bears, 514
Bees, *see* Apis, Bombus, Diadasia, Eulaema, Xylocopa
Bergmann's rule, 434–35
Biochemistry, in biological theory, 20 ff.
Biotic community, 418 ff., 524–25
Biston, 95–96, 100, 203, 220–21
Blood group genes, in man, 291 ff., 309–10
Blue-green algae, 68
Bombus, 108, 110, 359, 365

Bonplandia, 557–58, 562
Bos, 337
Bradytely, 544, 546
Branta, 460
Brassica, 115, 365, 376, 379, 471
Breeding systems, in relation to speciation, 451–52, 465 ff., 478–80, 482–83, 490 ff.; *see also* Inbreeding
Bufo, 99, 103–4, 381
Buttercup, 409–10

Camarhynchus, 520–21
Cantua, 557
Capra, 370
Causal theory, 6, 27 ff., 38 ff.
Causation, 3, 14 ff., 20 ff.
Ceanothus, 508, 513, 547
Celerio, 358, 416
Cepaea, 172, 289 ff., 438
Certhidea, 520–21
Character combinations, 114 ff., 448, 508 ff., 531, 540, 569
Chemical evolution, 41–44, 51 ff., 230–31
Chlamydomonas, 408
Chromosomal rearrangements, 167 ff., 246 ff., 471 ff., 490 ff., 497 ff., 509–10
Chroococcus, 68
Circumcision 128
Ciridops, 521–23
Clarkia: breeding systems and speciation, 468, 483; competition, 268–69, 505; evolutionary trends, 561; hybrid sterility, 441, 471; pollination races, 324; population structure, 287, 438, 460; reproductive isolation, 365, 377, 381, 386, 441, 505, 507, 513
Classification, 5, 33 ff., 83 ff., 424–25
Clupea, 325
Cobaea, 557–58, 562
Coccinella, 105
Colias, 123–24
Collomia, 562
Colon bacterium, 67
Columbine, *see* Aquilegia

601